T0297756

CAMBRIDGE LIBRARY COLLECTION

Books of enduring scholarly value

Physical Sciences

From ancient times, humans have tried to understand the workings of the world around them. The roots of modern physical science go back to the very earliest mechanical devices such as levers and rollers, the mixing of paints and dyes, and the importance of the heavenly bodies in early religious observance and navigation. The physical sciences as we know them today began to emerge as independent academic subjects during the early modern period, in the work of Newton and other 'natural philosophers', and numerous sub-disciplines developed during the centuries that followed. This part of the Cambridge Library Collection is devoted to landmark publications in this area which will be of interest to historians of science concerned with individual scientists, particular discoveries, and advances in scientific method, or with the establishment and development of scientific institutions around the world.

The Concise Knowledge Astronomy

The Concise Knowledge Astronomy, published in 1898, was one a series of popular reference books by experts. Agnes Clerke (1842–1907) was a successful author of books on astronomy and its history (three of her other works being reissued in this series), and her co-authors were astrophysicist Alfred Fowler, an internationally renowned expert in spectroscopy, and J. Elland Gore, a Fellow of the Royal Astronomical Society and expert on variable stars. Illustrated by over 100 photographs and drawings, the book aimed to provide the educated non-specialist reader with an understanding of contemporary astronomical knowledge. The application of new technologies, such as photography and spectroscopy, to astronomy in the nineteenth century had led to many new discoveries, and public interest in the subject had greatly increased. The book is divided into four parts – the history of astronomy, astronomical movements and instruments, the solar system, and sidereal astronomy.

The Concise Knowledge Astronomy

Agnes Mary Clerke
Alfred Fowler
John Ellard Gore

CAMBRIDGE UNIVERSITY PRESS

Cambridge, New York, Melbourne, Madrid, Cape Town, Singapore,
São Paolo, Delhi, Dubai, Tokyo, Mexico City

Published in the United States of America by Cambridge University Press, New York

www.cambridge.org
Information on this title: www.cambridge.org/9781108023887

This edition first published 1898
This digitally printed version 2010

ISBN 978-1-108-02388-7 Paperback

The Concise Knowledge Library

ASTRONOMY

Edited by Alfred H. Miles

Photograph of the Moon taken at Paris
by M.M. Loewy and Puiseux
with the great Coude Equatorial, February 14, 1894.

The Concise Knowledge Astronomy

BY

AGNES M. CLERKE, AUTHOR OF "A POPULAR HISTORY OF ASTRONOMY DURING THE NINETEENTH CENTURY";
A. FOWLER, A.R.C.S., F.R.A.S., DEMONSTRATOR TO THE ROYAL COLLEGE OF SCIENCE; AND
J. ELLARD GORE, F.R.A.S., M.R.I.A., AUTHOR OF "THE SCENERY OF THE HEAVENS," "THE WORLDS OF SPACE," ETC.

With many Illustrations from Photographs and Drawings

BY

MM. LOEWY AND PUISEUX (PARIS), E. E. BARNARD, GEORGE E. HALE, J. E. KEELER, W. W. PAYNE, J. E. ELLARD GORE, AND OTHERS

LONDON

Hutchinson & Co.

34 PATERNOSTER ROW, E.C.

1898

PREFACE

THIS work aims to present in concise form a popular synopsis of Astronomical Knowledge to date.

In Section I., Miss Agnes M. Clerke, author of "A Popular History of Astronomy during the Nineteenth Century," gives a brief historical sketch of the science from Hipparchus to the present time. In Section II., an attempt is made by Mr. A. Fowler, A.R.C.S., F.R.A.S., Demonstrator of Astronomical Physics to the Royal College of Science, to briefly outline the general principles of spherical and gravitational Astronomy, and to describe the instrumental means now at the command of observers in the various branches of Astronomical research. The author begs to record his indebtedness to Loomis' "Treatise on Astronomy," and Young's "General Astronomy," which have been frequently consulted, especially for memorial data; to Mr. W. Shackleton, for assistance in proof-reading; and to Mr. C. P. Butter, for valuable help in preparing the diagrams. Section III., contributed by Miss Agnes M. Clerke, deals with the Solar System; and Section IV., written by Mr. J. Ellard Gore, F.R.A.S., M.R.I.A., treats of the Sidereal Heavens.

The work is illustrated by a large number of diagrams and other illustrations, prepared expressly for its pages, as well as by a number of photographic and other reproductions of photographs and drawings made by distinguished astronomers in Europe and America. In this connexion numerous acknowledgments are due.

The Editor begs to express his sense of indebtedness to the following astronomers and publishers, for kind permission to reproduce original photographs and drawings from their works:

To M. Loewy, Director de l'Observatoire, Paris, for permission to reproduce the photograph of the Moon, which forms the frontispiece of this volume; to Professor Edward S. Holden, Director of the Lick Observatory, for permission to reproduce drawings and photographs of the Observatory at Nice, p. 2; the Thirty-six Inch Reflector of Lick Observatory, p. 40; the Meridian Circle of the Paris Observatory, p. 203; the Spectroscope adapted to the eye end of the Lick Telescope, p. 221; and Jupiter showing the Red Spot, p. 322; to Dr. Isaac Roberts, for permission to reproduce his photograph of the photographic telescope used by him, p. 196; to Messrs. Trichnor & Co., of Berlin, for permission to reproduce two illustrations of Donati's Comet, pp. 228 and 363; and one of Sun-spots and Magnetic Variations, p. 246—all from Langley's "New Astronomy"; to Messrs. Witherby & Co., for permission to reproduce the photograph of a Sun-spot by Janssen, p. 243; the photograph of Jupiter, p. 328; the photographs of Swift's Comet, pp. 374 and 375, Brooks' Comet, p. 381, and the Milky Way, p. 557, from *Knowledge;* to Messrs. Taylor & Francis, for the diagram of curves showing the development of Sun-spots, p. 257; to Professor George E. Hale, of Kenwood Observatory, Chicago, for his illustrations of Eruptive Prominences photographed at Kenwood, March, 24th, 1896, pp. 264 and 265, reproduced from the *Astrophysical Journal;* to the Council of the Royal Society, for the illustration of the Eclipsed Sun, p. 267, reproduced from "Philosophical Transactions"; to Professor Barnard, for the photographs of the Corona, reproduced on p. 269; and the drawings of the Transit of Jupiter's Satellite, on p. 330, reproduced from the *Monthly Notices* of the Royal Astronomical Society; the Eye of Mars, p. 302; and of Saturn and his Rings, p. 335; to the editor of the *Astronomische Nachrichten*, for the map of Mercury, by Schiaparelli, reproduced on p. 276; to the editor of *Nature*, for the drawing of Venus by Mascari, on p. 280; to Messrs. George Philip & Son, for the map of the Moon, given on p. 291, from Fowler's "Telescopic Astronomy"; to Messrs. Longman, Green & Co., for the Chart of Mars, p. 300, and the diagram of the Apparent Orbit of the Companion of Sirius, p. 439, from Proctor's "Old and New Astronomy"; to Professor W. W. Payne of Goodsell Observatory, for the use of the drawing of the Oases of Mars, p. 304, from "Popular Astronomy"; and the photograph of Holmes' Comet, p. 378, from the same work; to Messrs. A. & C. Black, for the illustrations of the Great Comet of September, 1882, p. 361, reproduced from Miss Clerke's "History of Astronomy"; to Messrs. Crosby, Lockwood & Co., for permission to reproduce the illustration of the Nebula in Andromeda 31 Messier, p. 398, from the frontispiece of Mr. J. E. Gore's "Visible Universe"; and also for the same authority, regarding the diagrams, showing the Stars visible in the Northern Hemisphere, p. 401; the Stars visible in the Southern Hemisphere, p. 403; the drawing showing the position of the Solar Apex, according to different computers, p. 429; and the photographs of the Spiral Nebula 51 Messier, p. 533; and the Milky Way in Sagittarius, p. 555, all from the same work; to Messrs. A. D. Innes & Co., for the use of the diagram, Apparent Orbit of Zeta Herculis, p. 436; Triple Stars, p. 451; and for permission to reproduce the photographs, 37 Messier, p. 505; the star cluster, Omega Centauri, p, 512; the Nebulæ of Orion, p. 521; and the Magellanic Clouds, p. 537, from "The Worlds of Space"; and to Messrs. Sutton & Co., for the use of the diagram of the Apparent Orbit of 70 Ophiuchi, p. 443; the photographs of the Double Cluster of Perseus, p. 503; the Star Cluster in Gemini, p. 504; the Star Cluster in Hercules, p. 507; and the drawing of the Trifid Nebula, Sagittarius, p. 525, from "The Scenery of the Heavens"; and the drawing of the Temporary Star of 1572, p. 481, from "Planetary and Stellar Studies," both published by them.

A. H. M.

CONTENTS

SECTION I.—HISTORY . . . . By Agnes M. Clerke.

SECTION II.—GEOMETRICAL ASTRONOMY AND ASTRONOMICAL INSTRUMENTS.

By A. Fowler, A.R.C.S., F.R.A.S.

SECTION III.—THE SOLAR SYSTEM.

By Agnes M. Clerke.

SECTION IV.—THE SIDEREAL HEAVENS.

By J. E. Gore, F.R.A.S.

LIST OF ILLUSTRATIONS

SECTION I.—HISTORY.

SECTION II.—GEOMETRICAL ASTRONOMY.

SECTION III.—THE SOLAR SYSTEM.

SECTION IV.—THE SIDEREAL HEAVENS.

ASTRONOMY

SECTION I.—HISTORY

By AGNES M. CLERKE

THE OBSERVATORY AT NICE.

[*See page* 192.

ASTRONOMY

SECTION I.—HISTORY.

By Agnes M. Clerke.

CHAPTER I.

FROM HIPPARCHUS TO LAPLACE.

In the year 134 B.C., a temporary star blazed out in the constellation Scorpio. It was observed by a man of extraordinary genius, and furnished the incentive to one of his most memorable works. This was the construction, on essentially modern principles, of a catalogue of 1,080 stars. Hipparchus thus, with deliberation and singular prescience, furnished a standard by which future changes in the heavens might be detected. He was a native of Rhodes, but belonged to the school of Alexandria; and at Alexandria, after three centuries, he found an able and ambitious successor.

Claudius Ptolemæus was one of the many "inheritors of unfulfilled renown." He combined, completed, and preserved what his predecessors, eminent or obscure, had done. Gathering materials from all quarters, and adding much of his own he reared an astronomical edifice so imposing, coherent, and substantial, that the lapse of fourteen centuries left it virtually unassailed, and, to a superficial judgment, unassailable. Fitly, then, this monument of industry and ingenuity kept the title bestowed upon it by the Arabs of "Almagest," signifying "the Greatest." It bears, nevertheless, perennial witness to the possibility of satisfying the human mind with the truth of appearances, apart from the truth of things. For although the Almagest embodies a large amount of real knowledge, that knowledge is throughout falsely interpreted. The Ptolemaic system was constructed on the principle of "saving the phenomena"—that is, of providing expedients geometrically valid, even if physically inadmissible, by which to represent the apparent movements of the heavenly bodies. That they might, to a great extent, be apparent only, was obvious to the cultivated Greek mind. The rotation of the earth on an axis was a familiar Pythagorean doctrine; it was adopted by Plato, and Aristarchus of Samos went to the length of ranking our green world as a planet revolving yearly round the sun. The idea, however, was too recondite for vulgar apprehension; it was tainted with a suspicion of impiety, and its development would, besides, have proved extremely embarrassing to the nascent science of that age. So Hipparchus chose the prudent alternative of treating astronomy from the purely mathematical standpoint; he submitted to the restrictions imposed by the hypothesis of equable circular motion; and, with wonderful skill, fitted the Apollonian eccentrics and epicycles to expound celestial wanderings. Ptolemy inevitably followed suit. He set some five dozen spheres in motion, while leaving the earth at rest; and at rest it remained until, in long meditations by the foggy shores of the Baltic, a grave-browed ecclesiastic elaborated certain cogent arguments in favour of its motion.

During the interval between Ptolemy and Copernicus, astronomy kept in the Alexandrian groove. Early in the eighth century, the seat of learning having been transferred to Baghdad, the charge of its crystalline machinery devolved upon Arabs and Jews, men of fine technical acquirements, but small originative power, men of the kind described in the

"Vicar of Wakefield," who, "had they been bred cobblers, would all their lives have only mended shoes, but never made them." Not but that they executed the necessary repairs with uncommon ingenuity, modifying the cumbrous structure given into their keeping to suit the fresh inequalities brought to light by their patient watchfulness. But their improvements consisted in adding to already intolerable complications —in piling orb on orb, in devising "trepidations" and oscillations, of which nature took small heed; so that the better they observed, the worse their system became.

The science was diligently cultivated. Al-Mamûm, son and successor of "good Harûn-al-Raschid," founded at Baghdad a school of astronomy, of which Albategnius, called "the Ptolemy of the Arabs," was the brightest ornament. He discovered, early in the tenth century, the movement of the "sun's apogee"—that slow revolution of the longer axis of the earth's orbit, regarded by astronomical glacialists as a factor in the production of recurring Ice Ages. The Persian grandee, Al-Sûfi (903-986) belonged to the same group. His "Description of the Stars" was a revised edition, not a simple reprint, of the Alexandrian list, and has the value derived from personal consultation with the skies. Thus, Algol, now purely white, is recorded in it as a decidedly red star. About a century later, Aboul Wefa detected the moon's "variation," independently noted, after five centuries, by Tycho Brahé. Then the Tartars had their turn. Nasir Eddîn (1201-1274) was a native of Khorassan; but his love of learning drew him to the city of the Khalifs, where he assembled a band of experts for the construction of new planetary tables, the old ones having lapsed into hopeless disaccord with the heavens. Last came Ulugh Beigh, grandson of the furious Tamerlane. He founded at Samarcand a kind of astronomical Solomon's House, built a grand observatory, and worked in it assiduously. His vigorous and ennobling reign of forty years was terminated by one of those domestic catastrophes which ordinarily fix the chronology of Eastern dynasties. He was murdered by his son in 1447, and the sands of the desert thereupon closed, so to speak, over his civilising efforts. His star-catalogue, edited by Francis Baily in 1843, is the outcome of fresh observations made in the old way. A Tartar prince, he ranks as an Arab astronomer.

Mohammedan science had already fulfilled its appointed task. A torch, still alight, had been handed on from East to West. Its extinction would have been a calamity. A total break in the cultivation of astronomy, for instance, would have cost ages to repair. The Ptolemaic system, it is true, disguised rather than revealed nature; yet it constituted a regulated body of knowledge, only looked at from a wrong point of view. An unbiassed spectator had merely to shift his position and open his eyes, in order to perceive the simplicity of the real celestial mechanism. No better illustration could be adduced of Bacon's aphorism that "truth emerges more easily from error than from confusion."

It was from the Moors in Spain that Christian Europe took its first lessons in antique science. The Alphonsine Tables were due to Oriental industry. They were compiled at Toledo about 1270 by an assemblage of Arab experts directed by Hassan, the Jew delegate of Alfonso X. of Castile. But they caught Western attention, and drew Western intelligence towards the abstruse art they exemplified. Thus a little treatise on the Sphere composed about 1230, by John Holywood, a Yorkshireman, known to cosmopolitan fame as Johannes de Sacrobosco, obtained astonishing popularity; at least sixty-five Latin editions of it appearing between 1472 and 1647, besides French, Italian, German, and Spanish versions, and endless commentaries. With the revival of classical learning, the Almagest, previously known in blundering Latin translations from the Arabic, came to be read in the original Greek, and thus re-emergent, roused fresh enthusiasm. Inspired by the afflatus, George Purbach (1423-1461) and his brilliant pupil, Johannes Müller of Königsberg in Franconia (Regiomontanus), successively professors of mathematics at Vienna, applied themselves to burnishing up the ancient epicyclical apparatus; while in Italy, the seductive opinions of the

Pythagorean school gained ground, as evidence came to light, that there had been astronomers before Ptolemy no less than kings before Agamemnon. The orthodox doctrine naturally continued to be taught at the universities; but some of the professors held esoteric opinions of a different cast, which they freely imparted to privileged disciples. The earth's rotation was spoken of as a matter of common knowledge by Lionardo da Vinci; it was inculcated in rhyme, before the close of the fifteenth century, by Girolamo Tagliavia, a Calabrese poet; it was debated by scholars and pedants; on all sides influences wrought to shatter the integrity of Ptolemaic convictions.

True progress, however, consists less in destruction than in re-organisation. And this demands powers of a high order. They were brought into play just at the right moment. Nicholas Copernicus was born at Thorn on the Vistula, February 19, 1473. At the age of twenty-three, having exhausted the teaching resources of the university of Cracow, he crossed the Alps in quest of instruction in Greek and mathematics. Towards the close of 1496, then, he was enrolled as a student at Bologna, and shortly afterwards became the pupil, assistant, and friend of the Ferrarese astronomer, Domenico Maria Novara. Here, beyond reasonable doubt, Copernicus adopted Copernican opinions. The question, *An terra moveatur?* was incessantly mooted at Bologna; advanced thinkers replied in the affirmative; Novara himself most likely took his intellectual beliefs from Plato and Aristarchus, while looking to Ptolemy for his daily bread. The transalpine scholar, at any rate, brought back with him to Poland in 1505, an unalterable persuasion that the heliocentric system belonged to the reality of things. He devoted eighteen years of his abode within the cathedral precincts of Frauenburg—from 1512 to 1530—to demonstrating its detailed conformity with the phenomena of the heavens; but allowed only a sketch of his results to be published. It was only at the earnest request of the Bishop of Culm that he finally delivered up to him the manuscript of "De Revolutionibus Orbium Coelestium," the first printed copy of which was laid on his deathbed, May 24, 1543.

The immediate effect was small. The new system of astronomy was admired, but not adopted. It indeed contradicted the evidence of the senses, and failed to compel assent from the understanding. For its author had not completely broken with tradition. He unfortunately retained the false supposition of equable circular motion, and thus greatly marred the simplicity of his scheme of the heavens. Orbs still kept rolling upon orbs, Mercury alone demanding a combination of seven to bear him over his course. But if seven, it might have been asked, why not seven times seven? The principle of representing appearances by transcendental means remained the same as before. Ignorance of the laws of motion raised other formidable objections. A whirling earth, it was thought, should leave behind all detached objects; absolute repose was taken to be the condition *sine quâ non* of stability. Then the seeming immobility of the stars implied for them a remoteness so extravagant, according to prevalent ideas, that even Kepler admitted it to be "a big pill to swallow." Copernicus was fully aware that the earth's orbital revolution must occasion stellar perspective displacements; indeed, he staked the truth of his theory upon future measurements of annual parallax. Nevertheless, four centuries passed before they were successfully executed.

Tycho Brahé was the last great mediæval observer. Like Hipparchus, he was summoned by a star—the marvellous "new star" of 1572; and, having obtained from Frederick II. of Denmark the grant of an islet in the Sound, he built upon it a mansion "royal, rich and wide," erected magnificent instruments, and used them, not only with consummate skill, but also with a certain princely pomp, donning robes of state before admitting the bright "populace of heaven" to audience. His stormy temper, however, led to disputes with the young King Christian IV.; he forsook Uraniborg, and died at Prague in 1601. Curiously enough, the very accuracy of his observations led him astray from speculative truth. For it enabled him to perceive the incompatibility of many facts with Copernican expedients for harmonising them, and intensified the difficulty raised to Copernican views by the absence of stellar parallax. So he devised a system of his own, in which the planets revolved round the sun, but the sun round the earth. It scarcely survived its contriver.

The invention of the telescope created descriptive astronomy. Without it, the mechanism of the solar system could have been laid bare, and the law of force regulating its action discovered; and in point of fact, Kepler's achievements owed nothing, and Newton's very little, to the optician's art. Inquiries, on the other hand, into the nature of the heavenly bodies were wholly inspired by it; it disclosed the amazing multitude of the stars, and opened endless vistas of research. No one could at first have divined the momentous character of the accident by which Hans Lippershey, a spectacle-maker at Middleburg in Holland, hit upon an arrangement of lenses serving virtually to abridge distance. It happened in 1608; and Galileo Galilei (1564-1642), hearing of it shortly afterwards at Venice, prepared on the hint a "glazed optic tube," and viewed with it, early in 1610, the satellites of Jupiter, the mountains of the moon, the star-streams of the Milky Way, and in 1611, the phases of Venus, the spots on the sun, and the strange appendages of Saturn. Thus, amid a tumult of applause, the telescopic revelation of the heavens began. It was brilliantly illustrative, although not demonstrative, of Copernican theory; and Galileo drove his own vivid conviction on the subject home to general apprehension by the literary skill with which he treated it in his famous "Dialogues" (1632). He most substantially promoted the new views, however, by his recognition of the laws of motion, and of force as the cause of motion. The problem of the heavens, stript thereby of metaphysical obscurities, was laid bare to the reason as one of pure mechanics; the planets came to be treated as ordinary projectiles, and distinct reasoning about the nature of their paths was rendered possible. Newton's great task was thus prepared and defined by Galileo.

Kepler's (1573-1630) three generalisations formed a still more indispensable prelude to its accomplishment. Their immediate effect was to sweep away the Copernican remnants of Ptolemaic lumber, and to disclose the harmonious plan upon which our system is ordered. But it was a geometrical plan only. Kepler indeed divined the influence of a central power, which he surmised to be of a magnetic nature; and he aspired towards the establishment of a truly physical astronomy. Yet he was far from perceiving the full implications of the laws he had himself, after half a lifetime of trial and failure, at last triumphantly discovered. These laws are:

(I.) The planets travel in ellipses of which the sun occupies one focus.

(II.) They travel at rates varying in such a manner that the "radius vector"—or imaginary line joining each to the sun—describes equal areas in equal times.

(III.) The cubes of their mean distances from the sun are proportional to the squares of their periods of revolution.

Now these are precisely the conditions under which planetary circulation should proceed if governed by a force emanating from the sun, and decreasing as the square of the distance from him increased. Moreover, Hooke, Halley, and Wren separately got so far as to perceive that it could be explained on this principle. But Isaac Newton alone could demonstrate what they divined, and even his supreme faculties were dangerously strained by the laborious process. This was not all. He showed that the earth exerts on the moon just the same kind of pull that the sun exerts on the planets; a pull identical with the familiar "attraction of gravitation," by which the globe we inhabit holds integrally together, retains its oceans in their beds, and bears with it through space its "cloud of all-sustaining air." Its domestic affairs are thus guided by the same unchanging rule that dominates its foreign relations.

The publication in 1687 of Newton's "Principia" marked an unprecedented advance in knowledge. The advance consisted in unification. A science of celestial physics, capable of indefinite future expansion, was founded on the sure basis of terrestrial experience. Canons of interpretation, derived from immediate perception, were proved applicable to the phenomena of the heavens. The line drawn in antique philosophy between the "corruptible" things under our feet and the "incorruptible" over our heads was forever rubbed out. Sublunary and empyreal regions were thrown together into one vast domain.

Although Newton's law is, in itself, of extreme simplicity, its actual workings are highly intricate. Because dependent upon a universal and unintermittent influence, they are self-modifying, so that each consequence becomes a cause, and to each cause is attached an endless train of effects. They can be dealt with only with the aid of the infinitesimal calculus, and then, not directly, but by successive and tedious approximations, or by arts and devices of almost superhuman ingenuity. Hence Newton's laurels would have remained comparatively barren had he not found successors in a group of men of extraordinary ability. What he had begun, Clairaut, D'Alembert, Euler, Lagrange, and Laplace carried on by showing the adequacy of a single law to account for every traceable deviation from undisturbed elliptical motion. In the course of a long and arduous campaign, they carried every position that they attacked. Over and over again, the principle of gravitation seemed to be compromised; over and over again, it was vindicated by these intrepid champions.

This process of gradual verification began in 1747, when Clairaut and D'Alembert sent to the Paris Academy of Sciences, on the same day, the first satisfactory solutions of the "Problem of three Bodies." The motions of the moon, nevertheless, did not at once fall in with the general theory; they were rendered amenable only after years of anxious toil. Barely the initial difficulties had been overcome when Euler, in 1753, published his "Theory of the Moon," from which Tobias Mayer of Göttingen constructed lunar tables. Now tables are the test of theories. Every row of figures they contain is a prediction, by the fulfilment, or non-fulfilment of which the underlying scheme must stand or fall. Through such comparisons, mathematical astronomers find out the shortcomings of their methods, or the insufficiency of their hypotheses, and are incited to refine the first, and correct the second. Demands for the application of the nicer criteria thus afforded suggest observational improvements, which seldom fail to bring to light minor discrepancies with theory, impelling to fresh efforts for their abolition. Such alternations of advance along the abstract and the practical lines result in a continual diminution in the *scale* of error, although not in its annihilation; absolute exactitude being, as it were, an asymptote, continually approached, but touched only at infinity—that is, never, under subsisting conditions. Even now the length of the moon's tether is four or five miles. To that extent, she may go astray from her computed path, not without occasioning disquietude to the responsible authorities.

So far as could be ascertained in the eighteenth century, her subjection to known law was completed by the dispersal of the mystery surrounding a slight, continuous acceleration of her orbital velocity detected by Halley in 1693. It had been in progress since the earliest recorded eclipse in 721 B.C., if not longer; there was no sign of its cessation or reversal, and the grave question arose, Was the principle of universal attraction, elsewhere unreservedly obeyed, here fatally complicated by the action of a resisting medium involving the eventual collapse of the earth-moon system? Laplace gave the answer, November 19, 1787, by proving the observed quickening of pace to be a necessary and simple consequence of a secular diminution in the ellipticity of the earth's orbit. This, however, will not go on for ever in the same direction; after many ages the tide of change will turn, and a complete restoration to the *status quo ante* will ensue.

Another master-stroke of Laplace's genius was his explanation, also in 1787, of the "long inequality" of Jupiter and Saturn. He demonstrated its strictly gravitational origin in the mutual disturbance of the two giant planets, rendered up to a certain point cumulative by the approximate commensurability of their periods. While Jupiter performs five circuits Saturn accomplishes nearly two, and the perturbation set up at their conjunction is hence both intensified and balked of compensation for 918 years.

The epoch of trial and confirmation immediately following the publication of the "Principia" lasted then a full century. During its course, difficulties had arisen only to be overcome; suggested qualifications of the single and simple law of gravity had proved unnecessary; at its close, recalcitrance had everywhere been overcome, and there was victory all along the line. And not only were the workings of the planetary system exhibited as depending upon an elementary principle, but they were further shown to be perfectly equilibrated. It contained within itself, so far as could be ascertained, no seeds of decay; its destruction could only come from without. This remarkable conclusion was established in a series of splendid treatises by Lagrange and Laplace. The special adaptation to permanence of the solar mechanism was demonstrated in them. Ruinous disturbances were shown to be excluded by the overwhelming disparity of mass between the central body and its attendants, no less than by the regularity and harmony of their movements and distribution. Thus only slight oscillatory changes can occur. Millions of years will elapse without producing any fundamental alteration. The machine is so beautifully adjusted as to right itself automatically through the mutual action of its various parts. And it is the force which perturbs that eventually restores.

The astronomical acquisitions of the century were embodied in Laplace's "Mécanique Céleste," published 1799-1805. This "Almagest of the eighteenth century," as it has been termed, is in a rare degree comprehensive and complete. It leaves nothing enigmatical. Every question propounded in it receives an answer, if not definitive, at least highly authoritative; and the range of these questions is very wide.

All the phenomena which the Greeks and Arabs had rightly observed, but wrongly interpreted, are not merely "saved" by geometrical artifices, but derived as a connected whole from one physical cause, absolutely prescribing that they should be thus, and no otherwise. The work is a record of unmixed triumphs. It seems as if the author, for want of more worlds to conquer, had laid down the sword of the calculus to take up the pen of the chronicler. With grave exultation, he proceeds from point to point, recounting the events of the campaign, commemorating the battles won by the brilliant staff of mathematical heroes to which he himself belonged, and expatiating in the broad subjugated plain. He scarcely looked beyond. There was indeed at that time no "beyond" where his methods of investigation were applicable. The "Mécanique Céleste" hints at no unsatisfied ambitions; it is a book of the *teres atque rotundus* sort—a world in itself well arranged and compact, to which outlying perplexities are allowed no access. Nor should this be counted a defect. As a monument to one of the greatest periods in the history of science, its fitting character was that of an ordered collection of acquired certainties.

The countrymen of Newton took no part in the striking series of operations by which the intricate consequences of the law of gravity were deduced and shown to correspond with reality. During the whole of the eighteenth century, they stood aside from the race towards verification. Their effacement was due to no lack of ability, but to a mistaken choice of means. Newton's synthetic method was a veritable Bow of Ulysses. It was too tough to be bent by other hands than his own. Thus, no sequel could be given to the "Principia." There was no possibility of following up the line of demonstration pursued in it. Newton himself would have vainly attempted to carry it much further. In order to advance, it was necessary, as Dr. Whewell remarked, to begin afresh. This, British mathematicians were unwilling to do. The easy and flexible analytical method brought to perfection on the continent remained strange to them. With inadequate strength, they persisted in wielding the cumbrous weapon of a giant—in using main force, so to speak, where skill and agility were required. Our insularity in this respect lasted until about 1816, when, by the joint efforts of the younger Herschel, Charles Babbage, and George Peacock (afterwards Dean of Ely), mathematical studies were revolutionised at the University of Cambridge.

The neglect in England of theoretical research was, however, partly compensated by the steady progress of practical astronomy. For a century and a half after its foundation in 1675, the Royal Observatory at Greenwich continued to be the main—almost the only source of information regarding the places of the heavenly bodies. Thence were obtained the data necessary for the correction of theory, since there alone the visible positions of the sun, moon, and planets were systematically determined. *Actual*, compared with *predicted*, movements gave so-called "tabular errors"; and tabular errors indicated theoretical shortcomings, the rectification of which led gradually, but surely, towards a higher plane of knowledge.

John Flamsteed (1646-1719), the first astronomer-royal, was, in Professor De Morgan's phrase, "Tycho Brahé with a telescope." By his diligence and insight he set on foot modern astronomy of precision. The "British Catalogue" of nearly 3,000 stars, was, in its day, an unique and most valuable work. His lunar observations were indispensable to Newton's calculations, which, indeed, through the insufficient supply of them, now and again came to a halt; he constructed new solar tables, and kept watch over the careers of planets and comets. His completion, in 1689, of a seven-foot mural quadrant, constituted a marked advance in the art of instrument-making. It was firmly fixed in the meridian, so that the distances from the zenith of the heavenly bodies at the moment of culmination could be read off on the limb, the time being simultaneously noted by a clock. Their positions in the sky relative to a set of forty otherwise known stars were thus completely determined, and they were determined essentially after the manner still in use.

On Flamsteed's death in 1719, Edmund Halley (1656-1742) succeeded to his place. An expedition to St. Helena in 1677, for the purpose of observing stars invisible in these latitudes, got him the name of the "southern Tycho." They were the very first so situated to be located on the sphere (except those few that came within Ptolemy's range), and a list of them, to the number of 341, was appended to the "British Catalogue." The purpose to which Halley devoted most sustained attention was, unluckily, that in which he was least successful. Early in life he formed the design of observing the moon through an entire revolution of its nodes, so as to bring lunar tables to the perfection required for solving the prize-problem of longitudes. But the *contumax sidus*—his opprobrious term for our satellite—proved more than a match for him. The eighteen years' watch was kept, notwithstanding that the watcher had reached the age of sixty-five before he was able to set about it; but in vain; nothing came of it. Halley's varied performances were, nevertheless, so considerable as to warrant Lalande in describing him as "the greatest of English astronomers"; and he ranked next to Newton among contemporary English men of science.

His cometary labours alone sufficed to perpetuate his name. He initiated the computation, on Newtonian principles, of the orbits traversed by such bodies—then a most toilsome process; and, among twenty-four, found three so much alike as to suggest the identity of the great comets of 1531, 1607, and 1682. A renewed apparition might then be expected in 1758, and he appealed to "candid posterity to acknowledge that this was first discovered by an Englishman." The prediction roused wide-spread interest, and as the epoch for its fulfilment drew near, Clairaut undertook the formidable task of determining to what extent it might be postponed by the retarding influence of Jupiter and Saturn. Many times he despaired of its execution, even with the efficient aid of Lalande and Madame Lepaute, the wife of a Paris clock-maker; and at last, after months of wearisome calculation, having succeeded in forming the differential equations representing the comet's disturbed motion, he threw down the paper on which they were written, with the exclamation, "Now, integrate them who can!" Eventually this, too, was done; and the comet, caught sight of on Christmas Day, 1758, by Palitzsch, a rustic star-gazer in Saxony, passed the sun within the month's "law" permitted to it by the French geometer. This signal triumph laid the sure foundation of cometary astronomy.

In 1679, Halley drew attention to the importance of transits of Venus for measuring the sun's distance; and developed later a method extensively used in observing the eighteenth century pair of transits in 1761 and 1769. But the accuracy actually attained in determining the instants of contact between the limbs of the sun and planet fell far short of what he had anticipated as attainable. The "black drop" interposed its pernicious effects, and occasioned wide discrepancies. The margin of uncertainty regarding the value of the great unit was, none the less, diminished, although it still remained uncomfortably wide; while the public interest excited by such rare events, the adventurous character of the expeditions sent to the uttermost parts of the earth for their utilisation, and the combined efforts of various nations towards the same end, served to popularise astronomy, and to give it something of that cosmopolitan stamp now borne by it.

Besides the discovery of the secular acceleration of the moon's motion, that of the long inequality of Jupiter and Saturn was due to Halley; he ascertained, in 1718, the proper movements of Sirius, Aldebaran, and Arcturus, thereby virtually demonstrating the non-existence of "fixed" stars; he associated auroræ with terrestrial magnetism; noted the globular star-clusters in Hercules and Centaur; and divined nebulæ to be composed of "a lucid medium shining with its own proper lustre," and filling "spaces immensely great." Yet, in spite of the comprehensiveness of his genius, his administration at Greenwich was a failure. He was a better astronomer than astronomer-royal.

James Bradley (1693-1762), who came after him, gave a narrower scope to his abilities, yet was of unsurpassed sagacity in connecting effects with their causes. Robert Hooke (1635-1703) had observed, in 1669, annual displacements of γ Draconis, a star nearly crossing the zenith of London, which he took for results of parallax; and Flamsteed, in 1694, had similarly interpreted a similar affection of the pole-star. They had both been misled by an "aberration," due to the progressive transmission of light combined with the advance of the earth in its orbit. Bradley determined to sift the matter thoroughly, and observed Hooke's star continuously from 1725 until 1728, first at Kew with Molyneux, then at Wanstead in Essex. It evidently described a small ellipse in the sky with a period of one year; yet its place in the ellipse was not what it should have been on the parallactic hypothesis; so he remained for some time in the dark about it. During a water-party on the Thames, however, in September 1728, he noticed that the slant of the pennant varied with changes in the boat's course, the wind remaining steady throughout. This gave him the clue he wanted; and his discovery of the "aberration of light" was communicated to the Royal Society in the month of January following. That of the nutation of the earth's axis followed in 1748. Both, setting aside their importance in themselves, were indispensable as preliminaries to accuracy in fixing the places of the heavenly bodies. For they are vital elements in the process of "reduction," by which the ore of truth contained in observations is extricated from the dross of casual circumstances. The raw material, collected by timing transits and reading circles, must be so refined and purified that the facts contained in it become mutually comparable. Before Bradley's time allowance was indeed roughly made for refraction in our atmosphere, and for the precession of the equinoxes; and, in the case of the moon, for parallax; but the effects of aberration and nutation had remained mixed up with a mass of disguising errors. Their elimination constituted an inestimable improvement.

In the immediate art of observation Bradley was a master. He did not live to possess an achromatic telescope; neither astronomical circles nor equatorial mountings were at his disposal. His leading instrument was an eight-foot quadrant, by John Bird, certainly of admirable workmanship; although of a type long since, and for good reasons, superseded. He amassed with it, nevertheless, a treasure of high-class observations. The bulk of them remained in manuscript until 1798, so that it was reserved for this century to turn them to account; but their value has only developed with the efflux of years. Those relating to the moon and planets, reduced by Sir George Airy, lent efficient aid towards perfecting the theories of those bodies. Those of 3,222 stars formed into a catalogue by Bessel were published in 1818 with the proud, but not unmerited title of "Fundamenta Astronomiæ." The same original data, again in 1886 reduced with the utmost nicety of care by Dr. Auwers of Berlin, afforded a splendid accession to knowledge of stellar proper motions. Acquaintance with Bradley's stars now extends over 144 years; and the amount and direction of their progress across the sphere during that long interval have, for the most part, become defined with tolerable certainty.

Nathaniel Bliss (1700-1764), the fourth astronomer-royal, filled the post only two years. Yet the observations made under his care form a sequel to Bradley's well worth having. The reign of his successor, Nevil Maskelyne (1732-1811), extended over forty-six years. His determinations of the sun, moon, and planets, were in great demand abroad for the correction of tables, and as criteria of theories; while, of the stars, he paid attention only to thirty-six, catalogued as reference-points in 1790. Their proper motions served Herschel for his second investigation, in 1805, of the sun's translation through space. By the close of the century, Maskelyne's instruments had lapsed into decrepitude; and only the stimulus supplied by Pond's strictures roused him to order one of Troughton's improved circles. But he died before it was mounted, and its employment fell to the share of his critic, John Pond (1767-1836), the sixth astronomer-royal. Maskelyne's most enduring title to fame is his foundation, in 1767, of the "Nautical Almanac."

English observers were ably seconded by English artists. Graham, Sisson, Cary, Bird, Ramsden, had, from the beginning to the end of the eighteenth century, no foreign competitors of note. Their quadrants and sectors were distinguished both for stability and for refinement of execution. The mechanical skill displayed in their construction was no less necessary for the promotion of practical astronomy than the subtlety of eye and hand needed to employ them to the best advantage. Bradley's work was conditioned by the performances of Graham and Bird. Without Graham's sector he could not have discovered the aberration of light; without Bird's quadrant the perennial worth of his Greenwich observations would have been impaired, if not destroyed. Observatories all over the continent were furnished in the latter half of the eighteenth century with instruments of English make; the art of accurately dividing circular limbs was invented in England, and nowhere else successfully practised. The innovation of substituting entire circles for quadrants was effectively introduced by Ramsden; and Piazzi came from Palermo in 1788 for the purpose of securing from him a five-foot altazimuth, at that date the finest sky-measuring machine in the world. Edward Troughton (1753-1835) ably carried on the tradition of his predecessors, and brought the altazimuth, transit-circle, and equatorial up to the modern standard of efficiency. But they were no longer in exclusive demand. The foundation, in 1804, of Reichenbach's Institute at Munich finally abolished the British monopoly in supplying astronomers with their exquisite and ingenious tools.

The improvement of refracting telescopes ran a somewhat similar course. The essential step of combining flint and crown glass, so as to bring differently-coloured rays to one focus, was taken in 1733 by Chester More Hall, a gentleman of fortune in Essex; but he published nothing, and the re-invention of the "achromatic" lens was left to John Dollond (1706-1761) a Spitalfields weaver. "I obtained," he wrote in 1758, "a perfect theory for making object-glasses, to the apertures of which I could scarcely conceive any limits." The excise duty on glass, however, which was repealed only in 1845, drew these limits very narrowly in this country; and it was through the extraordinary perseverance of a Swiss artisan named Guinand, in overcoming the difficulties connected with glass-making, and the genius of Joseph Fraunhofer (1787-1826) in moulding the material thus placed at his disposal, that refractors began at Munich to rise towards their present power and perfection.

The history of the reflecting telescope is British throughout. It was invented by Newton, made practically effective by John Hadley (1682-1744), and brought very near to theoretical perfection by James Short of Edinburgh (1710-1768); yet it is remarkable that not a single observation of lasting interest was made with any of his instruments, a few of which have survived, and are regarded with admiration to this day. The career of reflectors as engines of discovery began, but did not end, with William Herschel.

CHAPTER II.

A CENTURY OF PROGRESS.

On March 13, 1781, an event occurred without precedent in the history of astronomy. A new member of the sun's immediate retinue was disclosed. A hard-worked music-teacher at Bath performed this startling—indeed, according to antique notions—impossible feat; and the name of Herschel became known *urbi et orbi.* It was far from being by chance that the "new planet swam into his ken." The Octagon Chapel organist was no ordinary lucky amateur. He had, some time previously, made two notable resolutions. The first was to push the improvement of telescopes to the furthest verge of what was possible; the second, to leave no corner of the starry heavens unexplored. And he applied himself with marvellous energy, in despite of accumulated professional engagements, to carry them into execution. He thus rapidly grew to be an adept in the art of constructing specula, and a master in the art of using them.

Two lines of effort, accordingly, converged, in his case, towards celestial discoveries. With all his diligence in "reviewing" the heavens, he could not have distinguished at sight Uranus from a fixed star, but for the uncommon excellence of his seven-foot reflector; nor would the reflector, had it been used in the ordinary erratic fashion of casual star-gazers, been at all likely to have encountered the little bluish disc of the remote orb then slowly wending its way through the constellation of the Twins. The direct, and a momentous result of the discovery was to secure for astronomy the undivided powers of the extraordinary man who had made it. George III. attached him to his Court, delivered him from the drudgery of teaching, and gave him the means of carrying out his grand designs.

Their fulfilment involved the construction of great light-gathering machines. Herschel ardently desired to see as far and as much as the conditions of mortality permitted; he was the first to connect depth of penetration into space with extent of reflective surface; and he accordingly strained every nerve to secure the means by which to compass the end he had mainly in view. Nor was he content with mere size. His mirrors were as remarkable for beauty of figure as for breadth of aperture. They bore, on proper occasions, enormously high magnifying powers, and the precise roundness of the star-images formed by them excited the incredulous wonder of contemporaries. The quality of some of his largest instruments was guaranteed by the heavens themselves. Their approval was signified to the seven-foot reflector through the detection with it of Uranus; the "large twenty-foot," with a speculum of eighteen inches, revealed in January 1787, two Uranian moons, Oberon and Titania; and the monster forty-foot, through the tube of which George III. promenaded with the Archbishop of Canterbury, brought into view, within three weeks of its completion, Enceladus and Mimas, the innermost and hardest to observe of Saturn's numerous family of satellites.

The forty-foot was "Herschel's furthest"; he fully recognised that with it he had touched the line which divides failure from success. If, indeed, he had not overpassed it; for the subsequent career of the great telescope hardly bore out the promise of its start. It was an unwieldy engine, demanding vastly more time and labour to bring into play than the twenty-foot; and Herschel took such account of minutes as few men do of hours or days. His fiftieth birthday had in fact gone by before his optical ambition was satisfied; while his appetite for exploration was only whetted by what he had already accomplished. He estimated, however, that a "review of the heavens" with the forty-foot would have occupied 800 years; hence it was used only on special occasions. The Orion nebula was the last celestial object upon which, January 19, 1811, "its broad, bright eye" rested; and it was then, with due honour, placed on the retired list.

Two years before his death, which occurred August 25, 1822, the elder Herschel initiated his son into the secrets of speculum-building. The pupil was worthy of the master. John Herschel (1792-1871) aimed only at producing generally available instruments, and his success was easy and unqualified. His eighteen-inch mirrors seem to have been all but faultless. They certainly afforded him better views of the nebulæ than had been obtained by his father. Thus he first saw the "Dumb-bell" in its true oval shape; and his remarks upon annular lines of structure in elliptical nebulæ prove that features unmistakably imprinted upon Dr. Roberts' photographs had been antecedently visible to him, and probably to him alone.

The next stride in the enlargement of reflectors was made by an Irish nobleman, the third Earl of Rosse (1800-1867). His leviathan telescope, six feet in aperture, and fifty-four in length, has, in point of actual size, never been surpassed. Distinguished rather for light-grasp than for precise definition, it found its appropriate field in the nebular realms of the sphere; and the discovery of spiral nebulæ, with which it made its début, was one of high and wide significance.

William Lassell (1800-1881) of Starfield, near Liverpool, set the example, in 1840, of mounting reflectors equatorially, so as to enable them, by the application of clock-work, to follow automatically the diurnal movement of the heavens. His specula were of almost unrivalled perfection in form and finish. One twenty-four inches in diameter, now at Greenwich, left a splendid record. With it Lassell detected, October 10, 1846, the satellite of Neptune; September 18, 1848, simultaneously with W. C. Bond of Cambridge, U.S., Hyperion, the seventh in order of distance and last in order of discovery of Saturn's eight moons; and October 24, 1851, Ariel and Umbriel, the inner pair of Uranian satellites, of which Sir William Herschel had possibly, although not very probably, caught transient glimpses. He erected a similar instrument of fourfold capacity at Malta in 1861, registered with its aid 600 new nebulæ, and delineated the complex structure of many others, previously less well seen.

The four-foot reflector built in 1870 by Thomas Grubb of Dublin for the Melbourne Observatory disappointed expectation. An apparatus so delicate that the abrasion of $\frac{1}{20,000}$th of an inch makes all the difference between good and bad definition, is ill-fitted to endure the rough-and-tumble experiences of an ocean-voyage; and that it in some way "suffered a sea-change" is scarcely doubtful. It was the last great telescope of its kind, metallic specula, having, in the seventies, been superseded by mirrors made of glass upon which a thin layer of silver has been chemically deposited. These have many advantages over their predecessors. They are considerably more reflective; they are more easily constructed; their shape is less liable to injury; their brilliancy, although more evanescent, can be readily restored. They have the drawback, however, of being extremely sensitive to changes of temperature. A three-foot mirror of this description by Calver, was employed by Dr. Common at Ealing with surprising success, early in 1883, for the purpose of photographing the Orion nebula. It was mounted at the Lick Observatory, California, in 1896. Dr. Common has since himself constructed a similar instrument of five feet aperture, which is the most potent light-collector ever yet turned to the skies. It is curious to learn that the silver spread over its surface weighs less than one of the "fourpenny bits" some time ago withdrawn from circulation; the reflecting film is in fact only $\frac{1}{280,000}$ inch thick.

Reflectors are perfectly, and *naturally*, achromatic, rays of all colours being thrown back at the same angle, and consequently meeting at the same focus. This gives additional brilliancy to the images formed by them, compared with those given by object-glasses, the colour-correction of which has hitherto been so imperfect that much light has to be "thrown away" as worse than useless. New kinds and combinations of optical glass have, however, of late been invented, by which this grave defect may be cured. Reflecting telescopes, on the other hand, are less manageable, and suffer more from distortion through change of position. Their cheapness recommends them to amateurs; but they should, on principle, be reserved for special departments of work, such as nebular photography and the chemical delineation of stellar and nebular spectra.

The growth of refractors, like that of reflectors, has obtained from time to time the sanction of unexpected disclosures. Thus a superb fifteen-inch, turned out at Munich in 1847, for Harvard College, Cambridge, U.S., showed Hyperion to Bond, September 16, 1848, and on November 15, 1850, surprised him with a view of Saturn's dusky ring. This telescope was surpassed, after fifteen years, through the energy and genius of Alvan Clark, the famous self-taught American optician, originally a portrait-painter at Cambridge-port, Massachusetts. Before it had left the workshop, an eighteen-inch achromatic, now the leading instrument at the Dearborn Observatory, Evanston, Illinois, won maiden honours by disclosing to Alvan G. Clark, one of the maker's sons, January 31, 1862, the dim companion of Sirius, which, before being seen, had made itself *felt* by gravitational disturbances of its radiant primary. The Washington twenty-six-inch, by the same firm, was rendered illustrious by Professor Hall's discovery, in 1877, of a pair of Martian moons; the Lick thirty-six-inch, by bringing within the range of Professor Barnard's keen eyesight, September 9, 1892, Jupiter's tiny "fifth satellite." The diploma performance of the Yerkes forty-inch, mounted in 1896 at the Chicago University Observatory, is yet to come. Meanwhile, several very perfect refractors, up to thirty-two inches of aperture, have been built on this side of the Atlantic by Sir Howard Grubb of Dublin, and the MM. Henry of Paris; and a twenty-five-inch, finished so long ago as 1868, and at the cost of his life through the labours which it entailed, by Thomas Cooke of York, after having lain for upwards of a score of years choked by the fog and smoke of Gateshead, has recently begun a promising career at Cambridge, under the care of Mr. Frank Newall, son of the original owner.

And now we cannot but ask ourselves, has the *ne plus ultra* in telescopic magnitude been attained? There is no reason to suppose that it has, provided that due allowance be made for inexorable conditions. Climate is one of these. The largest instruments are those most readily crippled by atmospheric hindrances. The greater their powers, the fewer are the nights on which they are likely to be available. If they are to "shine in use," and not "rust unburnished," they must then be erected in exceptionally favourable localities, such as the summit of Mount Hamilton (the site of the Lick

Observatory), or the Harvard College southern station at Arequipa in Peru. In South Africa, too, but "up country"—not in the Cape peninsula—splendid facilities for astronomical observation are to be found.

From Professor Keeler's report it can readily be gathered, and he indeed explicitly states, that the Yerkes forty-inch marks the limit of useful size in equatorials. For the character of the star-images formed by it slightly change their character when it is directed to different parts of the sky; and this implies that its lenses become, as it moves, infinitesimally deformed through the effects of their own weight. No larger instrument, accordingly, can safely be permitted to swing in mid-air. The huge light-concentrating machines of the future will lie in wait for the objects to be observed, instead of pursuing them. They will either be supported horizontally, or mounted in the "Coudé" fashion invented by M. Loewy. In either case, the necessary movement will be performed vicariously by a plane mirror.

Thus, the optical and mechanical outlook is decidedly better than the atmospheric. The question, How to build giant telescopes? is more easily answered than the question, Where to place them when built? The ultimate barrier to seeing indefinitely far into space is the rigid circumstance that we live on an air-girt globe. The prospects of astronomy are deeply involved in the forecast of its hampering effects. The dependence of those prospects upon telescopic improvements became obvious when Herschel took the whole contents of the sphere "for his province." These are indefinitely numerous, indefinitely far-off, indefinitely faint. The task of their correlation undertaken by Herschel, and inherited from him by modern astronomers, can at no time be more than approximately fulfilled; but for each successive approximation more light is needed. Those who would investigate the universe can never get enough of that too scarce commodity.

Until Herschel conceived the novel idea of a comprehensive science of the stars, they had been chiefly regarded as convenient sky-marks, by which to track the wanderings of our nearer neighbours in space. When it was perceived that the sky-marks were not fixed, it became necessary to determine their movements; and this was very roughly done for fifty-seven stars by Tobias Mayer of Göttingen, in 1757; and more accurately for thirty-six by Maskelyne, a third of a century later. But if the stars were travelling, the sun could not be supposed to stand still; and the possibility of laying down his line of march through space, by extricating a common element from the confused network of mutually-crossing stellar paths, occurred to Mayer, and was actually realised by Herschel in 1783. His inquiry, with the scanty materials then at command, was a wonderful stroke of audacity, which very nearly hit the mark; yet few believed in his result until it was confirmed by Argelander in 1837.

The various attempts made, prior to 1782, to measure the parallaxes of some of the brighter stars were instigated by the wish to find a demonstrative argument in favour of the Copernican theory of our system. They had no reference to sidereal structure. Herschel, however, took up the subject simply for the purpose of fixing the scale of that vast edifice. Before sounding the skies, he sought to ascertain the length of his fathom-line. He never ascertained it. To the end of his life, he could only make plausible assumptions as to the distances of the stars. Their real parallaxes were insensible with his instrumental means. But he fortunately chose for his experiments Galileo's "double-star method." This consisted in determining the relative positions of two close stars, one of which, taken to be indefinitely remote, was designed to serve as a standard of reference for the perspective shiftings of the other. It was thus that Herschel's attention was directed to double stars. He found them to be astonishingly numerous—far more numerous than could have been anticipated by the doctrine of probabilities. In January, 1782, he presented to the Royal Society a catalogue of 269 star-pairs, and he had collected 434 more by December, 1784. From their abundance alone, the Rev. John Michell inferred their character of binary systems; and Herschel, after twenty years of observation, was able, in 1802, to announce the fact of their mutual revolutions. Thus was taken the second great step towards the unification of the Cosmos. Newton proved that terrestrial gravity dominates the solar system; Herschel showed that a law of attraction, presumably (and assuredly) identical in its mode of operation, extends through sidereal space.

One cannot reflect without amazement that the special life-task set himself by this struggling musician—originally a penniless deserter from the Hanoverian Guard—was nothing less than to search out the "construction of the heavens." He did not accomplish it, for that was impossible; but he never relinquished, and, in grappling with it, laid deep and sure the foundations of sidereal science. No one before him had thought of approaching the subject otherwise than by way of speculation; he alone had the boldness to attack it experimentally. Having invented for the purpose an ingenious method of "star-gauging," based upon the hypothesis that the stars are, on an average, scattered evenly through space, he concluded in 1784, from its application, that the Milky Way is the visual projection of a disc-shaped stellar aggregation, within which our sun is somewhat excentrically placed. The progress, however, of his telescopic studies convinced him that the continued action of a "clustering power" had long ago drawn the stars into many separate allotments, and annulled the original uniformity of their distribution. So the disc theory was given up, and the Milky Way came to be regarded as a collection of genuine clusters, arranged into an irregular ring encircling the solar system. This view, implicitly held by the elder Herschel from 1802, was explicitly stated by his son in 1847. The results that Herschel expected from star-gauging may, in the future, be derived from the more elaborate process of star-gauging by magnitudes, photographically executed; and the sky-charting work, rapidly progressing in all parts of the world, will at least supply ample materials for sounding the star-depths.

These are stored besides with the curious objects called

"nebulæ." They were little noticed until Herschel, on March 4, 1774, made

> "That marvellous round of milky light
> Below Orion,"

the subject of his earliest recorded observation. Except, indeed, as impediments to comet-hunting. Thus, Messier, one of the keenest sportsmen in that line who have ever scanned the sphere, tried to eliminate by enumerating them, and drew up in 1771 a list of 45 such misleading objects, enlarged in 1781 to 103. And Lacaille, during an expedition to the Cape in 1752-1755, picked up 42 more. So far this department of knowledge had been cultivated when Herschel began to "sweep the heavens." To *sweep* them, be it remembered. Not merely to gaze at hap-hazard, or to look out for show specimens, but to gather in the celestial harvest methodically, zone by zone, so as to "leave no spot of the heavens unvisited." The fruits were proportioned to his diligence. The nebulæ discovered by him amounted, in 1802, to 2,500. And he did not merely discover; he investigated them as well. He separated them into classes, noted the mode of their distribution, and searched out their relationships. To begin with, he believed them to be of a purely stellar nature—to be, in fact, independent galaxies. Miss Burney was informed by him in 1786 that he had "discovered fifteen hundred universes." A few years later, however, he reasoned out for himself the gaseous nature of a great many nebulæ, such as that in Orion, and those of the "planetary" sort; and published in 1811 a complete theory, strikingly illustrated with examples taken from his telescopic experiences, of stellar development out of nebulous stuff. The supposition that they included the revelation of "exterior universes" was thus rendered, to say the least, superfluous; yet it was not perhaps, even by him, wholly abandoned. It was, moreover, revived in consequence of the performances of the great Rosse reflector, from 1845 onwards, in resolving apparent nebulæ into "bee-like swarms" of stars. Meanwhile Sir John Herschel's examination of those wonders of the southern heavens, the Magellanic Clouds, had virtually decided nebular standing. For they contain within a limited compass, as Dr. Whewell argued in 1853, "stars, clusters of stars, nebulæ, regular and irregular, and nebulous streaks and patches. These, then, are different kinds of things in themselves, not merely different to us." That stars and nebulæ co-exist in every part of the heavens, has since been fully established; while the laws respectively governing their distribution over the sphere are related in such a manner as to leave no doubt that these two classes of sidereal objects unite to form the grand galactic whole. Hence, to all reasonable apprehension, "island universes" have vanished into the inane.

Sir John Herschel accomplished the unparalleled feat of sweeping the heavens from pole to pole. Having, within eight years from 1825, revised his father's work at Slough, he conceived the noble idea of rounding it off in the southern hemisphere; and, in 1833-4, transported his instruments from Slough to Feldhausen near Cape Town. During the four years of his residence there, he not only executed his proposed survey, registering 1,790 nebulæ—300 of them for the first time—and discovering and measuring 2,100 double stars, but carried out a number of special researches. He catalogued the miscellaneous contents of the Magellanic Clouds—systems *sui generis*, as he justly termed them—made a detailed and laborious study of the Argo nebula, applied pretty extensively the paternal method of star-gauging, observed Halley's comet at its second predicted return, measured the sun's heat-emissions, carefully watched the spot-maximum of 1837, and finally, struck with a sudden rise in magnitude of η Argûs, brought to general knowledge that star's extraordinary character. These varied results were embodied in a monumental volume, published in 1847.

One of the greatest triumphs of modern science has been the establishment of an "Astronomy of the Invisible." It was primarily due to Bessel's inquiries into the disturbed proper motions of the "Dog-stars," Sirius and Procyon. They convinced him that each of these brilliant orbs is attended by a massive satellite, round which it revolves as it advances, its path in the sky being thus not straight but wavy. Telescopic verification of his forecast was, nevertheless, delayed until 1862 in the case of Sirius, until 1896 as regards Procyon. The earliest, and still the most memorable result in this line is the discovery of Neptune. Bessel knew that the thing was to be done, and in 1840 planned the doing of it. But his powers began, soon afterwards, to be crippled by deadly illness, to which he succumbed, March 17, 1846. *Uno avulso, non deficit alter.* Adams and Leverrier separately undertook the enterprise he had relinquished, and each with perfect success. It was a formidable one. The *direct* problem of perturbations taxes the highest mathematical resources; the *inverse* problem is not only more arduous, but was then untried. Laplace and Lagrange had shown how to determine the perturbations produced by a known disturbing body; it was left for Adams and Leverrier to find an unknown body through its disturbing effects. Irregularities in the movements of Uranus betrayed the presence of Neptune, and by the powerful analysis brought to bear upon them, were made to serve as an index to his actual place in the heavens at a given epoch. This was done by Adams in September, 1845; but his calculations, deposited at the Royal Observatory in the hope that they would incite to a telescopic search for the new planet, remained there buried in a drawer. Sir George Airy had no faith in them, and he unaccountably received no reply to a test-question addressed to their author. In the following June, however, he was roused by the intelligence of Leverrier's advance towards the goal already attained by Adams, to arrange an exploratory campaign with the Cambridge "Northumberland equatorial." But here again, disbelief—reinforced by the absence of a detailed star-map—stepped in to retard proceedings conducted by Professor Challis in so leisurely a fashion that the object "wanted" was found before he had sifted his observations, September 23, 1846, by Galle of Berlin, acting under Leverrier's precise directions. It proved on inquiry to have been twice observed at Cambridge during the previous couple of months.

Gravitational astronomy won its crowning distinction by the discovery of Neptune. It afforded the first instance of a body made known as an unseen power previously to being visually detected. Many stellar systems, however, have since then been ascertained to include members which can only be *felt*, owing to their partial, if not total obscurity. Again, the spectroscope tells of the existence of others entirely beyond the range of direct vision with the most powerful optical appliances; not because they do not shine (although this is sometimes also the case), but because they revolve so close to their primaries as to form with them single and indissoluble telescopic objects.

The spectroscope and the photographic camera have been mentioned as aids to astronomy. Their adoption has profoundly modified the science, widening its borders, inviting it to undertake novel tasks, endowing it with previously undreamt-of powers. Realms of knowledge deemed inaccessible to human faculties have, as if at the touch of a magician's wand, been thrown open; and of the many paths leading into the interior, only a few have yet been pursued, and that for a short distance. The prospects of exploration are hence unlimited, and of bewildering variety.

Spectrum analysis is essentially a chemical method. It depends upon the principle firmly established in 1859 by Kirchhoff and Bunsen, two professsors at the university of Heidelberg, that different kinds of glowing vapour give out distinctive rays of variously coloured light, commonly called "lines," simply because, for the purpose of getting rid of overlapping images, and for convenience of measurement, they are transmitted through a narrow slit. Thus, the presence of a familiar, and almost ubiquitous deep-yellow line, named by Fraunhofer "D," and shown by a moderately powerful apparatus to be double, *infallibly* testifies to the presence of sodium; iron, rendered gaseous by heat, gives out several thousand lines ranging from end to end of the spectrum, not one of which is common to any other substance; hydrogen shows a radiant sequence exclusively its own; and so of all the remaining elements. To apply this mode of detection, the light from the source to be studied must be analysed, or dispersed into its various component colours through the unequal action upon them of a prism, or train of prisms. Dispersion can also be effected by "diffraction"; and since the spectrum thus produced is "normal," or dependent wholly upon wave-length, it is always employed where a high degree of exactitude is aimed at. The coloured fringes of shadows originate in this way, through the interference of ethereal undulations; while the rainbow is a prismatic phenomenon, drops of water performing the refractive office of actual prisms.

The rainbow exemplifies too—although less perfectly than the electric light—what is called a "continuous spectrum." Its tints merge one into the other insensibly, without any sensible dark interruption. Now, incandescent liquids and solids of every kind and quality give rainbow-like spectra; they emit light which *rolls out* into an unbroken band of colour. Hence there is nothing characteristic about them. They are to the chemical enquirer absolutely uncommunicative. Vapours and gases alone can be induced to show the *badge* of their particular nature.

Celestial spectrum analysis began with the sun. The solar spectrum is furrowed transversely by a multitude of fine dark lines, known as "Fraunhofer lines," because Fraunhofer brought them within scientific cognisance by carefully mapping and measuring them. Their significance remained a standing puzzle until Kirchhoff, in 1859, furnished the key to it, by demonstrating the correlation of radiation and absorption. In other words, vapours and gases have the faculty of arresting those precise rays of light which they are in a condition to emit. Hence, the ignited, although relatively cool vaporous envelope of a white-hot body like the sun, or the carbons of the electric arc, acts predominantly as an intercepting medium, stopping more than it sends out of its peculiar rays. There results a continuous spectrum crossed by dark lines of the same chemical significance as if they were bright. They would, in fact, show as bright if the brilliant background, upon which they are seen projected, could be withdrawn. The interpretation, upon this principle, of the Fraunhofer lines, proved the sun to be surrounded by hydrogen in vast quantities, by incandescent sodium, magnesium, iron, calcium, and a number of other metals. Spectrum analysis in this way assumed a double aspect. The hieroglyphics of coloured light were rendered legible, whether positively or negatively written. And the spectra of the heavenly bodies are actually found to be inscribed, some in one way, some in the other; not unfrequently, in both combined.

The new and marvellous power of investigation thus acquired was in 1864 applied to the stars by Dr. Huggins and his coadjutor, Professor W. A. Miller. They ascertained the presence in the atmospheres of Aldebaran and Betelgeuse, of nine or ten terrestrial elements, thereby setting on foot the science of stellar chemistry. Moreover, on August 29, in the same year, Dr. Huggins made the signal discovery of gaseous nebulæ. Admitting the dim rays of a "planetary" in Draco through the slit of his spectroscope, he perceived it to be composed of three bright green lines, one of them Fraunhofer's "F"—an emanation of hydrogen. This one observation verified after seventy-three years Herschel's inference of the existence in the heavens of a "fiery haze," destined, according to his long forecast of creative processes, eventually to "subside into stars."

By the discovery of celestial spectrum analysis, a third stadium of progress towards the unification of the sciences was reached. The first step was taken with the demonstration that the force retaining the planets in their orbits is no other than that which causes rivers to flow, and apples to fall upon the earth. The extension of the same law to the stellar universe through the discovery of binary stars, showing that matter, wherever existing, possesses at least one unchanging quality, constituted the second. It was now learned that the sun and stars were composed of the identical *species* of matter scattered in the dust of the earth, dug up from its bowels, condensed to make its oceans, entering into the very framework of our own bodies. An universal chemistry was established, based upon the relations of light to material molecules, and of material molecules to the ether filling space; and, as an inevitable consequence, the new branch of knowledge, termed "astro-physics," made its ardently welcomed advent. By it astronomy has entered into close alliance with the rest of the sciences. No laboratory experiment is any longer indifferent to her; and laboratory experiments, on the other hand, derive from the connexion vastly augmented importance. The youth of learning seems renewed. Secrets of nature, formerly believed to lie beyond the scope of investigation, have been penetrated; *nil desperandum* is the motto which astrophysicists have earned the title to adopt as their own.

The old art of direct observation has, during the latter half of the present century, developed in sundry novel directions. By the use of auxiliary appliances, the telescope has gained a wonderful increase of subtlety and power. Modern astronomical work may be divided into four classes:—telescopic, spectroscopic, photographic, and spectrographic or spectrophotographic. Daguerre's invention was almost immediately tried with the sun and moon; J. W. Draper and the two Bonds in America, Foucault and Fizeau in France, and Warren de la Rue in this country, being among the pioneers of celestial photography. But it was not until after the introduction of the collodion process that really useful results were obtained. With the regular employment at Kew, from 1858 onwards, of De la Rue's "photoheliograph," began the daily self-registration of sun-spots, suggested by Sir John Herschel in 1847; and pictures of the eclipsed sun, obtained with the same instrument at Rivabellosa in Spain, July 18, 1860, terminated a prolonged dispute as to the nature of the red prominences by exhibiting them as undeniably solar appendages. Lunar photography was meanwhile successfully prosecuted, and Henry Draper's picture, of September 3,

1863, remained unsurpassed for a quarter of a century. Star-prints were first secured at Harvard College, under the direction of W. C. Bond in 1850; and his son, G. P. Bond, made, in 1857, a most promising start with double-star measurements on sensitive plates, his subject being the well-known pair in the Tail of the Great Bear. The competence of the new method to meet the stringent requirements of exact astronomy was still more decisively shown in 1866 by Dr. Gould's determination from his plates of nearly fifty stars in the Pleiades. Their comparison with Bessel's places for the same objects proved that the lapse of a score of years had made no sensible difference in the configuration of that immemorial cluster; and Professor Jacoby's recent measures of Rutherfurd's photographs, taken in 1872 and 1874, enforced the same conclusion. To the "collodion period" also belongs the earliest spectrograph, taken by Dr. Huggins in 1863; but the analysed light of Sirius left an uncharacteristic, although a strong impression. No lines were visible in it; a "virgin page" was presented. Before prosecuting the subject, fresh developments had to be awaited.

The invention of gelatine dry plates was the decisive event in the history of celestial photography. Dr. Huggins turned it to account with marked success for depicting the spectrum of Vega, December 21, 1876, and was able, three years later, to exhibit to the Royal Society photographs of the spectra of six white, or Sirian stars, stamped with the ultra-violet series of hydrogen-lines, then for the first time recognised, whether on the earth, or in the sky. The uses of the camera have since then multiplied at a prodigious rate. Its versatility appears unbounded. There are very few departments of astronomy left in which the eye has the advantage over it. A volume might be written on its successes; its comparative failures would scarcely fill a page. Its extraordinary power of penetrating space would have amazed and delighted William Herschel. This is due to the indefinitely prolonged exposures rendered practicable by the employment of dry plates; and these exposures can be interrupted and resumed at pleasure. Three-night photographs are now quite commonly taken, following the example given by Dr. Roberts in 1889. Now every additional minute of exposure brings intelligence from further and further sky-depths, owing to the happy faculty of sensitive plates for accumulating impressions. The eye sees at once, or not at all; the chemical retina sees by degrees, storing up insensible effects until they become sensible, and this without definable limit. This is its most essential prerogative. For the portrayal of nebulæ and comets, it is inestimable; and by its means the boundaries of the sidereal system may be laid down before the twentieth century is far on its way. A picture of the great comet of 1882, standing out from a richly spangled background, taken at the Cape Observatory under Dr. Gill's direction, was the object-lesson by which the advantages of photographic star-charting were effectually learnt. They have been practically illustrated in the *Cape Durchmusterung*, a southern continuation, by photographic means, of Argelander's corresponding telescopic work at Bonn; and are being turned to account on a magnified scale, in the International Survey of the heavens, now in progress at seventeen observatories scattered over the face of the globe. Special problems have, meanwhile, been investigated with striking success, by the chemical method, and its fresh applications are innumerable. Hitherto, performance has usually outrun promise; but promise has now so quickened its pace as to make the issue of the race dubious. We can only be sure that the future will be full of surprises.

ASTRONOMY

SECTION II.—GEOMETRICAL ASTRONOMY AND ASTRONOMICAL INSTRUMENTS

By A. FOWLER, A.R.C.S., F.R.A.S.

THE LICK REFRACTOR OF THIRTY-SIX INCHES APERTURE.

SECTION II.—GEOMETRICAL ASTRONOMY AND ASTRONOMICAL INSTRUMENTS

BY A. FOWLER, A.R.C.S., F.R.A.S.

CHAPTER I.

THE EARTH AND ITS ROTATION.

IT is a common remark that we are creatures of circumstances, and in no sense is this truer than in its application to the conditions under which we view the heavenly bodies. At the commencement of a study of astronomy it is accordingly important to first ascertain as far as possible the nature of the earth on which we are situated, and to determine in what way our observations are affected by our local conditions.

THE HORIZON.—When we look at the sky we see a vast hemispherical vault of which we seem to occupy the centre. If we are at sea, the water and sky appear to meet at a certain distance, in whatever direction we look. Where these meet we have what is called the visible horizon. On land, the horizon is usually broken up by terrestrial objects, such as hills, buildings, or vegetation, but otherwise the appearances are the same as at sea.

SHAPE OF THE EARTH.—When we observe the horizon, whether from land or sea, our eyes are at a certain elevation above the level of the ground or water, as the case may be, and the higher we are situated, the greater is the distance of the visible horizon, although the circular outline is retained. No matter where we may be, the same appearances are noted, and we are thus led to infer that the earth is a globe, as no other shape could appear circular from all points of view.

There are other considerations which lead to the same conclusion with regard to the shape of the earth. One of the most familiar proofs that the earth cannot be flat is found in the aspects of a ship putting out to sea or coming into port, when observed from a somewhat elevated position on shore. A ship does not become visible in its entirety, as it would if diminishing distance were the only cause affecting its visibility; the masts are seen first, and then the lower parts of the vessel gradually make their appearance. This finds a simple explanation in the curvature of the surface of the sea, and as similar appearances can be seen in all parts of the world, a globular form is indicated.

The fact that one may continue to travel westward and yet return to the point of starting, is quite in harmony with the supposition that the earth is globular, but it does not furnish a proof. This facility would evidently be equally afforded by a cylindrical earth, or even by a flat earth of which the Pole occupied the centre.

Still another indication of the rotundity of the earth is given by the phenomena of an eclipse of the moon. On these occasions, as will appear later, the moon passes through the shadow of the earth, and as this shadow is always circular, nothing but a spherical, or nearly spherical, body can be in question.

SIZE OF THE EARTH ROUGHLY MEASURED.—Granting then that the earth is spherical, a measurement of its curvature will enable us to determine its size. To do this it is necessary to measure the distance of the visible horizon from the eye at a known elevation. Then it can be shown that if the height of the eye is only a small fraction of the diameter of the earth, the diameter is as many times larger than the distance of the horizon as that distance is greater than the height of the eye. Thus, to an observer whose eye is 5 feet above sea level, the horizon is $2\frac{3}{4}$ miles distant, while from the top of a lighthouse 66 feet high the sky would appear to meet the sea at a distance of 10 miles. One way in which an approximate measurement may be made is illustrated in Fig. 1. Three posts are placed in line, with their tops at the same height above the surface of some calm stretch of water such as is afforded by a canal. A telescope fixed to the first post, so that its centre is at the top, is directed to the upper end of the third post, and it is seen to sight the middle one at some distance from the top. When the posts are a mile apart, the line joining the two extremes turns out to be 8 inches below the top of the middle one.

In our diagram this 8 inches is represented by the distance *b d*, and if we imagine an arc of a circle *d e* concentric with

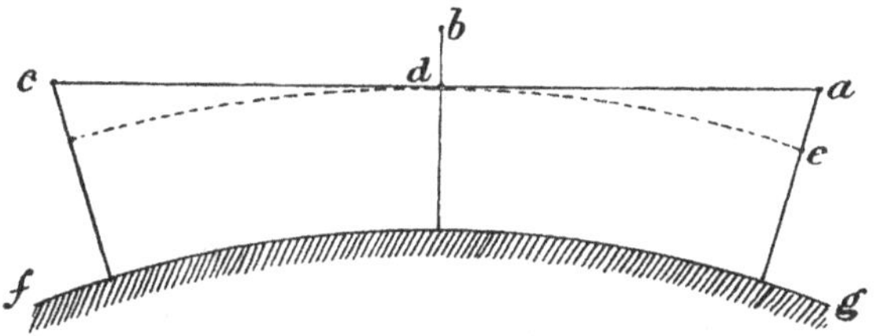

FIG. 1.—*Rough Measurement of Earth's Diameter.*

the surface of the water, the part which it intersects on the end post, namely *a e*, will also be 8 inches. This means that to an eye at *a*, 8 inches above the surface represented by *d e*, the visible horizon at *d* would be a mile distant. Applying the proportion named above, it results that the earth is 7,920 miles in diameter.

Owing to various causes, this method only furnishes a rough indication of the dimensions of our globe; but, if we had no other evidence, the result would suffice to explain that the irregularities of the earth's surface, though seeming so large to us who dwell upon it, are not inconsistent with the idea that the surface forms part of a sphere. The highest mountains with which we are acquainted do not exceed $5\frac{1}{2}$ miles in height, and this is only $\frac{1}{1400}$th part of the earth's diameter. On a globe 14 inches in diameter, representing the earth, the highest mountains would be less than a hundredth of an inch on the same scale; so that, taking the earth generally, it is practically a smooth globe.

DIFFERENT HORIZONS AT DIFFERENT PLACES.—So far then we have learned that the earth is a globe about 8,000 miles in diameter. This enables us to understand that persons in different parts of the earth will see the sky in different ways. At any given place we can see only what is above our horizon, and it results from the spherical form of the earth that no two observers have precisely the same horizon. If we consider a section of the earth, such as is shown in Fig. 2, an observer at the point *a* will have a horizon represented in section by the line *b c*, while the horizon of an observer at *d* will be represented by *e f*. It is clear then that an external distant object, such as the sun or a star, which may appear on the horizon in the direction *a b*, as seen from the point *a*, will be at a considerable angle above the horizon when seen from the point *d*.

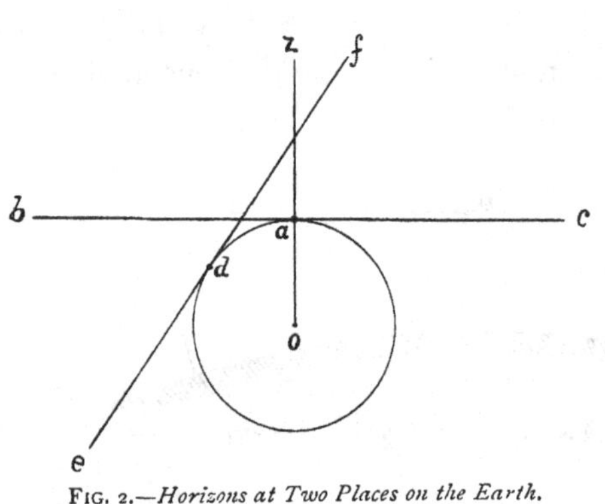

FIG. 2.—*Horizons at Two Places on the Earth.*

SENSIBLE AND RATIONAL HORIZON.—Having this conception of the horizon as a thing terrestrial, we may consider its astronomical relationships a little further. If we imagine the plane of the horizon prolonged until it cuts the distant sphere on which the stars and other celestial bodies seem to lie, it will meet that sphere in what is called the *sensible horizon*. A parallel plane passing through the centre of the earth is called the *rational horizon*, but as the starry sphere is at an almost infinite distance, the rational and sensible horizons coalesce into one celestial horizon.

Closely associated with the horizon is the point vertically

overhead which is called the *zenith*, and the point vertically below which is called the *nadir*. As the plane of the horizon is tangential to the earth's surface at the point of observation, the zenith is simply the prolongation into space of the line joining the centre of the earth with the place of observation; at the point *a* in Fig. 2, for example, the zenith is in the direction *o a z*.

The zenith as thus defined, however, is not the astronomical zenith, but what is called the geocentric zenith. As will appear later, the earth is not truly spherical, so that the direction of gravity does not pass exactly through the earth's centre, and the astronomical zenith is overhead in the direction of gravity.

DIURNAL MOTION OF THE HEAVENS.—In the day-time, when the sky is clear, we see the sun; at night, we sometimes see the moon, always some stars, and occasionally a comet. If we continue our observations, even for a few hours, we begin to recognise that the heavenly bodies have an apparent movement towards the west, very similar to the daily motion of the sun, with which everyone must have been familiar from childhood.

Continuing such observations, it is found that the great majority of the stars do not appear to change their positions relatively to each other, although their apparent places in the sky are different at different times. These have consequently been called the "fixed stars," but in the light of our present knowledge, the name is not to be taken too strictly. On account of this seeming fixity, the stars have been divided from very remote times into *constellations*, or groups, which enable us to name and identify individual members of the starry host. Other bright objects having the appearance of stars, when they are viewed merely by the naked eye, may be seen to change their positions with regard to the stars in that part of the sky in which they appear. These are the *planets* —the "wandering stars" of the ancients, to whom five were known, namely, Mercury, Venus, Mars, Jupiter, and Saturn.

Comets also are seen to share in the general westward

movement of the heavenly bodies, but, in addition, they have another movement relative to the stars situated in the same part of the sky.

If we closely observe the stars in Europe, we shall find some of them rising due east, and setting due west; others, again, will be found to rise in the north-east, and to travel nearly overhead; still others will be seen to rise south of east, attain only a small elevation above the horizon, and pass from our view as far south of west as they rise south of east. One point in the heavens appears stationary, and all the stars seem to traverse their daily courses round this as a centre. This stationary point is the north *celestial pole.* It is marked by no star, but a fairly conspicuous star is at present only about a degree and a half removed from it. The name given to this star is the Pole Star, or Polaris. As seen from London, stars within $51\frac{1}{2}°$ from the celestial pole never set, and such stars are said to be *circumpolar.*

When our place of observation is changed from one of middle latitude to one very near the Equator, these appearances are modified. We still see the stars rising and setting daily, but there will be *two* points which do not seem to move, one on the northern and the other on the southern horizon. One of these stationary points is identical with that seen from higher latitudes, and the other, which is called the *south celestial pole*, is diametrically opposite to it. What is more, stars which were not visible at all at our first place of observation will be seen in the south. All the stars will rise and set, and will alike be above the horizon for twelve hours.

If we could see the stars from the North Pole, the Pole Star, which is on the horizon of places at the Equator, would be found overhead, and all the stars visible to us would be ever above the horizon. Not only this, not one of the glittering stars which adorns the southern heavens would ever be seen at all.

In place of the rising and setting of stars, which lends such a great interest to their observation in other parts of the world, as seen from the poles the stars will simply travel round and round in circles parallel to the horizon.

To produce the apparent daily revolution of the heavens, and the changes in the appearances observed at different places, one of two causes must be at work; either the celestial bodies themselves must be performing a daily majestic movement from east to west round a motionless earth, or the earth itself must be whirling round from west to east, and so changing the situation of the observer's horizon with regard to external bodies. In the early days of astronomical observations this observed revolution of the heavens was thought to be real, but, with our present knowledge, we are no longer justified in regarding the earth as occupying a place of any such importance as that of the centre of the universe. By the earth's rotation, an observer, unless situated exactly at the North or South Pole, is carried round in a circle, and his horizon is gradually swept round so that on one side stars are setting and on the other side rising. The appearances at different places find a simple and sufficient explanation in the varying inclination of the observer's horizon to the earth's axis of rotation as the place of observation is changed.

A very simple experiment will assist one to comprehend the varying position of the horizon in different latitudes, and its effect upon the apparent diurnal movement of the heavens. Through the middle of an orange pass a knitting-needle, so that the two together may be taken to represent the earth and its axis. A circular piece of thin card pushed on to the needle at one end will represent the polar horizon, and, if the orange be rotated, it will be at once realised that such movement produces no change in the plane of this horizon, although different points on the visible horizon will be successively brought in line with different groups of stars or other external bodies.

Another piece of card should next be fixed on the orange by means of a pin at a point corresponding to the Equator. Again spinning the model earth on its axis, this horizon will be seen to constantly change its plane with regard to outside objects, and in a manner which perfectly accounts for the apparent movement of the heavens as observed from a point on the Equator.

A third piece of card touching the surface of the orange at an intermediate place will have an oblique movement, and as referred to this plane, the stars appear to traverse their daily rounds in oblique circles.

EXPERIMENTAL PROOFS OF ROTATION.—Not only does a supposed rotation of the earth accord perfectly with all that we can glean from observations of the heavens, but actual demonstrations of the reality of this movement are forthcoming. Sir Isaac Newton suggested one experimental method of setting the matter at rest. The further a thing is removed from the centre of the earth, the greater is the circle which it describes in a day, and the greater, consequently, the speed with which it must travel. Thus the top of a high tower moves more quickly than its base, and the surface of a mine than the bottom of the shaft. A stone let fall from the top of a tower thus starts with a greater forward velocity than that of objects at the base, and when it reaches the earth's surface, it will be a little east of the point where a plumb-line let down from its starting-point reaches the surface. This experiment has been tried, but there arc so many disturbing causes affecting the movement of the falling stone that the results are not very satisfactory, although generally confirming the earth's rotation from west to east. Evidently this method would fail at the Pole, and would be most effective at the Equator.

A much more beautiful and perfect proof is furnished by the celebrated Foucault's pendulum experiment. Again fancying ourselves at the North Pole, let us imagine a long and heavy pendulum, suspended in such a manner that the plane in which it swings is not affected by the earth's rotation. The trace of such a pendulum on a bed of sand placed beneath it would remain in a constant position if the earth were at rest. As the earth rotates, the bed of sand is twisted round, and the path of the pendulum apparently changes. The experiment was first actually carried out by Foucault in

1851, at the Pantheon in Paris, and it created a widespread interest. Since then, pendulums have been erected in various parts of the world, and all agree in essential results. The experiment can be seen in actual operation in the science section of the South Kensington Museum. The pendulum bob is a very heavy one, and before commencing the experiment, it is held out of the vertical by a loose band, which is fixed to the wall by a piece of string. On burning the string, the band falls off, and the pendulum starts its swing with little or no movement out of a plane. The pendulum bob is suspended by a long piano wire which is attached to a bracket carrying a conical pivot. The pivot rests on an agate plate at the end of a beam, and the weight of the bracket is compensated by an adjustable weight (Fig. 3). When swinging, the pendulum has a constant tendency to remain in one plane, and the turning of the beam beneath the pivot has no effect on the absolute direction of the plane of swing. Beneath the pendulum is a table divided into degrees, and the hourly apparent movement of the plane of swing at Kensington is observed to be nearly 12°.

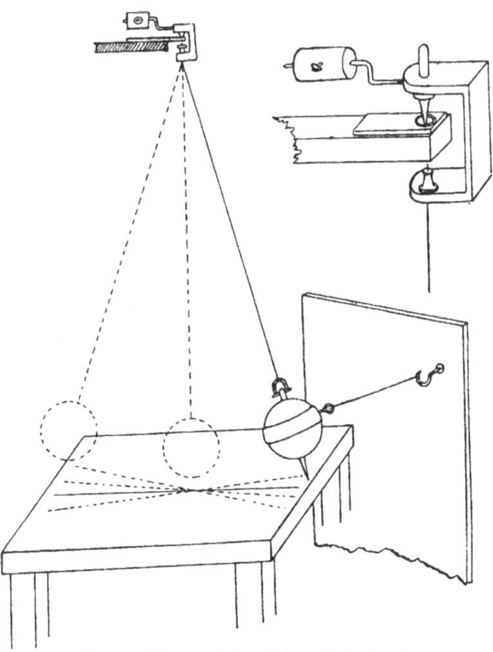

FIG. 3.—*Foucault's Pendulum Experiment.*

If the experiment could be performed at the North Pole, the

pendulum plane would apparently rotate from east to west, making a complete rotation once a day. At the South Pole the direction of movement would be reversed, but the rate would be the same as at the North Pole. The experiment, however, fails altogether at the Equator, while at places between the Poles and Equator the rate of movement varies with the latitude.

A more compact piece of apparatus for demonstrating the earth's rotation is the gyroscope, which we also owe to Foucault's ingenuity. The principle is exactly the same as in the case of the pendulum. A heavy disc is set in very rapid rotation, and is suspended in such a way that its points of support may be turned round without disturbing its plane of rotation. The results obtained with this instrument substantiate those derived from pendulums.

These experimental proofs of the rotation of the earth further teach us the same fact that we learn from observations of the stars, namely, that the earth makes a complete turn on its axis once a day.

LATITUDE AND LONGITUDE.—Having thus arrived at the conclusion that the earth is a globe turning on an axis once in twenty-four hours, the *North and South Poles* may be defined as the points where the axis of rotation meets the surface, while the *Equator* is the circle passing through places midway between the Poles. Imaginary circles passing round the earth through the Poles are called *meridians*, while circles parallel to the Equator are called *parallels*. These conceptions enable us to define very precisely the situation of any particular place upon the terrestrial sphere. We measure its angular distance from the Equator, as seen from the centre of the earth, and call this its *latitude;* London, for instance, is $51\frac{1}{2}°$ north of the Equator, and this is abbreviated to lat. $51\frac{1}{2}°$ N. All places on the same parallel have the same latitude, so that another measurement is required to designate the exact location of any one place. For this purpose the meridian passing through some place is agreed upon as a start-point, and we can then say that the place in question is so many degrees east or west; such a measurement represents the *longitude* of the place. At present there is no universal agreement as to the initial meridian, but in all British maps the meridian passing through the centre of the transit instrument at the Royal Observatory, Greenwich, is taken as the start-point. Longitudes are reckoned up to 180° E. and 180° W. New York, for example, is in long. 73° 58′ W., and Berlin in long. 13° 24′ E.

THE CARDINAL POINTS.—For general convenience in expressing the situation of an object, it is usual to say that it is towards the north, south, south-west, etc., as the case may be. A north or south line at any place, or a *meridian line*, as it is called, is in the direction of the terrestrial meridian passing through the place. The north point of the horizon is thus the point in which the meridian line meets the horizon towards the North Pole. The opposite point is south; while the east and west points lie in the directions at right angles. There are various ways in which a meridian line may be drawn. One of the simplest is to erect a vertical rod and to observe when its shadow thrown by the sun is shortest; at that moment the shadow marks the direction of north and south. This method is not very exact, as it is so difficult to tell when the shadow is shortest. A more accurate result may be obtained by drawing a circle round the stick as centre, and noting the points on this circle reached by the end of the shadow before and after noon; the point midway between these, marks the position of the shadow when shortest. By taking the average result of observations made with more than one circle, a good approximation can be obtained.

For a somewhat rough determination of the direction of the cardinal points, a watch showing the correct time may be utilised. Directing the hour hand to the sun, the south point will lie midway between that and XII. In the case of a watch having a dial marked up to XXIV., and reading XII. at mid-day, the latter figure would always point to the south when the hand indicating the hour was directed towards the sun. This will be easily understood if it be remembered that the sun is in the south at intervals of (approximately) twenty-four hours.

DAY AND NIGHT.—The succession of days and nights by which our daily arrangements are regulated is at once explained by the fact that the earth is round, and turns on its axis once a day. At any particular instant of time the sun can only shine on that half of the earth which is turned towards it. At all places included in the illuminated part the sun will be above the horizon, and it will be day. One half of the earth will be turned away from the sun, and to all places in that part it will be night. Under the conditions represented in Fig. 4, to a person situated at the point P it will be midnight; he will, however, be carried by the earth's rotation along the circle P Q R; when he arrives at a point on *a b*, the sun will be rising to him, and his day will commence. On reaching the point R the sun will be on the spectator's meridian, and it will be noon. After another interval he will arrive at the boundary of light and shade, and his night will commence.

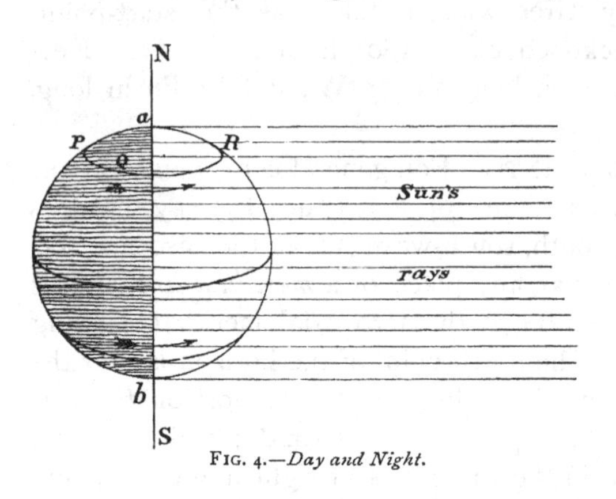

FIG. 4.—*Day and Night.*

ATMOSPHERIC REFRACTION.—In common with other substances through which light can pass, the atmosphere by which the earth is surrounded has the effect of bending rays of light out of their courses, and on account of this we do not see the heavenly bodies in their true positions. If the air were of uniform density the effect of this refraction would be as illustrated to the left in Fig. 5. The light from a star S will reach the observer at O after striking the atmospheric shell at *a* and being refracted along the line *a* O; consequently the observer will see it in the direction O S′, and not in the direction O S, which it would have if the air were absent. As a matter of fact, the atmosphere becomes less dense in passing upwards, so that the rays of light are subjected to a succession of small deviations ; two such refractions are illustrated at the right of Fig. 5. When a star is overhead there is no refraction, and the greatest displacements of a star's positions are produced on the horizon, where the light has to pass through a great thickness of atmosphere.

Refraction always makes the heavenly bodies appear higher in the sky than they otherwise would be, and some very curious effects can be traced to it. Thus the sun becomes visible on account of refraction some time before it has actually risen, and remains visible for a little while after it has really descended below the horizon. The amount of refraction varies with the temperature and pressure of the air, but the average amounts for different elevations above the horizon are as follows :

TABLE OF MEAN REFRACTIONS.

Altitude.	Refraction.	Altitude.	Refraction.	Altitude.	Refraction.
0°	34′ 54″	12°	4′ 23″	30°	1′ 38″
2°	18′ 9″	14°	3′ 45″	40°	1′ 8″
4°	11′ 39″	16°	3′ 17″	50°	0′ 48″
6°	8′ 23″	18°	2′ 54″	60°	0′ 33″
8°	6′ 29″	20°	2′ 35″	70°	0′ 21″
10°	5′ 15″	25°	2′ 2″	90°	0′ 0″

Refraction is responsible, among other things, for the curiously distorted appearances of the sun and moon, when they are very near the horizon.

TWILIGHT.—The atmosphere, or rather the solid and liquid particles which it always contains, has the property of reflecting light, and hence it does not suddenly become dark when the sun has set. Even until the sun has descended 18° below the horizon, the upper parts of the air continue to reflect his beams, and this is the origin of *twilight.* In the tropics the sun sets almost vertically, so that it gets below the twilight limit comparatively quickly, and this explains the short twilight which is remarked by all who have visited a tropical country. In our own country the sun has an apparent oblique

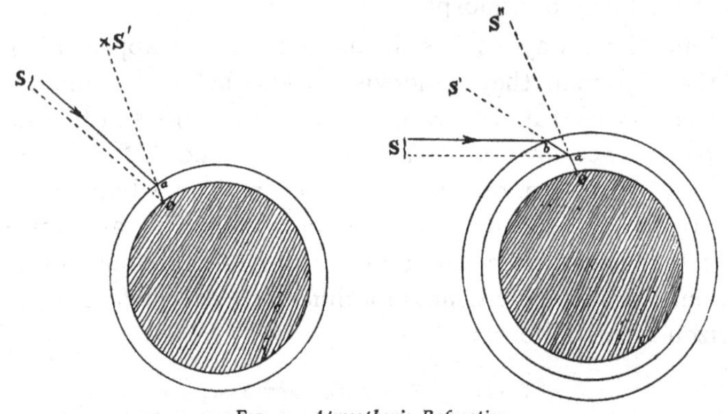

FIG. 5.—*Atmospheric Refraction.*

motion, and a relatively long period elapses before twilight ends. The increase in the duration of twilight is, indeed, very noticeable in merely travelling from London to the north of Scotland in summer-time.

Within the Arctic Circle, at places where the sun itself is never visible for months together, its reflected beams in the form of twilight may be seen for months.

CHAPTER II.

THE EARTH'S REVOLUTION ROUND THE SUN.

APPARENT MOVEMENTS OF THE SUN.—During any day on which we may observe the sun, it will be seen to rise at a certain place on the horizon, gradually ascend into the heavens to a certain point, then as steadily sink towards the west until it disappears at some point on the western horizon. If we watch the sun about the 20th of March, we shall find it to rise due east, and set due west; it will be above the horizon for exactly twelve hours, and below for the same length of time. When this happens, we have the *vernal* or *spring equinox*, as the nights are then equal in all parts of the world. From this time to the third week in June, we shall find the sun to rise more and more to the north of east, and to set gradually further north of west. This is accompanied by a daily increase in the apparent height of the sun at noon, and by increasing length of day and reduction of night. For some days before the 21st of June the change of the sun's place of rising and setting is very slow, and after this day the places of rising and setting begin to recede to the south. We then have the *summer solstice*, so-called because the sun seems to stand still, in so far as its northward travel is concerned. The point of rising or setting of the sun goes on moving nearer to the south point of the horizon, until about September 22, we again have the sun above the horizon for twelve hours, and below the horizon for an equal period; this is the *autumnal equinox*. The southward movement is continued until December 21, after which the rising begins to take place further towards the north. When furthest south, we have the *winter solstice* in the Northern Hemisphere, the sun being above the horizon for only a short time, and reaching only a small altitude at noon. From December 21 to March 20, the sun rises further to the north, at first very gradually, and afterwards more rapidly. These varying amounts of sunshine correspond to the short days of winter, and the long days of summer. A diagrammatic representation of the apparent path of the sun at the solstices and equinoxes for some place, such as London, is given in Fig. 6.

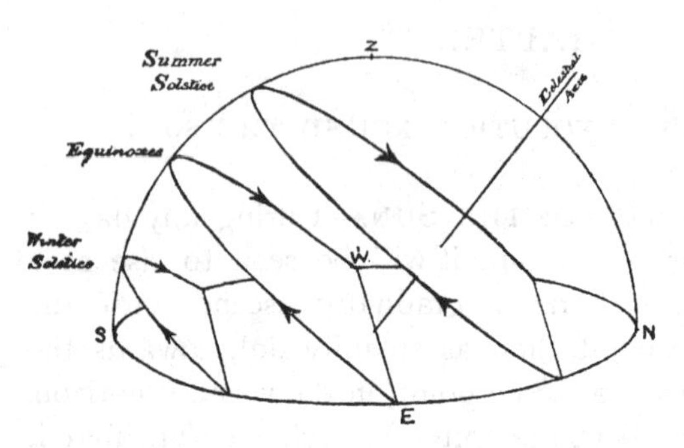

FIG. 6.—*Apparent Paths of Sun at Equinoxes and Solstices.*

It is clear, then, that our relations to the sun are very different from our relation to the stars, inasmuch as the apparent position of the sun, as projected upon the sky, is constantly changing, but returns to similar conditions at the end of a year. If our place of observation is changed, the apparent diurnal movement of the sun is affected in the same way as that of the stars.

To explain these annual changes of the sun, with regard to an observer's horizon, it is only necessary to suppose that the sun marches northwards towards the celestial pole from the winter to the summer of the Northern Hemisphere, and southwards from summer to winter. It is not to be imagined, however, that this apparent movement towards or from the north celestial pole is necessarily a real movement of the sun; we shall, in fact, very shortly see that it is only an apparent movement due to the changing situation of the earth with respect to the sun.

THE ECLIPTIC.—A very small amount of actual observation, without the aid of instruments, suffices to show that the changes in the sun's relation to any observer's horizon at different parts of the year are associated with a change in its situation among the stars. If we direct our gaze towards the south at midnight, we are looking towards that part of space which is directly opposite to the sun, as will be evident from Fig. 4, and if the sun's apparent movement were only in a polar direction, we should always see the same stars in the same part of the sky at the same hour. Such, however, is not the case. The stars are found more and more towards the west at the same hour as the year advances. Sirius, for instance, is due south about midnight on December 31 ; but at the end of January it will pass through the south point shortly before ten P.M. Similar changes are noted in the case of all the stars, and they indicate either an easterly movement of the sun among the stars, or a westerly motion of the stars with regard to the sun. If it were possible to see the stars in the immediate neighbourhood of the sun, this relative motion could be directly observed ; but under the actual circumstances, the apparent track of the sun amongst the stars must be determined indirectly. When we make observations at midnight, we know that the sun is opposite to stars which are due south at that moment ; and the height which it reaches above the horizon at noon indicates its angular distance from the celestial pole. It is thus possible to trace the sun's apparent path on a map of the stars, or upon a celestial globe; this is called the *ecliptic*, and it is found to be a great circle of the celestial sphere—that is, it is a circle contained in a plane which passes through the centre of the sphere.

The observed movement of the sun among the stars might be produced either by a revolution of the sun round the earth in a year, or by a revolution of the earth round the sun in the same period, the stars being supposed at rest at a greater distance than the sun. There are many phenomena which indicate that it is the earth which moves round the sun, but the most direct proof is found in what is known to astronomers as the aberration of light.

ABERRATION AS A PROOF OF THE EARTH'S REVOLUTION.—While engaged on an observation having for its object the determination of the distance of a star, Dr. Bradley made a discovery of very great interest and importance to astronomers. What he found practically amounts to this, that in order to see a star exactly at the centre of the field of view of a telescope we must direct the optical axis of the instrument at a small angle to the line joining the earth and star, irrespective of other deviations, such as that produced by refraction. The direction of this displacement is constantly changing throughout the year, but it is common to all the stars, and the fact that the original apparent position is regained at the end of a year at once associates aberration with a revolution of the earth round the sun.

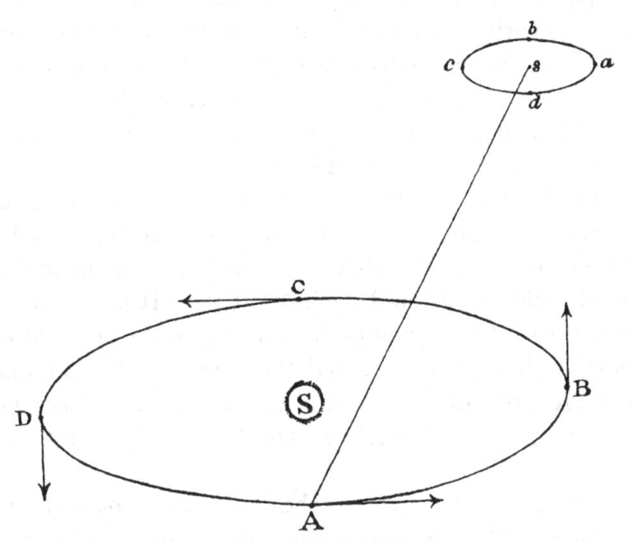

FIG. 7.—*Aberrational Orbit of a Star.*

In Fig. 7 we have a perspective view of the earth's orbit with the sun at S. A star *s* would appear in the direction A*s* when seen from the earth, supposed at rest at the point A; actually it is seen at *a*, ahead of its place, and in the course of a year it describes the *aberrational orbit*, *a b c d*, these points corresponding to positions A B C D of the earth in its annual path.

As a result of aberration, then, each star appears to revolve once a year in a small elliptic path about its average position.

The breadths of these ellipses vary according to their angular distances from the ecliptic, but all have precisely the same length of about 41″. Half the length of the ellipses, which amounts to 20″.5, is accordingly called the *constant* of *aberration*.

The fact that the earth's velocity in its orbit forms a sensible fraction of the velocity of light is the cause of aberration. If we let an object fall down the middle of a tube which is at rest, it will fall to the bottom without touching the side if the tube be held vertically. When the tube has a forward movement, however, it must be inclined at an angle in order that the falling body may pass clear to the bottom, and the greater the speed of the tube the more it must be inclined. So it is with light which comes from a star and traverses the tube of a telescope situated on a moving earth; the tube must be inclined to the actual path of the light rays.

Other proofs that it is the earth which moves round the sun are furnished by the parallaxes of the stars, and by spectroscopic measures of the earth's velocity.

APPROXIMATE SCALE OF EARTH'S ORBIT.—A very beautiful application of the constant of aberration is in the measurement of the distance of the earth from the sun. We have only to bear in mind that the apparent size of the sun does not change very much, in order to realise that the path of the earth must be very nearly a circle; if the distance changed very much there would be a correspondingly great change in the sun's apparent diameter. Now the constant of aberration is a measure of the relative velocity of the earth in its orbit and the velocity of light. There are several ways of determining the velocity of light, and it is known to be very nearly 186,300 miles per second. In a right-angled triangle having one angle equal to the constant of aberration, the side opposite to this angle would represent the velocity of the earth, if the longer side represented that of light. In such a triangle the proportion between these sides would be nearly as 1 to 10,000. That is, the velocity of light is about 10,000 times that of the earth in its orbit. The earth's velocity is thus found to be about 18½ miles per second, so that the distance which it traverses in a year is found by a simple multiplication. In this way the circumference of the earth's orbit is obtained, and it is easily deduced that the radius of the orbit, which is nothing more than the sun's distance, is not far from 93,000,000 miles.

THE ZODIAC.—The space about 8° above and below the ecliptic constitutes what is called the *zodiac.* The zodiac is of very great antiquity, and marks out the region traversed by the sun and all the planets known to the ancients. It is divided into twelve parts of 30° each, called signs of the zodiac, from the supposed outlines of animals, etc., marked out by the stars. The names of these signs are probably familiar to everyone from the well-known rhyme:

" *The Ram, the Bull, the Heavenly Twins,*
And next the Crab the Lion shines,
The Virgin, and the Scales,
The Scorpion, Archer, and the Goat,
The Man that bears the Watering-Pot,
And Fish with glittering tails."

The astronomical names and symbols corresponding to these are as follows:—

♈ Aries, ...	The Ram.
♉ Taurus, ...	The Bull.
♊ Gemini, ...	The Twins.
♋ Cancer, ...	The Crab.
♌ Leo, ...	The Lion.
♍ Virgo, ...	The Virgin.
♎ Libra, ...	The Balance.
♏ Scorpio, . . .	The Scorpion.
♐ Sagittarius, ...	The Archer.
♑ Capricornus, . . .	The Goat.
♒ Aquarius, ...	The Water-Bearer.
♓ Pisces, ...	The Fishes.

The sun enters the sign Aries at the vernal equinox in March, and the others in successive months. On account of the precession of the equinoxes (see p. 69), however, the sun no longer enters the *constellation* Aries at the vernal equinox, but it is still said to enter the *sign* Aries.

INCLINATION OF THE EARTH'S AXIS.—The revolution of the earth round the sun provides us with a very satisfactory explanation of the apparent easterly movement of the sun among the stars. There is, however, another very important point. We have seen that during a year the sun has a movement towards and from the Pole, as well as an easterly movement. The plane of the earth's orbit, therefore, cannot be coincident with the plane of the Equator; if it were, the sun would have the same apparent movement every day—it would always rise due east, and set due west, in all parts of the earth. The ecliptic, moreover, would be coincident with the celestial equator. When the ecliptic is determined by observations in the way already explained (p. 57), it is found to intersect the celestial equator in two points, and the plane containing it is inclined at an angle of very nearly $23\frac{1}{2}°$ to the equatorial plane. This inclination of the Equator to the ecliptic, or "obliquity of the ecliptic," indicates that the earth's axis of rotation is inclined to the plane in which the revolution round the sun is performed, the actual inclination being about $66\frac{1}{2}°$.

Further, the axis of rotation must remain parallel to itself during the revolution of the earth. Otherwise, the situation of the celestial pole would be seen to change, and the Pole Star would no longer serve to show us which way lies north.

It is precisely this inclination of the earth's axis which brings about the varying lengths of days and nights which we associate with different seasons.

THE SEASONS.—Let us in the first place contrast the conditions in summer with those which obtain in winter. Imagine that we can view the sun and earth from a very distant point lying in the plane of the ecliptic, and situated so that a line joining it with the sun is perpendicular to the line joining the sun and earth in summer or winter.

The sun will thus appear in some position represented by O in Fig. 8; in the summer of the Northern Hemisphere the earth will be in the position S, and in winter in the position W, since it travels half way round its orbit in six months' time. An observer situated at London will be $38\frac{1}{2}°$ from the North Pole, and he is represented by the point A in our diagram. The horizon at noon of such an observer is represented by the line H R, tangential to the surface of the sphere at the point A. At noon, then, the altitude of the sun is equal to the angle O A H. When it is winter in the Northern Hemisphere, the earth's axis is inclined away from the sun,

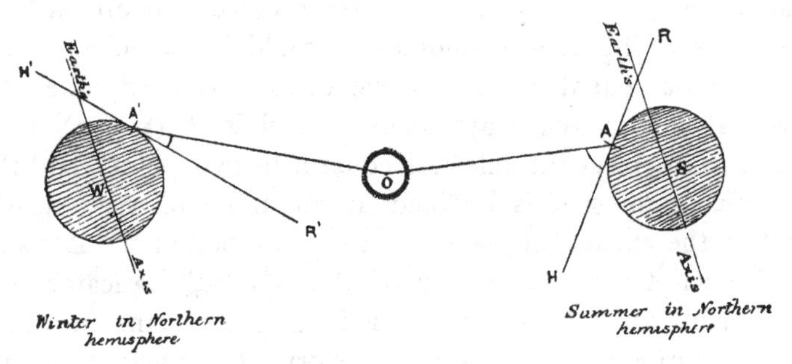

FIG. 8.—*The Sun's Altitude in Summer and Winter.*

and our observer at London is so situated that at noon his horizon is the line H′ R′, while the sun's altitude is the angle O A′ R′, which is no less than 47° smaller than in summer. People who dwell in the Southern Hemisphere enjoy the long days of summer at the time when our own days are shortest, and *vice-versa*, and the reason for this is clearly that when the position of the earth's axis presents the greatest part of the Northern Hemisphere towards the sun, the greater part of the southern half of our globe is turned away from the sun.

At the equinoxes, which occur very nearly midway between the solstices, the earth's axis is directed neither towards nor away from the great source of light and heat, so that both hemispheres are presented to the sun under exactly the same conditions. This state of affairs is shown diagrammatically in Fig. 9. The sun's altitude at noon at the commencement of spring is equal to that at the beginning of autumn, and depends only upon the observer's latitude. The half of our globe which is then flooded with the sun's rays comprises both the North and South Poles, and it is evident that as the earth turns round, every place upon it, whether in Arctic or equatorial regions, receives the benefit of twelve hours sunshine, and at the same time has a night of twelve hours duration.

THE MIDNIGHT SUN.—The facilities which are now offered

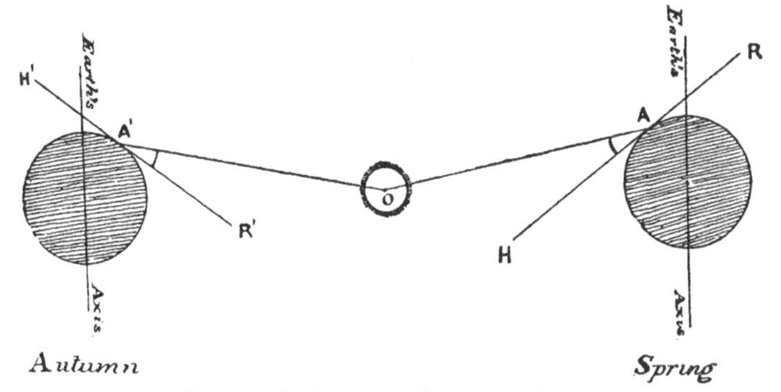

FIG. 9.—*The Sun's Altitude at the Equinoxes.*

for foreign travel have induced many people to pay a visit to the north of Norway, one of the objects in view frequently being to witness the so-called "midnight sun." It seems somewhat paradoxical to speak of night when the sun is above the horizon, but it simply means that in high latitudes the sun may be seen over the northern horizon when it is midnight at places further south which have the same longitude. We have seen that in our own country there are certain stars which never set, and when we get to the Pole itself, all the stars which are there visible will present this peculiarity.

In order to see the sun at midnight, then, what we have to do is to travel towards the Pole until we reach a latitude where the sun itself becomes circumpolar. At the Pole this would be the state of things during the whole of the northern summer, when the sun is north of the Equator, and since the sun never travels northward more than 23½°, it can only be circumpolar at places within that angular distance from the Pole, that is, within the Arctic Circle.

Let A in Fig. 10 be such a place, the sun being to the left. At noon the horizon of A is represented by H R, and the

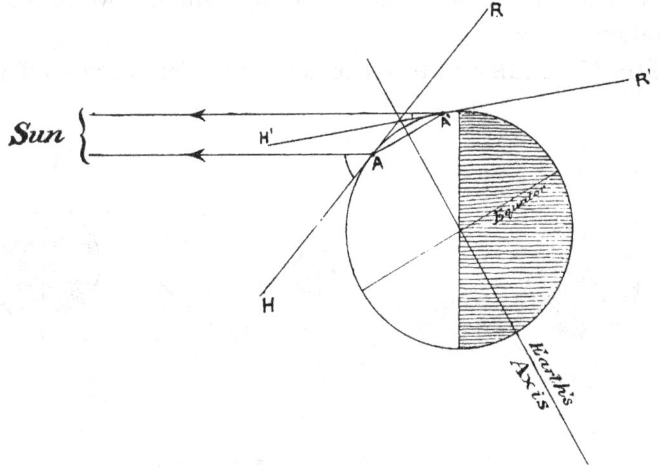

Fig. 10.—*The Midnight Sun.*

sun will appear in the south at a certain altitude, S A H. At midnight the earth's rotation will change the observer's position to A′ and his horizon to H′ R′, but it will not have taken him out of sunshine. The sun will then appear due north, but, except at the Pole, its altitude, S A′ H′, will be lower than at noon. At a place situated on the Arctic circle, latitude 66½°, the midnight sun would only be visible for one night at the summer solstice, were it not that refraction causes it to appear above the horizon when it is geometrically more than its own apparent diameter below.

At Tromsö the midnight sun is visible from May 19 to July 22, and at the North Cape from May 12 to July 29.

Nature, however, exacts compensation for this lavish share of summer sunshine in high latitudes, and there is a correspondingly number of dreary days in winter when the sun does not not rise at all.

CHAPTER III.

HOW THE POSITIONS OF THE HEAVENLY BODIES ARE DEFINED.

TWO MEASUREMENTS REQUISITE.—In order to make a more precise study of the movements of the heavenly bodies, it is essential that we should have some very definite means of specifying their positions upon the celestial sphere. To define the position of any object, at least two measurements are required. If, for example, one wishes to draw attention to a particular letter on the page of a book, it is only necessary to say that it is so many lines from the top, and a certain number of letters from the end of the particular line on which it lies. In the same way, latitude and longitude sufficiently indicate the situation of a place on the surface of the earth, and similar measures can be employed to indicate the places of the heavenly bodies.

ALTITUDE AND AZIMUTH.—The horizon and zenith at any place—being in a constant position with reference to the earth—may be utilised for indicating the positions of external bodies. We may say, for instance, that at noon on June 24, the sun, as seen from London, is 62° above the horizon, or 28° from the zenith. Technically, the former is called the *altitude* of the sun, being the angular distance above the horizon, while the latter measure is called the *zenith distance*.

We may next note that an object, besides having a certain altitude, is a certain number of degrees from the north, south, east, or west points, measured horizontally; if we reckon from the north point through E, S, and W, from 0° to 360°, such a horizontal measurement is called *azimuth;* if reckoned north or south of the east or west points it is called the *amplitude* of the body Fig. 11 illustrates these terms. In this diagram the observer is placed at O, N S and E W respectively representing a north and south, and an east and west line in the horizon; the point Z is the zenith, and S a heavenly body. A vertical circle drawn from Z through S will meet the horizon at a point A. The azimuth of S is thus the angle N O A, and its amplitude is the angle E O A, while the altitude of S is simply the angle A O S. Measurements of altitude and azimuth are made by means of an instrument called the altazimuth, an account of which will be found on page 202.

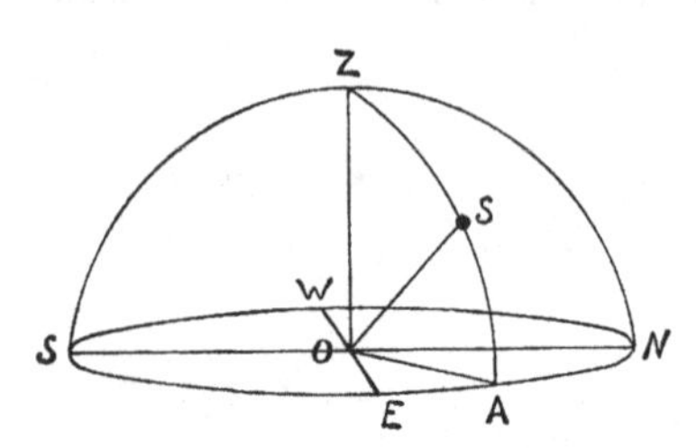

FIG. 11.—*Altitude and Azimuth.*

DECLINATION.—Altitude and azimuth only specify the position of a star for a particular place at a particular time. A better system is evidently one which is independent of the observer's situation on the earth. Of the two measurements required, one is readily decided upon; we can say that the sun, or star, or other heavenly body is a certain number of degrees from the north celestial pole; or, what is just as good, we can state the number of degrees north or south of the celestial equator, which lies midway between the poles. The former measurement gives what is called the *north polar distance* of the star, and the latter its *declination.*

RIGHT ASCENSION.—Just as the latitude of a place on the earth does not tell whether it is in Europe or North America, so declination alone fails to locate a heavenly body. We must have some measurement equivalent to terrestrial longitude, and it is therefore necessary in the first instance to select a start-point, which shall do for stars what Greenwich does for our geographical maps. By universal consent the fundamental point for the stars is a point situated on the celestial equator where it is crossed by that part of the ecliptic occupied by the sun at the vernal equinox. This zero mark is called the *First Point of Aries*, and is frequently denoted by the symbol ♈ identical with that employed for the corresponding sign of the zodiac.

The location of this reference point being thus determined, the *right ascension* of a celestial body may be defined as its angular distance from the First Point of Aries, as measured along the celestial equator. Like terrestrial longitude, it may be stated in degrees, but it is more usually expressed in hours, minutes, and seconds of time, for the reason that in general the measurement of a right ascension consists of an observation of the time at which the body in question comes to a certain position.

The right ascensions and declinations of stars are best determined when they are on the meridian of the place of observation, and such measurements are made by means of a transit instrument. When a star is on the meridian, its declination is estimated by the angle at which the instrument is inclined to the celestial equator when directed to the star. The fact that the earth is turning on its axis furnishes us with a simple method of finding the right ascensions of the heavenly bodies. Imagine a plane passing through the observer's position on the earth and through the earth's axis. This, prolonged indefinitely, cuts the celestial sphere in his meridian, and it is evident that on account of the earth's rotation it will turn completely round every twenty-four hours. It may therefore be regarded as the hour-hand of a clock, which is provided with figures ranging from I. to XXIV. When this gigantic clock hand sweeps past the First Point of Aries, all stars then seen in the plane—that is, all stars which are on the meridian—will have zero right ascension. After a complete rotation it will again sweep through the First Point of Aries.

USE OF STAR TIME.—Meanwhile, suppose we have a clock regulated so that it marks twenty-four hours between these two meridian passages of the First Point of Aries. Evidently, then, the time by this clock at which any object in the sky is seen on the meridian will depend upon its angular distance from the celestial meridian passing through the First Point of Aries. As the earth is rotating through 360° in twenty-four hours, reckoned by our clock, the meridian plane will travel at the rate of 15° per hour, so that, for example, a star 60° from the celestial meridian passing through the First Point of Aries, will appear to cross the observer's meridian at IV. hours by the clock. A clock so regulated to keep time with the stars is called a sidereal clock, and the sidereal time at which a celestial body crosses the meridian, or "souths," is the right ascension of that object. Such a time measurement can be converted into angular measure by allowing 15° per hour, 15′ per minute, and 15″ per second of time.

CELESTIAL LATITUDE AND LONGITUDE.—In some astronomical questions it is often convenient to adopt a different system of co-ordinates to indicate the situation of a celestial body. Just as the earth's equatorial plane serves as a basis for the measurement of declination, the earth's plane of revolution—that is, the plane of the ecliptic—is used as the term of reference for *celestial latitude,* which may be defined as the angular distance of an object above or below the plane of the ecliptic. *Celestial longitude* is the angular distance from the First Point of Aries measured along the ecliptic.

A diagram such as that in Fig. 12 may assist the comprehension of these co-ordinates. Here the observer is supposed to be situated at the point O, at the centre of the celestial sphere. To him the north and south celestial poles will appear in some such positions as N and S, and the celestial equator will be represented by a great circle at right angles to the line joining these two points. The apparent path of the sun—the ecliptic—will be indicated by another great circle, which is inclined to the Equator; and the poles of the ecliptic will be represented by P and P′.

The Equator crosses the ecliptic at the First Point of Aries, marked ♈. Considering now a star which the observer sees in the direction of the line O S, its position would be reckoned as follows in the two systems:—

Right Ascension = Angle ♈ O R }
Declination = " S O R }

Celestial Longitude = Angle ♈ O L }
" Latitude = " S O L }

Either pair of co-ordinates can, by a mathematical process, be expressed in terms of the other.

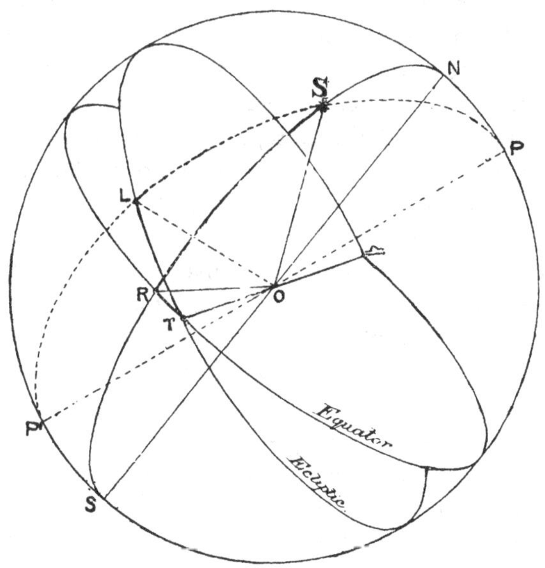

FIG. 12.—*Right Ascension, Declination, Celestial Latitude, and Celestial Longitude.*

PRECESSION OF THE EQUINOXES.—It is not too early to remark that the First Point of Aries is not absolutely a fixed point on the celestial equator. This is on account of the precession of the equinoxes, which consists of a backward movement of the First Point, due to a change in the position of the earth's equator. As a point common to the ecliptic and equator, it is conveniently retained as the starting-point of right ascensions and celestial longitudes, but in consequence of precession, these co-ordinates are subject to a constant change. The amount of precession for a point on the Equator is $50''\cdot2$ per annum, and this movement requires 25,800 years for a complete revolution.

GEOCENTRIC AND HELIOCENTRIC POSITIONS.—When observing objects at a very great distance, they will appear in the same direction to a spectator on the earth as they would if he could by some means be transferred so as to be able to see them from the sun. If, for instance, one sees the Peak of Teneriffe from a distant ship, its apparent direction will be very slightly affected by a change of a mile in the ship's position. But a similar change of place would produce a greater difference of direction when a nearer body was under observation. If an object is relatively near to the sun and earth, its direction, and, therefore, its apparent position on the celestial sphere, will be different, as seen from the earth and sun. Such will be the case with planets and other bodies which lie in our immediate neighbourhood, speaking astronomically. Hence, it is often convenient to distinguish between the *geocentric* position of a celestial body—referring it to the position it would occupy if it could be seen from the centre of the earth—and the *heliocentric* position, representing it as it would appear to an observer occupying the centre of the sun. We thus have geocentric and heliocentric latitudes and longitudes of the nearer heavenly bodies.

STAR CATALOGUES.—The problem of constructing catalogues showing the positions of the stars is one of considerable practical value, as well as one of great scientific importance. In the first instance, such catalogues were of necessity compiled from data acquired by naked eye observations, so that the ancient catalogues comprise only a small number of stars.

As far back as 295 B.C., the positions of stars were determined by Timocharis with sufficient accuracy to lead Hipparchus to his great discovery of the precession of the equinoxes about 170 years later. From observations at Rhodes, Hipparchus drew up a catalogue of 1,022 stars, giving their latitudes and longitudes; this is preserved for us in Ptolemy's "Almagest," where the positions are corrected for precession, and reduced to the epoch 150 A.D. The next catalogue of importance was due to the industry of Tycho Brahé (1546-1601), who gave the positions of 1,005 stars with greater accuracy than had been previously obtained; indeed, notwithstanding his want of optical aid, it has been estimated that the probable errors of his measures were not more than 24″ and 25″ in right ascension and declination respectively. The last of the naked eye catalogues is that of Hevelius, giving the positions of 1,553 stars.

Coming to more recent times, in which the employment of telescopes has vastly increased the power of accurate observation, there are the catalogues of Flamsteed, Halley, Lacaille, Lalande, Argelander, the British Association, and catalogues of the stars in particular parts of the sky which have been published by all the leading national observatories. Eighteen observatories are now taking part in the construction of an international star catalogue by means of photography, and this is intended to record with great accuracy the positions of nearly 3,000,000 stars. A modern star catalogue usually places the stars in the order of their right ascensions, and, in addition to the two co-ordinates, furnishes the necessary data for determining the exact situations of the stars at any particular time.

CHAPTER IV.

THE EARTH'S ORBIT.

EXACT SHAPE OF THE ORBIT.—It will be clear that if we made our annual journey in a circle we should always be at the same distance from the sun, and the apparent size of that luminary would never vary. This, however, is not the case. Exact measurements, which are best made by means of the transit instrument, indicate variations which, though not perceptible to the unassisted eye, establish a want of circularity. The observations bearing on this point consist of a measurement of the time required for the sun to cross the meridian—the larger its apparent diameter, the longer it will obviously be in passing the meridian. An observation of the sidereal time at which the centre of the sun passes the meridian determines the right ascension, and from this one can calculate the sun's longitude.

If such observations be made at intervals during a year, we can utilise them for determining the shape of the earth's orbit independently of a knowledge of the actual size. In Fig. 13 let us suppose the sun to be situated at the point S; from S we draw a line, S A, representing the line joining the earth and sun at the vernal equinox when the sun's longitude is zero. If our observations include a measure of the sun's diameter on that day, let S A be drawn on some convenient scale. To plot the observations for other days, we must

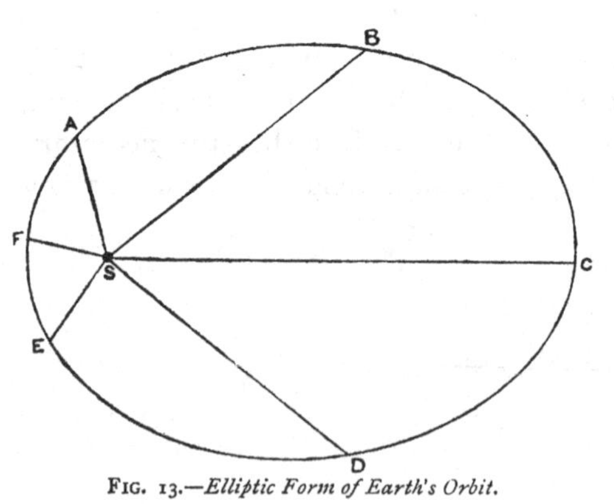

FIG. 13.—*Elliptic Form of Earth's Orbit.*

draw S F, S E, etc., at angles A S F, A S E, etc., equal to the sun's longitude, and make the lengths inversely proportional to the apparent diameters, on the same scale as S A. The other observations can be plotted in the same way, and the earth's orbit is then found to be an ellipse with the sun in one of its foci. Actually, the earth's orbit is much more nearly circular than is shown in Fig. 13, and in illustration of this the following numerical data may be given :—

1896.	Jan. 1	Greatest	apparent	diameter	of sun	= 32′ 35″·2	in long.	281°
	July 3	Least	,,	,,	,,	= 31′ 30″·6	,,	102°
	March 29	Mean	,,	,,	,,	= 32′ 4″	,,	9°
	Oct. 5	,,	,,	,,	,,	= 32′ 4″	,,	193°

It thus appears that in 1896 we were nearest to the sun on January 1, as on that day the sun's apparent diameter was greatest, while we were furthest removed on July 3.

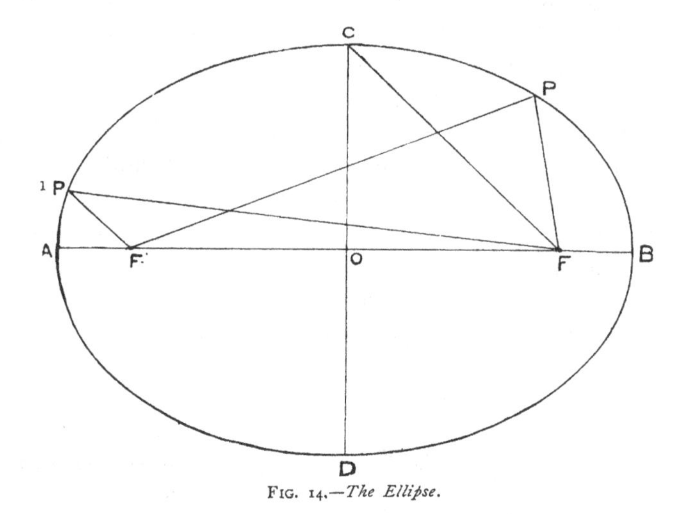

FIG. 14.—*The Ellipse.*

The ellipse is a curve of such importance in astronomy that an understanding of some of its properties is essential for further progress. This beautiful closed curve lies in one plane, and its figure is such that the sum of the distances of any point upon it from two fixed points within the curve is constant. These two fixed points, F F′ (Fig. 14), are called the foci of the ellipse, and we have, for example, the sum of the lengths P F and P F′, equal to the sum of P′ F and P′ F′. The line A B passing through the foci is the greatest distance across the ellipse, and is called the major axis; at right angles to this is the minor axis C D.

Following our definition of the ellipse, we see that as B is a point upon its circumference, B F + B F′ must be equal to the sum of the distances of any point P from the foci. But since B F is of the same length as A F′, the sum of the distances of the point B from the foci, and therefore of all other points, is equal to the major axis. Hence the average or mean distance of the focus F from all points on the ellipse is half the length of the major axis. It follows also that C F is equal to the semi-major axis O B.

At the point O, where the axes intercept each other, we have the centre of the ellipse, and the ratio between the distance from the centre to either of the foci and the semi-major axis, *i.e.*, $\frac{OF}{OB}$ is called the eccentricity of the ellipse. Thus, in an ellipse of eccentricity 0·5, the foci would lie mid-way between the centre of the ellipse and the extremities of the major axis. The eccentricity is always less than unity; if it become unity, the two foci merge together, and the curve becomes a circle.

To draw an ellipse, two pins may be stuck into a piece of paper at the points intended as foci. A loop of thread is then made and thrown over the pins. A pencil placed inside the loop, so as to stretch it, and traced completely round, will outline an ellipse. The size and shape of the ellipse may be varied by changing the length of the thread and the distance between the pins. Such, then, is

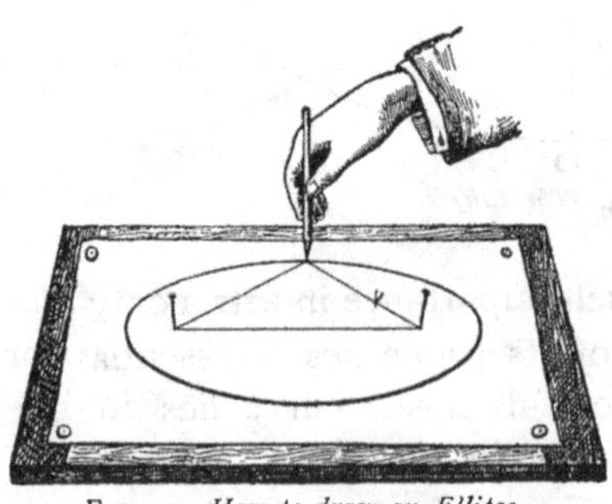

FIG. 15.—*How to draw an Ellipse.*

the curve in which our earth performs its annual journey round the sun, the sun being relatively fixed in one of the foci.

APHELION AND PERIHELION.—When the earth is in that part of its orbit where it makes its nearest approach to the sun, it is said to be in *perihelion;* when at the point furthest removed from the sun it is in *aphelion.* The line joining these two points is obviously the major axis of the earth's orbit, and when this is imagined to be prolonged indefinitely into space it is called the *line of apsides*, or *apse line.* When the earth is in perihelion, the sun's apparent diameter will be the greatest possible, and when in aphelion it will be at a minimum. A knowledge of these limiting values of the apparent solar diameter enables us to determine the eccentricity of the orbit of the earth. The sun's apparent diameter when the earth is in perihelion amounts to 32′ 35″·2, and to 31′ 30″·6, when the earth is in aphelion, from which it results that the value of e is 0·0167.

UNEQUAL SPEED OF THE EARTH.—The observations by which we are enabled to determine the true form of the earth's orbit are not quite exhausted of their usefulness; we can utilise them still further for studying the varying rate of the earth's motion. If the earth moved through equal angles every day, the apparent movement of the sun would always be uniform, and in that case the sun's daily increase of longitude would be constant.

The following figures, however, prove that this uniformity does not exist :—

1896.	Sun's daily motion in longitude.
Jan. 1	1° 1′ 8″·5
Mar. 29	1° 0′ 6″·7
July 3	0° 57′ 12″·1

Facts such as these led Kepler in 1609 to the discovery of his famous second law of planetary motion, namely, that the

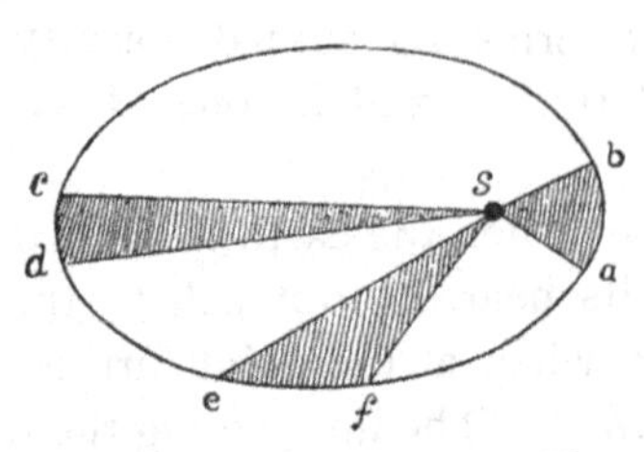

FIG. 16.—*Illustrating Kepler's Second Law.*

radius vector (the line joining the sun and earth in the case of the earth's orbit) describes equal areas in equal times. For the sake of clearness, imagine the earth's orbit to be represented by the elongated ellipse in Fig. 16, with the sun in the focus *S*. When the earth is near perihelion, it will move over a certain distance, *a b*, in a given time; some time afterwards it will be in another part of the orbit, and in the same interval as before it will traverse the distance *c d;* again, in another equal interval of time, it will move from the point *e* to the point *f*. The law tells that the areas *S a b*, *S c d*, and *S e f*, are equal so long as equal times are in question; in different parts of its path, then, the earth's rate of motion must vary, *c d*, for example, being smaller than *a b*. It will be seen that the motion is most rapid when the earth is in perihelion, and least rapid when in aphelion.

CHANGES IN THE EARTH'S ORBIT.—Owing to disturbances caused by the proximity of other bodies, the earth's orbit is not always of the same shape. The eccentricity is steadily diminishing, and in about 24,000 years the orbit will be very nearly a circle; it will afterwards become more elliptical again, until in another 40,000 years or so the eccentricity will be about 0·02. So far as our knowledge goes, the eccentricity will never exceed 0·07.

The direction of the major axis of the earth's orbit, that is, the line of apsides, moves forward at the rate of about 11″ per annum, so that at this speed a whole revolution will be made in a period of 108,000 years.

On account of precession, the equinox moves backwards along the orbit at the rate of 50″·2 per annum, so that the movement of the apse line with regard to the equinox is 61″ in a year; or, in other words, the perihelion point of the earth's orbit makes a complete revolution with respect to the equinoctial point in a little over 20,000 years. The earth at present passes through perihelion in our northern winter, but owing to this motion of the apse line it will in 10,000 years time be at aphelion in winter. Northern winters will then be somewhat colder than at present. The plane of the orbit itself is subject to changes, with the result that the obliquity of the ecliptic is variable in amount. In the course of ages the obliquity may oscillate between the limits 24° 35′ 58″ and 21° 58′ 36″. The mean value during 1896 was 23° 27′ 9″·9.

THE EARTH'S REAL PATH.—In this and preceding chapters, we have had occasion to consider various features of the earth's orbit, but it must now be pointed out that what we call the orbit of the earth is not quite the same thing as the earth's actual path in space. The earth, as we know, is accompanied by the moon, and these two bodies are bound together in such a way that it is really the centre of gravity of the earth and moon which describes an elliptic orbit round the sun; the moon is so small in relation to the earth that the centre of gravity of the two companions lies within the earth's surface, but, nevertheless, an oscillatory displacement of the earth's centre in space is produced by the moon's monthly circuit round the earth. We judge of the earth's movement by the apparent movement of the sun, and we actually find a monthly inequality in the sun's apparent motion. A very good illustration of this may be found in the varying celestial latitude of the sun. It will be clear that if the earth always moved in the plane of the ecliptic, the sun's latitude would always be zero. If, on the other hand, the earth has a motion round the common centre of gravity, it will be above the ecliptic when the moon is below, and *vice versa*; the sun will, therefore, not always appear to be in the ecliptic, and its latitude will depend upon that of the moon. The following figures from the "Nautical Almanac" will illustrate this point:

	Sun's apparent latitude.	Moon's latitude.
1896, April 1	0″·70 S.	5°9′ S.
" 10	0″·01 N.	1°41′ N.
" 16	0″·39 N.	5°6′ N.
" 22	0″·07 S.	0°48′ N.
" 29	0″·74 S.	5°1′ S.

The displacement in right ascension amounts to a little over 6″, and is, therefore, large enough to be directly measurable.

On account of this association with her satellite, the earth's centre moves some hundreds of miles above and below the plane of the ecliptic.

The so-called "perturbations," or disturbing effects of the other planets, also cause the earth to depart more or less from the plane of the ecliptic and from a geometrical elliptic path. Nevertheless, these disturbances can be calculated and allowed for, so that when we speak of the earth's orbit we really mean the path which the centre of gravity of the earth and moon would traverse if subject only to the influence of the sun.

CHAPTER V.

MEAN SOLAR TIME.

SUN-DIAL TIME.—The changing directions of shadows thrown by the sun have been utilised from very remote periods for the measurement of time, the instrument usually employed being a sun-dial. On account of the varying declination of the sun, it is necessary to employ as a time-measurer the shadow of a line which lies parallel to the earth's axis, that is, if we wish the same hour marks to be permanently useful. Such a rod must lie in the plane of the meridian, and be inclined to the horizon at an angle equal to the latitude of the place. If the shadow be received on a horizontal dial, hours may be marked upon it corresponding to the duration of the longest day at the place where it is set up. Sometimes, as on old churches, one sees a vertical sun-dial, the rod, or *style*, as it is called, being still parallel to the earth's axis, but as a dial facing the south is only serviceable for twelve hours, another on the north wall is necessary for times before six in the morning and after six in the evening. As indicated by the sun-dial, it will always be noon when the sun is on the meridian, that is, when it is due south.

The time indicated by sun-dials is distinguished astronomically as *apparent time*, and an *apparent solar day* is the time which elapses between two successive southings of the sun. It is longer than the sidereal day, for the reason that the sun moves eastward among the stars.

NECESSITY FOR MEAN TIME.—The varying speed of the earth in its orbit, or what comes to the same thing, the variable rate of the sun's apparent eastward movement, prepares us for the discovery that the intervals between successive noons as indicated by sun-dials are unequal. That is, the apparent solar day is not of uniform length, and our clocks could not be regulated to indicate noon at the same moments as the sun-dial unless they were rated afresh every day. All our daily actions are regulated by the sun, and our timekeepers must also be controlled by its movement if they are to be as convenient as is necessary for purposes of everyday life. Our clocks and watches are therefore regulated to measure twenty-four hours in the time corresponding to the average duration of the apparent solar day throughout a year. In other words, they are controlled by the movements of an imaginary sun, called the *mean sun*, which is supposed to come to the meridian after equal intervals, and in order that it may do this while having a uniform motion, it must of necessity move along the celestial equator. In this way the time shown by our clocks and watches never departs very greatly from that shown by sun-dials, the maximum discrepancy being little more than a quarter of an hour. A *mean solar day* is thus the average length of the apparent solar days throughout a year.

THE EQUATION OF TIME.—The difference between apparent and mean solar time is called the *equation of time*, and a knowledge of its amount enables us to determine mean time from an observation of apparent time.

One of the causes of this difference we have already seen to be the varying speed of the earth in its orbital movement; this produces a correspondingly irregular motion of the sun amongst the stars, and in consequence the true sun comes to the meridian after unequal intervals. Neglecting for a moment another cause of the varying length of the day, the relation of the apparent and mean solar days would be somewhat as follows:—Let us suppose that when the earth is at perihelion, we set our clocks to the same time as the sun-dial. In the interval which elapses before noon next day the true sun will have moved faster than the mean sun, because the earth, which produces the apparent eastward movement of the sun, is then travelling at its greatest speed. Consequently, our meridian will overtake the mean sun before it comes up to the true sun, and mean noon will occur before apparent noon; the difference will be the equation of time for the day, and it must evidently be added to apparent time in order to give mean time. This will go on for a certain period, when, in consequence of the reduced rate of the earth's orbital velocity, the sun's eastward motion will be less than that of the mean sun, and the two will again come to the meridian at the same time when the earth reaches its aphelion point; clocks and sun-dials would then give identical times. After aphelion passage, the earth is moving slowly, and the apparent eastward velocity of the true sun will be less than that of the mean; our meridian will therefore come to the true sun before it overtakes the mean sun, so that apparent noon will precede mean noon, and the equation of time will have to be subtracted from apparent time to give mean time. The two suns would again come together when the earth reached perihelion, and the equation of time, so far as this cause was concerned, would vanish. As the earth's orbit is only slightly elliptical, the equation of time due to this cause alone would never amount to more than seven minutes.

This, however, is by no means the whole cause of the equation of time; a still greater source of variation is the obliquity of the ecliptic. To investigate the part played by this inclination of the fundamental planes, let us now suppose that the true sun has a uniform angular motion in the ecliptic, while the mean sun moves uniformly along the Equator. Both these fictitious suns would have the same rate of movement along their respective paths, since they come back to the same places after the lapse of a year. If, then, these two suns start together at the equinox, both would indicate noon at that time, and there would be no equation of time. The "ecliptic sun" would then be moving at an angle of 23½° to the Equator, as along *a b* in Fig. 17. If the distance *a b* represents the average daily movement of the "ecliptic" sun, and *d c* the equal movement of the mean sun, it is clear that our meridian will overtake the true sun at *b* before the mean sun at *c*, so that apparent noon will precede mean noon, and the equation of time must be subtracted from apparent time to give mean time. The difference becomes greater up to a certain limit, and then since both suns will traverse 90° in the same time, they will pass the meridian together at the solstice.

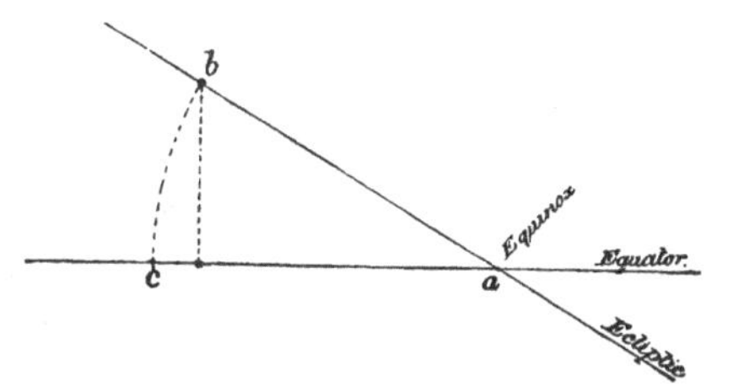

FIG. 17.—*Effect of Obliquity of Ecliptic upon the Equation of Time.*

In the next quarter of a revolution, from solstice to equinox the difference is similar, but in the opposite direction, and the same applies to successive quadrants described throughout the year.

The net amount of the equation of time at any moment is thus the added effects due to two causes.

In 1896 the greatest and least values of the equation of time at Greenwich mean noon were as follows:—

	M.	S.	
Feb. 11	14	27	to be added to apparent time.
April 14	0	7	" " "
May 13	3	50	to be substracted from apparent time.
June 13	0	6	" " "
July 25	6	17	to be added to apparent time.
August 31	0	0	" " "
Nov. 2	16	20	to be substracted from apparent time.
Dec. 24	0	7	to be added to apparent time.

A somewhat notable effect, owing its origin to the equation of time, is seen in the times of sunrise and sunset given in our almanacs. On November 8, for example, the sun rises at Greenwich at 6h. 58m., and sets at 4h. 31m., thus apparently making the afternoon about half an hour longer than the morning. As reckoned by the sun-dial, however, the morning and afternoon would differ only by a few seconds, and the peculiarity noted arises simply from the fact that our clocks keep time with the mean, and not with the true sun.

DETERMINATION OF TIME.—Although the sun-dial may be used to indicate the time of day with sufficient accuracy for some purposes, its use is limited by the fact that it can only be employed when the sun is visible at the place of observation. Other modes of measuring the flow of time have, therefore, long been adopted. In early days, the rate at which a candle burned, or at which water or sand escaped through a small aperture, was employed as a time-measurer. Coming to more recent times, clocks and watches serve a similar purpose, but from what has already been stated, it is evidently necessary to regulate them according to the results of astronomical observations.

The most precise determinations of time are made by means of a transit instrument, that is, an instrument by which the exact moment at which a celestial body passes the meridian can be observed. The positions of certain fundamental stars called "clock stars" have been determined with great accuracy, and it is therefore known to within a very small fraction of a second at what sidereal time one of these stars will pass the meridian. If the sidereal clock does not indicate this time when the star is observed on the meridian, its error can be noted and corrected. In this way the sidereal time is ascertained, and its equivalent in mean solar time is only a matter of simple calculation.

Another method is to observe, by means of a sextant, or an altazimuth, the time, by a clock, at which the sun or a star has a certain altitude before noon, and the time at which it has the same altitude after noon. Midway between these times marks the time at which the body passed the meridian; the true sidereal time of passage is furnished by the known right ascension, and the corresponding mean time can therefore be calculated.

At sea, time is most frequently determined by observing the altitude of the sun in the morning or evening, when it is nearly in an east or west direction. The time by the chronometer corresponding to a certain altitude of the sun is noted, and by spherical trigonometry the apparent solar time is deduced; mean solar time is then obtained by correcting for the equation of time. The nearer the sun is to due east or west, the more accurate are the results obtained by this method.

TIME AT DIFFERENT PLACES.—In all these methods of finding the time, *local time* is alone determined, whether it be sidereal or solar. When solar time is in question, we have seen that mean noon is determined by the passage of the mean sun across the meridian. All places on the same meridian will thus have equal times; but at places on different meridians, the local times will be different. When it is noon at Greenwich, it will be something before noon at places to the west of Greenwich (for the reason that the sun has not yet crossed their meridians), while at places to the east it will be afternoon, because the sun has already passed the meridian. As the earth rotates through 360° in a day, it will turn 15° in an hour, or 1° in four minutes. Hence at places 15° east of Greenwich the time will be an hour in advance of Greenwich time, while at places 15° west it will be an hour earlier. For places in other longitudes, the difference of time is in the same proportion. The following are the local times at several places when it is noon at Greenwich:—

	A.M.		P.M.
Dublin	11.35	Paris	0.9
New York	7.4	Berlin	0.54
Toronto	6.42	Calcutta	5.53
Vancouver	3.38	Melbourne	9.40

Throughout the whole of England and Scotland, Greenwich mean time is exclusively employed in preference to local times. This has the very practical advantage of uniformity, and as in no case does local time differ more than half an hour from Greenwich time, there is little inconvenience in regard to the beginning and end of day.

Until recently, the time systems of other countries have been mainly based on the times corresponding to their various national observatories. At present, what is called "zone time," in which the hours alone differ from Greenwich time, has been adopted in several European states, as well as in other parts of the world.

The present state of time reckoning on this much improved plan is indicated by the following table :—

Country.	*Standard time.*
England Belgium Holland	Greenwich time.
Denmark Germany Italy Switzerland Norway and Sweden	Mid-European time, 1 hour fast on Greenwich.
Colony of Natal	2 hours fast on Greenwich.
United States Canada	4, 5, 6, 7, or 8 hours slow on Greenwich, according to longitude.
Japan	9 hours fast on Greenwich.
Western Australia	8 ,, ,, ,,
South Australia	9 ,, ,, ,,
Victoria, New South Wales, Queensland, and Tasmania	10 ,, ,, ,,

TELEGRAPHING TIME.—An important part of the work of

the chief national observatories is the determination of correct time, and its communication to the public at large. Railways have especially created a demand for a uniform and accurate system of time-reckoning, and to meet this need there is usually an organised service providing an automatic distribution of time-signals by means of the electric telegraph. The transmission of such time-signals was first established on a large scale in connection with Greenwich Observatory, and at the present time signals are sent to the General Post Office, whence they are distributed automatically to post offices and subscribers throughout the kingdom. In addition, signals are sent direct to Westminster for the regulation of the great clock on the Houses of Parliament, and time-balls are dropped at certain hours at Greenwich and Deal, in order that navigators may have the opportunity of rectifying their chronometers.

THE YEAR.—The day is too small an interval of time to be conveniently employed as a unit for chronological purposes, so that at present the count of time by days is practically limited to the number of days in a month. A greater unit, but still too small, is supplied by the month, and the necessity for a more serviceable unit early led to the adoption of the length of the year. This is at once a natural division of time, corresponding to the recurrence of the seasons, and sufficiently answers all requirements for measuring extended intervals.

If we determine the exact time required by the sun to pass from one fixed point in the heavens to the same point again, we shall find the time in which the earth makes a complete revolution round the sun, that is, the time in which a line joining the earth and sun sweeps through an angle of 360°. This interval, which is called the *sidereal year*, amounts to 365 days 6 hours 9 minutes 9 seconds of mean solar time. It will be clear, however, that the most useful year is that which will give us the same day of the month at the same season in all years. If there were no precession of the equinoxes, this would be of the same length as the sidereal year, but on account of precession the passage of the sun from the vernal equinox to the same equinox again occupies less than a sidereal year. In fact, this equinoctial, or *tropical year* amounts to 365 days 5 hours 48 minutes 46 seconds; that is, about 20 minutes less than the sidereal year. This is the year which is always understood, unless it is otherwise stated. If our calendars were regulated according to the sidereal year, the same day of the month would in time run through all possible changes of seasons, the 25th of December, for instance, occurring at one time in winter, and gradually changing through spring, summer, and autumn.

THE CALENDAR.—The earlier calendars with which history acquaints us were mainly based on the lunar month of about 29½ days, twelve of which made up a lunar year of 354 days. The calendar year was thus more than 11 days shorter than the actual year, and in order to bring the dates into agreement with the seasons, arbitrary intercalations were occasionally made by the authorities.

In the year 45 B.C. a great reform was introduced by Julius Cæsar; 365¼ days was adopted as the length of the year, and it was prescribed that ordinary years should be reckoned as consisting of 365 days, while every fourth year divisible by 4 without remainder should be a *leap year* of 366 days. Matters were so much simplified by this arrangement that the Julian calendar remained unaltered until 1582, and is even now retained throughout Russia.

The tropical year, as we have seen, is less than 365¼ days, so that the Julian calendar does not quite keep course with the seasons. Although the difference is only 11¼ minutes, it amounts to an entire day in 128 years, so that if the vernal equinox occurred on the 21st of March at one time it would occur on the 20th after 128 years. If, then, it be desired to bring the existing dates of any particular year into agreement with dates at a previous period, as regards the seasons, a correction in addition to that ordained by Cæsar must be introduced. In the time of Pope Gregory, in the year 1582, the vernal equinox fell on the 11th of March, and the necessity of a new calendar came to be recognised. The astronomer Clavius, with the authority of the Pope, devised our present "Gregorian" calendar. This arrangement, first of all, altered the actual date of the equinox from the 10th to the 21st of March, that is, to the day on which it occurred in the year of the great Council of the Church at Nicæa, 325 A.D. To bring about this alteration it was necessary to drop 10 days from the calendar, and it was therefore decided that the day following the 4th of October, 1582, should be called the 15th instead of the 5th. To prevent subsequent changes in the date of the equinox the Julian rule for leap year was slightly modified. If the date number of a year is divisible by 4 without remainder it is still to be a leap year, unless it be a century year, in which case it must be divisible by 400 without remainder if it is to be called a leap year.

It was not until 1752 that the Gregorian calendar was adopted in England, and as 1700 was a leap year according to the Julian rule the old style date was 11 days behind the Gregorian date. An Act of Parliament decreed that the day following September 2, 1752, should be called the 14th. The Act was carefully planned so as to prevent injustice in the collection of rents and the like, but it was only accepted after considerable opposition.

It has lately been pointed out that if we wish to make the day of the year correspond with the seasons for all time, a modification of the Gregorian calendar must be adopted. By the Gregorian rule, three leap years are omitted every four centuries; but Mr. W. T. Lynn has drawn attention to the fact that if one were dropped every 128 years instead, the calendar would be sensibly perfect, and the seasons would always commence on the same dates.

CHAPTER VI.

THE MOVEMENTS OF THE MOON.

THE MOON'S REVOLUTION.—Apart from the changes in the appearance of the moon due to the ever-varying phases, the first fact which strikes the attentive observer is that the moon has an eastward movement among the stars, and that this motion is much more rapid than that of the sun. Indeed, the moon gains a whole revolution upon the sun in a period of about $29\frac{1}{2}$ days, this being the interval between two successive new or full moons. As referred to the stars, however, it is found that the moon and any particular star which cross the meridian together at a certain time will again do so after the lapse of only $27\frac{1}{3}$ days. Besides this eastward movement among the stars, the moon moves towards and away from the Pole; the full moon, for instance, is sometimes seen high in the heavens at midnight, and at other times very low. Indeed, the moon's apparent movements resemble in a very general way those of the sun, but they cannot be attributed to a revolution of the earth round the moon, as those of the sun are to a real movement of the earth round the sun. We have seen that there are direct proofs of the earth's revolution round the sun, and a revolution round the moon, even in a smaller orbit, would not be consistent with the observed movements of the greater luminary. Being convinced of the reality of the moon's movements around the earth, we can next proceed to investigate the circumstances of its varied motions.

Just as we learn the conditions of the earth's movements by observations of the sun's apparent movements which are their natural consequence, we can determine the moon's motions by studying its varying situations with regard to the much more distant stars. We can measure the moon's right ascension and declination at different times with the transit instrument, and, if desired, we can mark out the apparent path on our star charts or celestial globes. In this way it is found that the moon moves in a plane which is inclined at $5^\circ\ 9'$ to the plane of the ecliptic. As to the shape of the orbit, we have only to observe the changes in the moon's apparent size; when it is nearest to us it will appear largest, and when furthest removed its apparent diameter will be least. Actual observations show that, like the orbit of the earth, the moon's orbit is an ellipse, with the earth in one focus. Owing to various causes, the orbit is somewhat variable in shape, and its eccentricity ranges from 0·07 to 0·045. When the moon is at the point of its orbit nearest to the earth, it is said to be in *perigee;* and when at the most distant part of its orbit, in *apogee.*

The earth's orbit, as we shall see by and by, is very small as compared with stellar distances, and the moon's apparent movement, with regard to the stars, is not affected by the revolution of the earth and moon round the sun; consequently the interval between its passing a star and overtaking the same star again is a measure of the time in which the moon's movement round the earth is performed—this is 27 days, 7 hours, 43 minutes, and is called the moon's *sidereal period.* The direction of the moon's motion is opposite to that of the hands of a clock, a movement which is said to be *direct* (motion in the reverse direction would be *retrograde*).

PHASES.—Two circumstances lead us to suppose that the light of the moon is borrowed from the vast store thrown out into space by the sun. First, the fact that it puts on *phases,* for if it were a body shining by its own light we should always see a full moon. Second, the fact that the phase we see depends absolutely on the moon's situation with regard to the sun and earth.

There is every reason to suppose that the moon is a dark globular body, so that the sun can only illuminate that hemisphere which is turned towards it. At new moon the illuminated part is turned directly away from us, and we are thus led to infer that when new the moon lies directly between the earth and sun. At full moon, on the contrary, the whole of the illuminated part is presented to us, and we therefore conclude that at this time the earth lies between the sun and moon. On account of the inclination of the moon's orbit to that of the earth, the sun, earth, and moon do not always come exactly in a straight line at new or full moon; when they do, the interesting phenomena of solar and lunar eclipses occur. (Chapter VIII.)

A diagram will help to elucidate the production of the moon's intermediate phases. Supposing the sun's rays to proceed from the left, the earth being at O, the moon will be at A when new. Proceeding towards B, a small portion of the illuminated side will be turned towards us, and the moon will be a crescent. On reaching the point C, exactly half of the sunlit hemisphere will be visible to us, and we have the moon's *first quarter.* Passing to the point D we see more than half of the bright part of our satellite, and it appears gibbous in form, until it reaches E, where it becomes full.

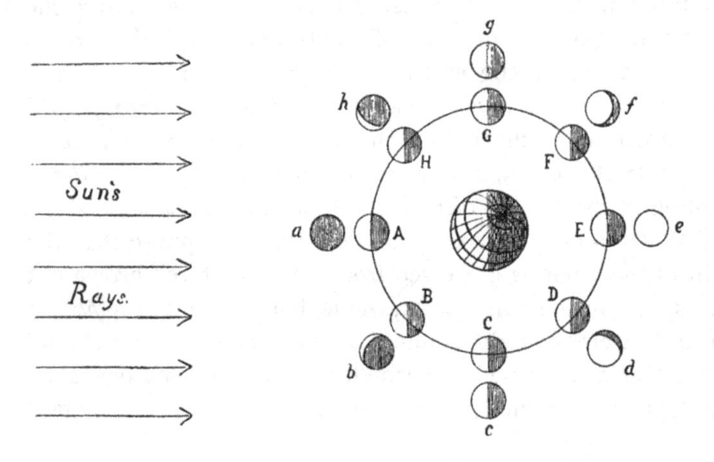

FIG. 18.—*The Moon's Phases.*

Similar phases occur in inverse order during the movement along the other part of the orbit.

Such would be the conditions as to the phases of the moon, if the earth were at rest.

THE MONTH.—If the earth were fixed in space with regard to the sun, the moon's phases would be repeated in the time corresponding to its period of revolution round the earth. This is 27 days 7 hours 43 minutes, and measures the length of a sidereal month.

It is much more useful, however, to refer the month to the phases actually observed. If in Fig. 19 we have the sun, earth, and moon represented at a full moon by S, E, and M respectively, the next full moon will not occur until the three bodies occupy the positions S, E′, and M′, the earth having travelled about 30° along its orbit. Between two full moons, then, the moon must make a complete revolution round the earth, and through an additional angle, A E′ M′, which will be equal to the earth's angular motion in the interval. This movement of the moon occupies 29 days 12 hours 44 minutes, and is the duration of a *lunar month.* It also determines the

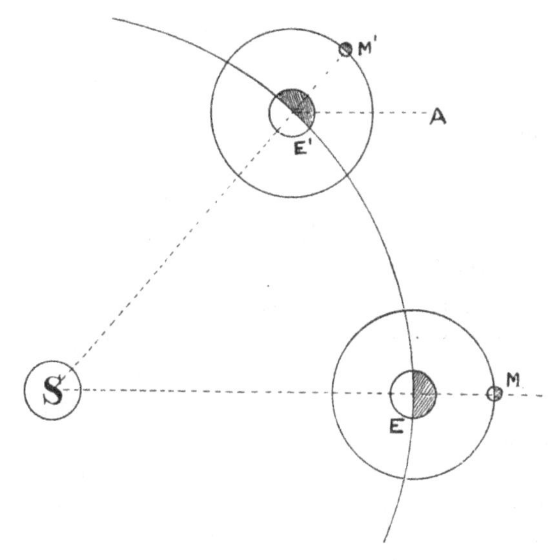

FIG. 19.—*The Lunar Month.*

synodic period of our satellite, a term which, taken generally, signifies the period in which a planet or satellite recovers the same position with respect to the sun when observed from the earth.

A calendar month, of which there are twelve in a year, must of necessity consist of a whole number of days, and the average duration of such a month is longer than that of a lunar month.

A remarkable relation exists between the synodic month and

the length of the year. In 19 Julian years of 365¼ days there are almost exactly 235 synodic months, so that after the completion of this period full moons again occur on the same days of the month. The discovery of this cycle is usually ascribed to Meton, a Greek astronomer, 433 B.C. It is accordingly known as the *Metonic Cycle*, and is still used in the calculation of the moveable festival of Easter.[1]

ROTATION AND LIBRATIONS.—Even observations made without instrumental assistance show that the surface of our satellite always presents the same face to us, and without further inquiry one might suppose that it had no axial movement corresponding to that of its primary. If there were no rotation, however, we should in turn see all parts of the moon, and the observed circumstances indicate that it must rotate on an axis, in the same direction as that of its orbital movement, and in the same time. In Fig. 20 let E represent the earth, and *a b c* the part of the moon which is turned towards us when it is at M. When the moon arrives at M′, observations show us that the same part is presented to our view, so that the part corresponding to that we saw in position M is represented by *a′ b′ c′*. Now, if the moon had not rotated in the interval, the line joining *a* and *c* would have retained the same direction, and would have been in the position *d e*; the part *c′ e* would thus have been carried out of sight, while another part which was not seen when the moon was at M would have come into view. In order that we may see the same part of the moon in two different positions, M and M′, the dividing line *a c* between the visible and invisible portions must turn through an angle equal to that between the lines *d e* and *a′ c′*; and since this angle is equal to that described by the moon in the same time, the period of the moon's rotation on its axis must be equal to that of its revolution round the earth.

On account of the elliptical form of its orbit, the angular movement of the moon is not quite uniform; like the earth, it

[1] There is a very complete paper on "How to find Easter," by Dr. Downing, in the *Journal* of the British Astronomical Association, vol. iii., p. 264.

is subject to the law of areas. Hence, as the rotation is equable, the foregoing explanation does not strictly hold. In fact, this varying velocity results in a *libration in longitude*, which means that we sometimes see a little more of the western edge and sometimes of the eastern edge. There is also a *libration in latitude* on account of the fact that the moon's axis is inclined to the plane of its orbit, so that at

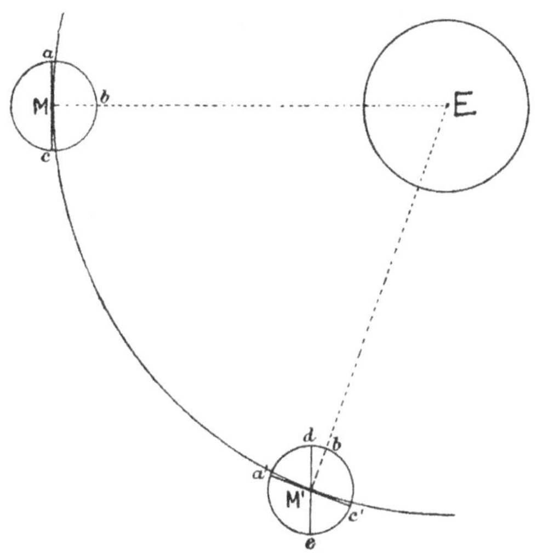

FIG. 20.—*The Moon's Rotation.*

different times we see more of the North or South Pole, as the case may be; in this respect the moon behaves to the earth somewhat as the earth does to the sun in regard to the seasons, but the inclination is not so great.

The moon is so near to us that the portion of it which we see depends to a slight extent upon our terrestrial location. When the moon is rising we see a little more of its western edge than will be seen by an observer to the east of us, where the moon is in the south, and more than we ourselves shall see when it has come to our own meridian. Just before the time of setting we get to see a little beyond the eastern edge. This is called the *diurnal libration*, and never amounts to more than a degree.

Thanks to these librations, we are enabled to make telescopic observations of 9 per cent. of the moon's surface which would not otherwise be open to our investigations.

CHANGES OF THE MOON'S ORBIT.—The moon's orbit is by no means to be regarded as a hard and fast geometrical figure. Indeed, it is subject to such great distortions in consequence of "perturbations" that the computation of the moon's position at any future time is one of great complexity. One of the most easily recognised changes in the orbit is the revolution of its *nodes*, that is, of the points where it crosses the plane of the ecliptic.

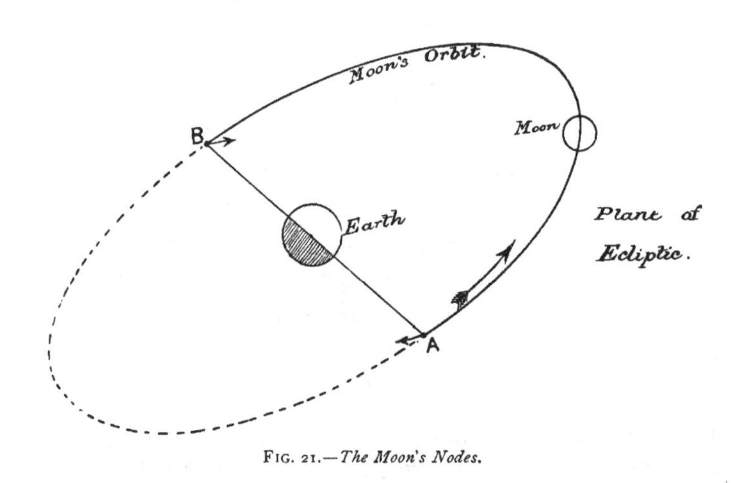

FIG. 21.—*The Moon's Nodes.*

The latter being a plane of indefinite extent, to which the moon's orbit is inclined at 5° 9′, the moon will be alternately above and below the ecliptic for about half its period of revolution. The point where it passes from south to north of the ecliptic, A in Fig. 21, is the *ascending node*, and the corresponding point on its southward path is the *descending node* of the orbit. Connecting these two points is the line of nodes (A B), and by observations of the points where the moon's path intersects the ecliptic at different times it is found that the line of nodes *regredes* or moves backwards.

The rate of this revolution of the moon's nodes is very irregular, but a whole revolution is made in 18·6 years.

This retrogression of the moon's nodes may be well illustrated by the following heliocentric longitudes of the ascending node as given in recent "Nautical Almanacs":

1892	January 1	53° 51′·56.
1893	"	34° 28′·69.
1894	"	15° 19′·00.
1895	"	355° 49′·31.
1896	"	336° 29′·61.

The line of apsides of the moon's orbit joins the perigee and apogee; the direction of this line in space changes in a very variable manner, but in the long run it makes a complete revolution in 8·9 years.

When the sun is passing through the moon's line of apsides it temporarily increases the eccentricity of the orbit; when at right angles to this line, the orbit becomes more nearly circular. This disturbance of the moon has accordingly a period equal to that required for two successive passages of the sun over the apse line of the moon's orbit.

Such are a few of the movements which come within the province of the *lunar theory*, a fuller treatment of which is beyond our scope.

THE HARVEST MOON.—The full moon which occurs nearest to the autumnal equinox is called the *harvest moon*, for the reason that it rises very nearly at the same hour for several nights together, and so gives us a greater share of moonlight, by which harvest operations may be extended. At that time the sun will be at the autumnal equinoctial point, and when it is setting in the west, the vernal equinoctial point, and the moon with it, must be rising due east. The part of the ecliptic then above the horizon will extend from the east to the west point, but will lie wholly below the celestial equator (Fig. 22). As the moon's path is very slightly inclined to the ecliptic, its movement will thus make only a small angle with the horizon, and for several nights together it will rise at nearly the same time.

In March, when the sun is near the vernal equinox, the full moon will be near the autumnal equinoctial point; when the sun is setting, the moon will be rising as before, but in this case the part of the ecliptic which is above the horizon lies wholly above the celestial equator. The ecliptic is thus inclined at an angle to the horizon greater by 47° than when the vernal equinox is rising in autumn; the moon's path being near the ecliptic, its movement during a day will at this

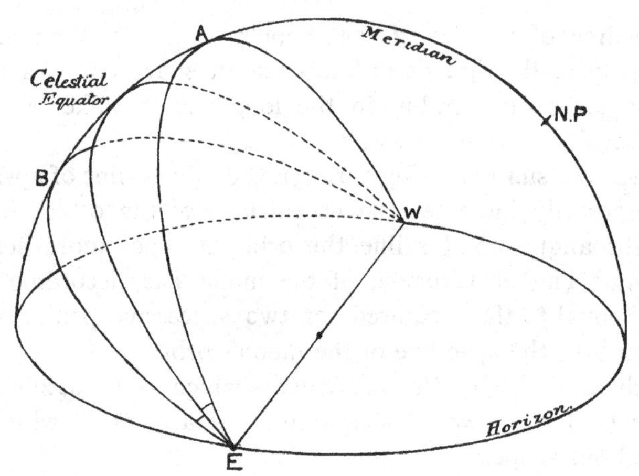

FIG. 22.—*Position of Ecliptic at Sunset at Vernal Equinox* (E A W) *and Autumnal Equinox* (E B W).

time carry it a long way below the Equator, and it will rise much later the following day.

In the Southern Hemisphere, the conditions are reversed, the harvest moon occurring at our vernal equinox, which, however, is the commencement of the southern autumn quarter.

The phenomena of the harvest moon recur, but are not so marked, in the month of October, and it is then called the hunter's moon.

It is important to bear in mind that this rising of the moon at nearly the same hour for several days occurs every month, but as the risings then occur either in daylight or after midnight, and the moon is not full, no special attention is drawn to them.

Again, since the phenomenon of the harvest moon depends upon the small inclination of the path of the full moon to the horizon when it is at the equinoctial point, the circumstances will be modified by the latitude of the place of observation. At the Equator, for example, there will be no harvest moon, as there the ecliptic is always greatly inclined to the horizon; in fact, it will be inclined at the same angle in spring as in autumn.

The moon's path being inclined to the ecliptic, the conditions as to the harvest moon will depend to a small extent upon the position of the moon's nodes, which, as we have seen, revolve in a period of a little less than 19 years. At times, then, the moon's path will be inclined 5° more, and 9 years afterwards 5° less, than is the plane of ecliptic, and under the latter conditions the harvest moon will be most pronounced.

HIGH AND LOW MOONS.—At the time of full moon, the moon is in the opposite part of the heavens to that occupied by the sun, sometimes being 5° above and other times 5° below. Manifestly, then, if the sun be high in the heavens at mid-day, it will be only a little below the northern horizon at midnight, and the moon, consequently, will be only a small distance above the southern horizon. In summer, then, quite apart from the fact that the nights are shorter, there is less moonlight. In winter, on the other hand, the sun descends far below the northern horizon at midnight, and the full moon has a high elevation in the southern part of the sky. By this happy arrangement, the full moon is longest above the horizon when its light is of greatest benefit to mankind.

CHAPTER VII.

MOVEMENTS OF PLANETS, SATELLITES, AND COMETS.

APPARENT MOVEMENTS OF PLANETS.—It has already been pointed out that like the sun and moon, the planets also have an apparent movement with respect to the more distant stars. Mercury and Venus are never seen very far from the sun, while other planets, among which are Mars, Jupiter, and Saturn, may be seen in the part of the heavens opposite to the sun.

One point, and that a very important one, which we notice from our observations is that the planets never depart very far from the ecliptic, so that the planes in which they perform their movements are nearly coincident with the plane in which our own annual journey round the sun is performed. The apparent movements of the planets are such that it is quite impossible to regard these bodies as circulating in regular orbits round the earth itself. If they revolve round any other body it is manifest that their apparent or geocentric motions will be compounded of the real movements of the planets and that of the earth. It is not necessary here to trace the steps by which it has been determined that the planets revolve in regular orbits around the sun. Suffice it to say that their observed movements are simply and sufficiently explained by supposing that, like the earth, which may now be regarded as a planet, they travel in elliptic orbits with the sun at one of the foci. Besides this revolution, the planets have a rotatory motion about their axes, but this question cannot be studied apart from the telescopic features, and will therefore be treated in Section III. of the present work.

The circumstance that the planets Mercury and Venus are never seen long after sunset or before sunrise, indicates that their orbits must lie between us and the sun. Hence, they are distinguished as the *interior planets*, while those outside the earth's orbit are called the *exterior planets*.

MOVEMENTS OF INTERIOR PLANETS.—Let us consider briefly the conditions under which we observe the interior planets. If such a planet be represented by M in Fig. 23, while the earth is represented by E traversing a larger orbit, the planet is said to be in *inferior conjunction* with the sun, when it lies directly between the sun and earth. The actual movements of the planets being direct — that is, anti-clockwise—the planet at M has an apparent westerly motion as seen by an observer situated on the earth, and from this we gather that it moves more rapidly than the earth.

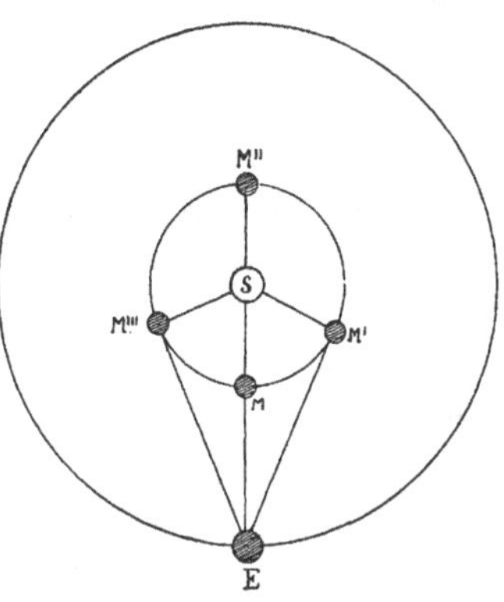

FIG. 23.—*Movement of an Interior Planet.*

For simplicity let us regard the earth as being at rest at the point E. Then, as the planet reaches the position M , where it is as far as possible to the west of the sun, it is said to be at its *greatest western elongation.* Proceeding in its orbit, the planet's apparent movement is direct, and it eventually comes in line with the sun on the further side as seen from the earth; it is then said to be in *superior conjunction.* From this point the planet moves to the east of the sun until it comes to the point M''', after which the motion becomes retrograde, and the planet proceeds to inferior conjunction again. When at its greatest distance to the east of the sun, as at M''', the planet is said to be at its *greatest eastern elongation.* Taking the term *elongation* in general, it may be regarded as a measure of the angular distance of a planet from the sun as observed from the earth.

If the orbits of the planets were perfect circles, the greatest elongation distances of an interior planet would always be the same; sometimes, however, we are nearer to the sun than at the other times, and the apparent separation of the planet from the sun would seem greater than at other times, even if there were no other cause at work. The variations of the elongation distances are greater than can be accounted for by our own varying distance, and are naturally attributed to the elliptical form of the orbits of the interior planets themselves. Mercury, for example, sometimes only departs 18° from the sun, while at other times it reaches as far as 28° east or west.

When we take account of the fact that the earth has also a movement along its orbit, it will be seen that the same conditions hold good with regard to elongations and conjunctions, except that the intervals between them will be longer.

MORNING AND EVENING STARS.—From superior to inferior conjunction an interior planet is to the east of the sun. It then rises after the sun, and sets after the sun, so that it is visible for a short time in the early evening; in other words, it is an *evening star* during this part of its path. Between inferior and superior conjunctions, the planet is conversely a *morning star.* This is illustrated in Fig. 24, where the position of an observer towards whom the sun is rising is shown at A. An

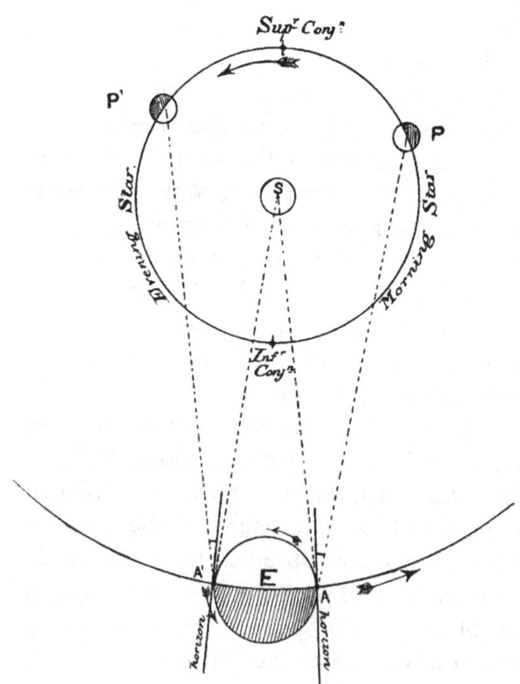

FIG. 24.—*Morning and Evening Stars.*

interior planet at P is above the horizon at sunrise, but will be below at sunset, the observer having been carried to A′ by the earth's rotation; it will thus be a morning star. When the planet occupies the position P′ it is below the horizon at sunrise, but will remain in sight after the sun has set in the evening, the observer then having been transferred to A′ by the earth's rotation.

PHASES OF INTERIOR PLANETS.—From the conditions which have been stated with regard to the movements of the interior planets, one is not surprised to find that telescopic examination reveals that these bodies put on phases similar to those of the moon. At superior conjunction the planets exhibit a fully illuminated disc, at greatest elongations they appear as a half moon, while at inferior conjunction their dark sides alone are presented to us. The apparent sizes of the planets, as measured with the aid of a telescope, are also found to vary according to their positions; when at inferior conjunction, the planet is much nearer to us than at other times, and it consequently appears larger. The apparent brightness of an interior planet also varies. At superior conjunction the whole of the disc is illuminated, but the planet is then so far removed from us that its light is very feeble. On the other hand, at inferior conjunction, when it is nearest to us, the dark side of the planet is turned towards us. The greatest brightness thus occurs at some intermediate point. In the case of Venus this is between the greatest elongations and inferior conjunction, when it is 40° from the sun. It is then bright enough to be seen with the naked eye in full sunshine, and has sometimes, on such occasions, been erroneously regarded by ignorant persons as the Star of Bethlehem.

TRANSIT OF VENUS.—If an inferior conjunction occurs when the planet is very near to a node—this term having the same significance as in the case of the moon (p. 94)—the planet, whether it be Mercury or Venus, will be seen projected as a dark spot upon the bright disc of the sun. Such an occurrence is called a *transit of Venus* or of Mercury, as the case may be. Just as we do not get an eclipse of the sun every month, so we do not get a transit of Venus every time the earth and that planet have the same heliocentric longitude, and for the same reason, namely, that the plane of the orbit is inclined to the ecliptic. As we shall see in another chapter, a transit of Venus has a most important application in the determination of one of the fundamental constants of astronomy—the sun's distance. The conditions as to the recurrence of transits are of great interest. In the case of Venus, the *synodic* period is 584 days, this being the time which elapses between two successive inferior conjunctions. Five synodic periods are thus very nearly equal to eight years, and 152 synodic revolutions are even more nearly equal to 243 years. As seen from the earth, the sun crosses the nodes of the orbit of Venus on June 5 and December 7, and since there can be no transit when the planet is more than $4\frac{1}{2}^{\circ}$ from the node, the transits will all occur about these dates. A transit will be followed by another after the lapse of 8 years, if the planet is not too far from the node; but there can be no other transit with the planet at the same node until 243 years have elapsed. There are, however, transits occurring at similar intervals when the planet is at the other node. The following dates on which transits have occurred, or will occur, will illustrate the foregoing statements :—

Interval	Date	Interval
8 years	December 7, 1631,	243 years (December 7, 1631 – December 9, 1874).
	December 4, 1639,	243 years (December 4, 1639 – December 6, 1882).
8 years	December 9, 1874,	
	December 6, 1882,	
8 years	June 5, 1761,	243 years (June 5, 1761 – June 8, 2004).
	June 3, 1769,	243 years (June 3, 1769 – June 6, 2012).
8 years	June 8, 2004,	
	June 6, 2012,	

MOVEMENTS OF EXTERIOR PLANETS.—The exterior

planets are at once recognised as such by their occasional appearance in the part of the sky opposite to that of the sun. They are then said to be in *opposition*. When in the same line as the sun, and on the remote side of it, as at P′ in Fig. 25, the planet is in *conjunction*. The apparent movements of such a planet are very complex. Neglecting for a moment the earth's motion, it is evident that the apparent rate of movement of the planet with reference to the stars will vary very considerably according as the planet is near opposition or near conjunction, the movement appearing to be most rapid when the planet is nearest to us. Upon this unequal rate of motion is superposed a varying direction of motion produced by the changing position of the earth. When the planet is at P, and the earth at E, both are moving in the same direction, but as the earth has the greater angular velocity, the apparent motion of the planet will be retrograde, that is, the planet will appear to go backwards in its path. If the earth be near the point E′, its orbital movement will be directed away from the planet, and will scarcely affect its apparent position ; accordingly, about this time the planet has a direct movement in the heavens. Between these two points the direction of the apparent movement of the planet has changed, so that at some intermediate position it would seem to have suspended its wanderings ; here we have a *stationary point*. For a certain time, before and after conjunction, the linear directions of movements of the earth and planet will be opposed to each other, and on this account the *direct* apparent motion of the planet will be accelerated. Presently, as the

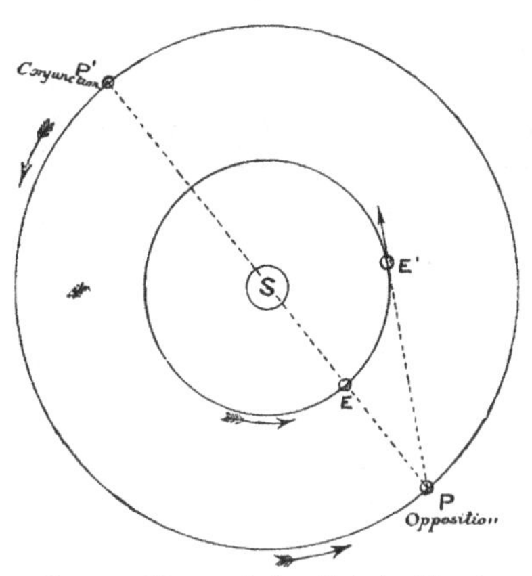

FIG. 25.—*Movement of an Exterior Planet.*

earth gains on the planet, another stationary point will be reached, and with the approach to opposition the planet will again retrograde.

If both orbits were in the same plane, these apparent movements would all be backwards and forwards along a great circle of the celestial sphere coincident with the ecliptic, the eastward movement predominating. The planes in which the planets perform their revolutions are, however, inclined to the ecliptic, and the result is that they appear to us to travel in loops, some of which are illustrated in Fig. 26.

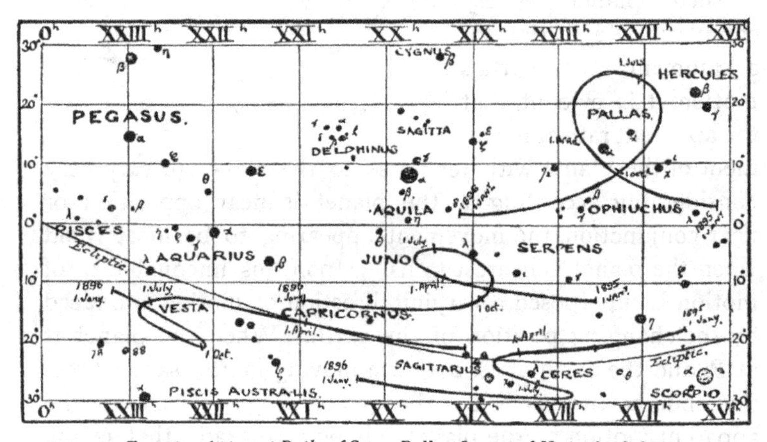

FIG. 26.—*Apparent Paths of Ceres, Pallas, Juno, and Vesta, in* 1896.

From the fact that we are constantly within the orbit of an outer planet, it is evident that we must always see more than half of the planetary hemisphere on which the sun is shining. Consequently, an exterior planet never puts on a crescent phase, or presents the appearance of a half moon. The nearer the planet the greater will be the dark area which it is possible for us to observe. In the case of Mars, for example, we sometimes see it gibbous like the moon about three days from full, but in the more distant planets this gibbosity is scarcely perceptible. The greatest phase of an exterior planet occurs when it is at *quadrature*, that is, when a line joining the earth and sun is perpendicular to one joining the earth with the planet.

FAVOURABLE AND UNFAVOURABLE OPPOSITIONS.—A little consideration of Fig. 25 will make it perfectly clear that an exterior planet is very much nearer to us at a time of opposition than at a conjunction. We are, in fact, then, nearer to the planet by the diameter of the earth's orbit, a matter of some 186 millions of miles. Accordingly, the planets, more especially our neighbour Mars, are best studied in the telescope about a time of opposition. Now, if we had to deal with circular orbits, the distance of a planet at opposition would remain constant, and we should see the planet equally well at all oppositions. It is found, however, that this is not the case, and the ellipticity of the orbits of the earth and planets supplies a simple and sufficient explanation. Sir Robert Ball illustrates this in the case of Mars by a diagram similar to Fig. 27. It will be seen that, when the opposition occurs in August, the earth is much nearer to Mars than when it happens at other times. The least favourable oppositions are those which occur in February, the planet then being nearly twice as far removed from us as at the nearest approach during an August opposition.

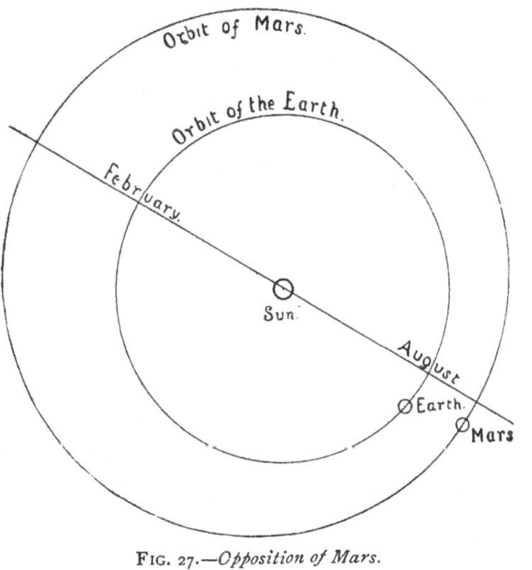

FIG. 27.—*Opposition of Mars.*

As regards the more distant planets, the diameter of the earth's orbit and the variations of opposition distance are of less importance, since they form a much smaller proportion of the distances of those planets from the sun.

ELEMENTS OF A PLANETARY ORBIT.—A complete study of the apparent movements of the planets with which we are acquainted shows that their real movements are performed round the sun in ellipses, the sun being placed at a focus. Each orbit, like that of the earth, has its perihelion and aphelion points, and its apse line; not being coincident with the ecliptic, it will have a line of nodes, and an ascending and descending node. Each planet will further have a particular inclination to the ecliptic, and a period of revolution peculiar to itself. Consequently, to systematise our knowledge of any particular orbit, certain conventions are adopted, and the seven things we must know, in order that we may specify the size of the orbit, its position in space, and the situation of the planet in its orbit, are as follows:—

a = Semi axis major of elliptic orbit.
e = Eccentricity.
i = Inclination to ecliptic.
Ω = Longitude of ascending node.
π = Longitude of perihelion.
P = Period of revolution. (u, the mean daily motion, sometimes replaces P.)
E = The epoch, giving the longitude of the planet at some particular time.[1]

The first two quantities indicate the size and shape of the orbit, the next three its position with regard to the ecliptic, and the last two are required to determine the situation of the planet in its orbit. Some of the elements are illustrated in Fig. 28.

[1] The application of Kepler's third law gives us $P = a^{\frac{3}{2}}$ years, but as this is not strictly true, both P and a must be given where the greatest possible accuracy is desired.

DETERMINATION OF A PLANET'S PERIOD.—Observations enable us to determine the synodic period of a planet, and knowing that the earth's period is a year, it is a simple matter to determine that of the planet. In the case of an exterior planet, the interval from opposition to opposition furnishes the best means of determining the synodic period. The exact moment of an opposition cannot usually be directly observed, and what one actually does is to measure the R. A. and declination of the sun on several days about the time of opposition, as also those of the planet; then, by reducing these co-ordinates to celestial longitude and latitude, it is not

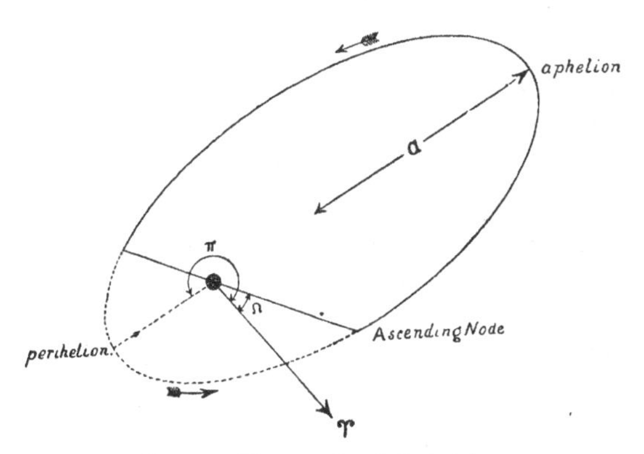

FIG. 28.—*Elements of an Elliptic Orbit.*

difficult to determine at what moment the longitudes differed by 180°, that is, the moment at which opposition took place. The problem of finding the planet's sidereal period, then, amounts to this: at what rate must the planet be moving in order that the earth may make a complete revolution, and move, in addition, through the same angle as the planet? In other words, what must be the period of the planet in order that the earth may gain a whole revolution in the interval corresponding to the synodic period? The daily movement of the planet will be $\frac{360°}{P}$, and that of the earth $\frac{360°}{365\frac{1}{4}}$, if P denote the number of days in the planet's sidereal period. The earth's gain per day will thus be the difference between these two quantities, and since a whole revolution is gained in the synodic period, the gain per day can be expressed as $\frac{360°}{S}$, where S represents the synodic period ; thus we get

$$\frac{360°}{365\frac{1}{4}} - \frac{360°}{P} = \frac{360°}{S}$$

or

$$\frac{1}{365\frac{1}{4}} - \frac{1}{P} = \frac{1}{S}$$

The synodic period of Mars is 780 days, and the application of the foregoing formula leads us to 687 days as the time of its revolution round the sun.

A single determination of a synodic period does not give precise results, for the reason that the orbits of the planets are elliptical, and the intervals consequently dependent upon whether the planet is near perihelion, or far removed from it when an opposition is observed. It is, therefore, necessary to determine the time of opposition at long intervals, and so reduce the errors in measuring the length of a single period.

MOVEMENTS OF SATELLITES. — Telescopic observations show that some of the planets are accompanied by *satellites*, which revolve round their primaries as the moon revolves round the earth. The apparent movements of these bodies, with regard to the planets, are very similar to those of the interior planets with regard to the sun, having similar points of greatest eastern and western elongations. The facts which have been collected show that each satellite, like our own moon, moves in an elliptical orbit, with the planet in one of its foci. With one exception, the satellites attending the planets of our system have a direct movement ; those of Uranus, however, have apparently a movement in the same direction as the hands of a watch, but this can be regarded as direct, if we consider the plane of the orbit to be inclined more than 90° to the plane of the ecliptic.

THE ORBITS OF COMETS.—Another class of bodies which circulate round the sun now claims our attention. These are the *comets*, some of which are never seen without the aid of telescopes, while others have been brilliant enough to excite a widespread wonder and interest. They usually have a very rapid movement relatively to the stars; and to learn something as to their real motions, we commence by measuring their right ascensions and declinations as frequently as possible. When such observations are plotted, they give us the geocentric movement of a comet, which generally seems very irregular, and gives one the idea that it is subject to no law. Unlike the planets, comets do not usually keep near the ecliptic, but move in planes inclined at all angles to it. Their rates of apparent movement also change very rapidly.

When the effect of the earth's movement upon that of a comet is eliminated, it is found that the movement of the comet is performed either in an ellipse, a parabola, or an hyperbola, the sun in each case occupying one of the foci.

From our definition of the eccentricity of an ellipse, it will be seen that, when the eccentricity is zero, we have a circle. When the eccentricity becomes unity, the ellipse becomes a parabola, so that the latter curve may be regarded as part of an ellipse, of which the foci are at an infinite distance apart. In the case of the hyperbola, the eccentricity is greater than unity.

Comets which move round the sun in ellipses are called *periodic comets*, for the reason that they return regularly into the sun's neighbourhood. Those which traverse parabolic or hyperbolic paths will pass once round the sun and continue to journey into the depths of interstellar space until their movements are changed by the proximity of other bodies into the neighbourhood of which their wanderings may take them.

When a new comet is observed, one of the things which astronomers endeavour to do is to determine its orbit, so that its path may be predicted with sufficient accuracy to enable it to be picked up readily with a telescope when it becomes so feeble that it is no longer visible to the naked eye. In the first instance, the motion is assumed to be parabolic, and any deviation from such an orbit forms the subject of a rigorous calculation by means of which the precise form is determined.

CHAPTER VIII.

ECLIPSES AND OCCULTATIONS.

ECLIPSES OF THE MOON.—As the various members of the solar system shine only by virtue of the light which they receive from the sun, they will cease to be visible if by any means they are deprived of the sun's rays. Each planet or satellite must evidently cast a shadow which is turned directly away from the sun, and any other body passing wholly or partially within such a shadow will be proportionately debarred from receiving the direct light of the sun.

Were the sun a mere point of light these shadows would be parts of cones, the apex always being at the sun, and they would be prolonged indefinitely into space. As a matter of fact, every individual point upon the sun's disc is competent to cast a conical shadow, and the net result is that

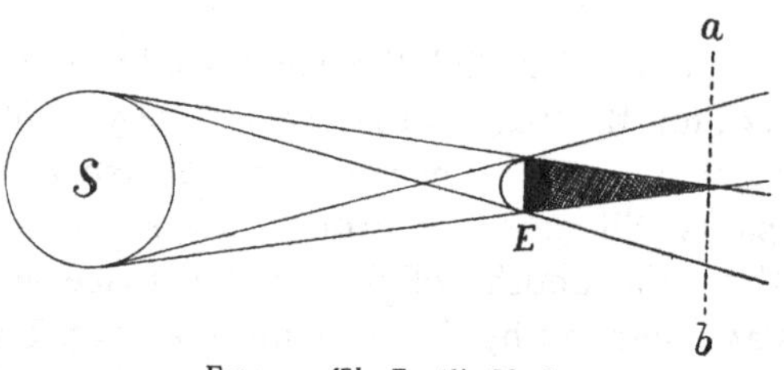

FIG. 29.—*The Earth's Shadow.*

only a relatively small space behind a planet or satellite is really in total darkness. This will be readily understood from Fig. 29, in which S is the sun, and E the earth. The total shadow now becomes a cone, with the apex turned directly away from the sun, but round this there is a region of partial shadow which is only illuminated by portions of the sun. If we imagine a section of the shadow across the line *a b*, we should find a central disc of total darkness called the *umbra*, and surrounding this a ring of half shadow called the *penumbra*.

From the known dimensions of the sun and earth, and the distance between them, it is easy to calculate the size of the earth's shadow cone, and its length is found to be greater than the distance of the moon. The axis of this shadow will, of course, always be in the plane of the ecliptic. If, then, at the time of opposition, the moon is sufficiently near the plane of the ecliptic, it will pass through the shadow, and we shall have the phenomena of a *lunar eclipse*. When the moon is wholly immersed in the umbra, the eclipse is total, and if it further passes quite symmetrically through the shadow, the eclipse is said to be central. This would always be the state of affairs if the moon performed its monthly journey in the plane of the ecliptic, and a total eclipse would occur every month. The moon's orbit, however, is inclined to the ecliptic, so that for a central eclipse, the moon must be simultaneously at opposition and at a node. If the moon be near the node when at opposition, a total eclipse may occur, but it cannot be central, and the duration of the total obscuration will be reduced. Still further from the node, the moon will be above or below the ecliptic, and will be only partially involved in the shadow cone ; such an eclipse is called a partial one. Beyond a certain distance from the node, the inclination of the moon's orbit will take the moon entirely out of the umbral shadow, and no eclipse will be possible.

The circumstances of an eclipse of the moon thus vary very considerably, and there is still another reason why we may expect them to be different. We have seen that the earth's distance from the sun changes throughout the year, and, in consequence, its shadow will be of varying length, and the diameter of the shadow at any specified distance will not be constant. The moon, again, is not always at the same distance from the earth, and it will, therefore, pass through varying depths of shadow in different eclipses, and with different velocities.

The breadth of the earth's umbral shadow at the point

where the moon passes through it is, on the average, about three times the moon's diameter, and the time taken for the moon to traverse this distance is about two hours. The duration of totality in a central eclipse may, therefore, amount to two hours, while an additional two hours may be occupied by the partial phases.

THE LUNAR ECLIPTIC LIMIT.—The greatest distance of the moon from a node at which a partial eclipse is possible, is called the *lunar ecliptic limit*, and is very easily calculated. In Fig. 30, let E N represent

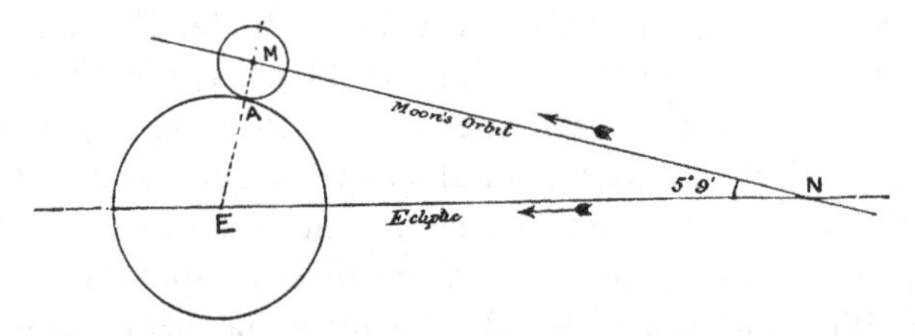

FIG. 30.—*The Lunar Ecliptic Limit.*

a part of the ecliptic, N being the node of the moon's orbit, and E the centre of the earth's shadow. As the orbit of the moon is inclined about 5° 9′ to the ecliptic, it may be represented by the line N M, inclined at an angle to N E. If E A be the radius of the earth's shadow, which, on the average, is about three-quarters of a degree, and M A the moon's apparent semi-diameter (about a quarter a degree), it is clear that the point beyond which no eclipse is possible is that in which the line M E, perpendicular to N M, is equal to the sum of the semi-diameters. All the quantities for solving the triangle N E M are thus known, and it can be readily calculated that N M, the greatest distance of the moon from the node at which an eclipse would be possible, under average conditions is about 11°.

Taking into account the varying distances between the sun, earth, and moon, it is found that an eclipse must always occur if the moon is within 9° of the node, and may occur if it be 12° from the node. These figures refer to the passage of the moon through the umbra, as the effect of its entrance into the penumbra is too slight to be observed.

The entrance of the moon into the earth's shadow is a definite phenomenon, which is independent of the observer's position on the earth, and the phases of the eclipse are seen at exactly the same moment from all places where the moon is above the horizon. The computation of the circumstances at a given place is accordingly a simple one.

When a lunar eclipse is not total at any of its phases, it is usual to specify its *magnitude* by the ratio of the greatest measurement of the obscured part to the moon's diameter. Thus the magnitude of the partial eclipse of February 28th, 1896, is given in the "Nautical Almanac" as 0·870, the moon's diameter being taken as unity.

The conditions of lunar eclipses which have been stated have reference to the moon's passage through the earth's geometrical shadow, but the actual conditions are greatly modified by the fact that the earth is surrounded by an atmosphere which refracts the sun's light so much that the moon is seldom quite obscured during totality. The commencement of the total phase is also rendered difficult of observation by the somewhat indefinite boundary between the umbra and penumbra.

ECLIPSES OF THE SUN.—If the moon performed its revolution in the plane of the ecliptic, it is evident that it must always come between us and the sun once in each month. This it does not do, but occasionally it happens to be in the ecliptic when in conjunction, and the moon is then seen to be projected upon the sun. In other words, there is an eclipse of the sun. Let us consider the circumstances, in the first instance, to an observer placed at the centre of the earth. If the centres of the moon and sun appear in the same straight line, the eclipse will be *total* or *annular*, according as the moon or sun has the greater apparent diameter. Both these forms of eclipses are possible, on account of the varying apparent diameters of the sun and moon consequent upon their variable distances from the earth. If the moon appear the larger it will evidently cover up the whole of the sun, but when it is the smaller, a ring of sunlight will be visible round the dark body of the moon, and the eclipse will be an annular one. These conditions are illustrated in Fig. 31, *a* and *b* representing a total and an annular eclipse respectively. If the moon and sun be not quite in the same straight line, the moon may still be seen partially projected

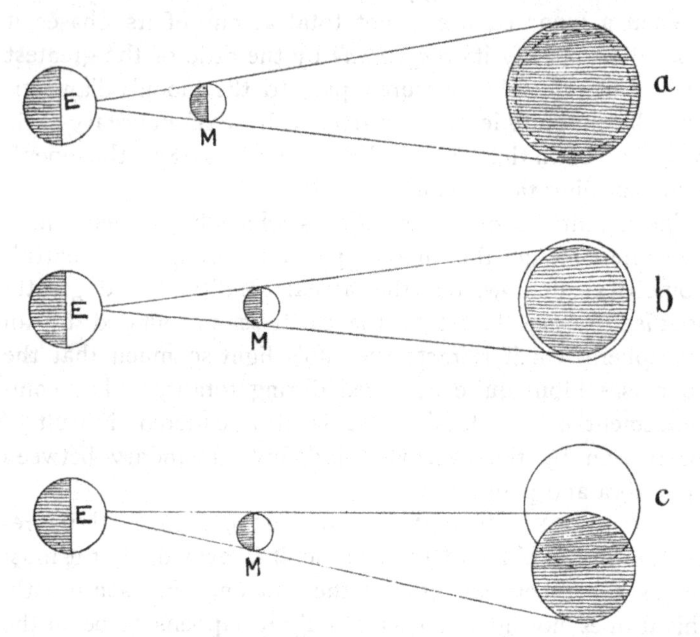

FIG. 31.—*Eclipses of the Sun.* (*a*) *Total Eclipse.* (*b*) *Annular Eclipse.* (*c*) *Partial Eclipse.*

on the sun's disc, in which case there will be a *partial eclipse* of the sun, as in Fig. 31, *c*.

In a total eclipse there are four so-called *contacts:* the first when the moon is seen to encroach upon the sun's disc, the second when the advancing edge of the moon reaches the opposite limb, the third when the following edge of the moon first touches the sun's boundary, and the fourth when the projected moon finally passes off the sun. The interval between the second and third contacts marks the duration of totality. As referred to our supposed observer at the centre of the earth, the duration evidently depends upon the apparent rate of the moon's eastward movement as compared with that of the sun, as well as upon the differences of the apparent diameters of the two bodies.

The production of eclipses of the sun may also be considered as arising from the immersion of an observer in the shadow of the moon. This shadow has its axis turned from the sun, but is so short that it does not always reach the earth. If an observer comes near the axis of the conical shadow, and within the apex, the eclipse will be total; if he is in the axis, but outside the apex, the eclipse will be annular.

The whole of the shadow of the moon is so small that only a few places on the earth's surface can be simultaneously immersed in it, and when we come to discuss the conditions of an eclipse with regard to a particular observer, the problem becomes a complicated one. At some places the eclipse may be total, at others it will be only partial, while at others no eclipse will occur at all. These differences are due to the fact that the sun is scarcely appreciably displaced by the change of locality, while the apparent position of the moon may be affected to the extent of nearly a degree. Again, the observer situated on the earth's surface has a movement of his own, produced by the earth's

FIG. 32.—*Duration of a Solar Eclipse.*

rotation, and his rate of motion depends upon the latitude in which he is situated. The effect of this movement upon the conditions of the eclipse are very pronounced. Suppose for a moment that the sun, moon, and earth, are fixed along the same straight line S M E in Fig. 32, a terrestrial observer a *a* on the earth's Equator would see an eclipse at noon ; if he were not in rotation, and the three bodies remained at rest, the eclipse would be a perpetual one. He is, however, carried onward by the earth's rotation, and even if the moon were at rest, it would appear to him to pass over the sun in the reverse direction. This retardation of the moon will be less in amount for observers away from the Equator, and also for observers to whom the sun is not on the meridian when eclipsed. The effect of rotation on an observer at *b* (Fig. 32), for example, is to move him almost in the direction of the line joining the moon and sun, and the backward tendency of the moon due to rotation is very slight. On account of the earth's rotation, then, the duration of a solar eclipse is lengthened, the greatest increase occurring at those places where the sun is on the meridian at the time of eclipse.

There is another source of gain of duration of an eclipse to the observer who sees the phenomenon about noon. The moon's apparent diameter is then augmented by a greater amount than at other places, because the observer is then nearest to the moon ; while the sun's apparent diameter is not appreciably affected. The greater the difference in the apparent diameters of the sun and moon, the longer will totality last.

These and other circumstances have all to be taken into account in computing the conditions under which an eclipse will be seen at any given place.

According to an eminent authority, Professor Young, the greatest possible diameter of the moon's shadow, where it strikes the earth, is 167 miles. It may, however, cover a larger space on the earth's surface, because the latter does not pass perpendicularly through the shadow. To all persons within the shadow, the eclipse will be total, but to those on its outer boundary the duration of totality will be for an instant only. The penumbral shadow has a cross section about 4,500 miles in diameter, covering sometimes a space on the earth's surface 6,000 miles across. To all persons within this area, but not in the central shadow, the eclipse will be partial. The shadow spot travels over the earth's surface, because of the moon's movement, but its track and speed are greatly modified by the earth's rotation. The movement of the shadow, as affected by the earth's rotation, would be along a parallel of latitude; but its ultimate direction of movement, though trending eastwards, depends upon this, combined with the direction of the moon's movement at the time of the eclipse. Thus, a portion of the track of the total eclipse of April 16, 1893, is as that shown in Fig. 33.

FIG. 33.—*Track of Eclipse of April* 16, 1893.

These considerations will suffice to explain the necessity for very precise calculations as to the position of the central line of an eclipse, when observers are sent out for the purpose of recording the phenomena.

Under the most favourable combination of conditions, that is, when the eclipse occurs at noon at a place on the Equator, an eclipse cannot be total for more than 7 minutes 58 seconds, nor be annular for a longer time than 12 minutes 24 seconds. From first to last contact may occupy as much as 2 hours, when all the circumstances are similarly favourable. (Loomis.)

THE SOLAR ECLIPTIC LIMIT.—In order that an eclipse of the sun may occur, the moon must be so near the ecliptic that it can be seen projected on the sun, either wholly or partially, from some point on the earth. It must therefore not be very far from the node, and the distance it may be from the node, while still being seen upon the sun, is called the *solar ecliptic limit.* As in the case of lunar eclipses, this distance is determined by the inclination of the moon's orbit, and the distances of the moon and sun from the earth. The latter being variable quantities, the limit is not always the same. It is calculated without much difficulty that an eclipse *must* occur if the new moon happens when it is within 15° 21′ of the node, and may occur within 18° 31′. These are called the minor and major ecliptic limits respectively. For total or annular eclipses, the limits are respectively 9° 55′ and 11° 50′.

NUMBER OF ECLIPSES IN A YEAR.—If the moon's nodes were fixed, the sun would pass through the line of nodes twice a year. At such times an eclipse of the sun must necessarily occur if the moon were within 15° 21′ of the node on either side. The sun requires more than a month to traverse this space of 30° 42′, and the moon must therefore pass through each node at least once while the sun is traversing these limits. It follows, then, that there must be at least two eclipses of the sun in a year. Since the line of nodes of the moon's orbit revolves backwards in a period of about nineteen years, the sun returns to the same node after an interval of 346·6 days, and there must accordingly be two solar eclipses in this interval. If, then, there be an eclipse early in January, there will be another about the middle of the year, and another at the end of the year, so that on this ground alone there is a possibility of three solar eclipses in a year.

Again, while the sun is passing through the ecliptic limits, it may happen that an eclipse occurs on its entrance, and then another will occur before it gets beyond on the other side of the line of nodes. In this way two eclipses may occur in the region of each node passage, and if the first of the series occurs early in January, five eclipses of the sun may occur in a single year.

The sun, however, is not a month in traversing the lunar ecliptic limit. Consequently, a whole year may elapse without the moon being sufficiently near the node to pass within the earth's shadow, and in many years there are accordingly no eclipses of the moon. Only one full moon can occur within the lunar ecliptic limits when the sun passes the node, but if there be an eclipse at one node, there may also be one six months later at the other node. As in the case of the solar eclipses, the "eclipse year" is one of 346·6 days, so that if there be an eclipse of the moon early in January, there may possibly be three altogether in the course of the year, but there could not be three lunar eclipses if the extra solar eclipse were possible. Altogether, then, there may be seven eclipses in the course of a year—five of the sun and two of the moon. Usually there are four or five, some particulars of which are furnished by all respectable almanacs. It will be observed that the number of solar eclipses is much larger than that of lunar ones, but as the latter are visible at all places having the moon above the horizon, while the former are restricted to small parts of the earth's surface, more lunar than solar eclipses are visible at any specified place.

RECURRENCE OF ECLIPSES.—We have seen that the sun requires only 346·6 days to travel from one of the moon's nodes back to the same node again, in consequence of the regression of the nodes, while the moon requires 27·2 days. Suppose, then, that the moon and sun are at a node, and there is an eclipse at new moon; after 346·6 days the sun will return to the same node, but the moon will not be at the node, nor will it be exactly new. It will not be until the sun has returned nineteen times to the node that the moon is also very nearly new at the same node again. Nineteen returns of the sun to the moon's nodes occupy a period of 6,585·78 days; 223 intervals between successive new moons (synodic months) cover 6,585·32 days, while 242 node passages of the moon require 6,585·357 days. In this period of 18 years 11⅓ days (or $10\frac{1}{3}$ days if there are five, and $12\frac{1}{3}$ if there are three leap years in the interval), the sun and moon thus return to nearly the same conditions as affecting the possibility of eclipses. This period was called the *Saros* by the Chaldeans, by whom it was employed in the prediction of eclipses. The adjustment of periods, however, is not quite precise, so that predictions based upon the Saros are only approximations, which serve as a guide for more accurate computations.

This eclipse period is still more remarkable from the fact that it almost exactly represents 239 passages of the moon through perigee, so that after the lapse of 18 years $11\frac{1}{3}$ days the moon is almost at the same distance from the earth, as well as nearly at the same phase and the same distance from a node.

As the Saros includes a fraction of a day, an eclipse is not necessarily repeated at the same place after the lapse of 18 years $11\frac{1}{3}$ days, for the reason that the eclipse will not occur at the same time of day, and the sun may be below the horizon. After three Saroses, however, the eclipse will be repeated nearly at the same hour, but even then it will not be seen under the same conditions, because the track of the shadow will be in different latitudes, for the reason that the moon does not return *exactly* to the node in the interval between 223 new or full moons, and eclipses can only occur when the moon is new or full.

Beginning as a partial eclipse, an eclipse of the moon will gradually become of greater magnitude at successive intervals of 18 years 11 days, until it becomes a total eclipse, and will again gradually become of smaller magnitude, until it ceases to be reproduced at all. Altogether, it would be repeated once in every 223 months for 865 years.

Since the solar ecliptic limit is greater than the lunar, a solar eclipse is repeated at similar intervals of 18 years for about 1200 years. Most of these eclipses would be partial, 27 would be annular, and 18 total. During this period, the track of the central eclipse would shift northwards if the moon were at a descending node, and southwards if at an ascending node, until finally it passed altogether clear of the earth.

It must be remarked, however, that, in the period corresponding to a single Saros, about 28 eclipses of the moon, and 43 of the sun, usually appear, so that altogether about 71 series of eclipses are in progress. Of the solar eclipses which occur during a period of 18 years, about 12 are total at some places upon the earth.

OCCULTATIONS OF STARS AND PLANETS BY THE MOON.—In its monthly round, the moon is constantly passing in front of some of the stars which lie in its apparent path, and these luminaries will, therefore, at times, be hidden temporarily by the moon's disc. Occasionally a planet may appear in the same line of vision as the moon, and that also will pass from view until subsequent motion again removes the intercepting body. These disappearances are closely allied to the phenomena of eclipses, and receive the name of *occultations.* On account of the moon's eastward movement, it is evident that the disappearance of stars or planets when occulted will take place on the eastern edge of the moon; but since the moon trends north or south in some parts of its orbit, the disappearance near the northern and southern edges may occur slightly on the western side of the north or south point of the moon's limb. Similarly, the reappearance generally occurs on the western side of the moon, but occasionally may occur on the eastern side—that is, when the northern or southern edge of the moon does not much more than appear to graze the stars.

The calculation of the circumstances of an occultation is very similar to that involved in the computation of eclipses. (A simple graphical method for working out the conditions of an occultation is described by Major Grant, R.E., in the *Geographical Journal* for June, 1896.)

ECLIPSES AND OCCULTATIONS OF SATELLITES BY PLANETS.—Just as we find the moon eclipsed by passing through the earth's shadow, we find the satellites of other planets to be at times invisible for a similar reason. We thus observe *eclipses* of the satellites. The satellites may also be invisible to us for the reason that they are behind the planet, and they are then said to be *occulted*. These satellite phenomena are especially remarked in the case of Jupiter, and their observation is one of great interest. When a satellite passes between the sun and the planet it throws a shadow on the surface of the planet similar to that of the moon upon the earth. This is visible to us as a dark spot, and from the centre of that dusky patch an inhabitant of Jupiter would undoubtedly see a total eclipse of the sun. To us on the earth the passage of such a shadow across the planet's disc is but a "transit of the shadow" with its "ingress" and "egress."

The times of all these appearances are computed from a knowledge of the movements of the satellites.

CHAPTER IX.

HOW TO FIND OUR SITUATION ON THE EARTH.

DETERMINATION OF LATITUDE.—In order that we may precisely define our situation upon the terrestrial sphere, we have seen that two measurements are necessary, namely, latitude and longitude. The first of these indicates the angular distance from the Equator, and the latter the angular distance east or west of an arbitrary initial meridian. It is necessary for us then to learn something of how these important co-ordinates can be determined.

In considering the apparent movements of the heavenly bodies in different latitudes, we have already seen that at places on the earth's Equator the north celestial pole is on the horizon, while at the North Pole it is in the zenith, and in other latitudes is elevated at different angles. If one sails from England to the Cape, for example, the Pole Star is seen to gradually get lower and lower in the sky, until, on crossing the Equator, it descends below the northern horizon and is no longer visible. Sailing northward, as to Norway, the Pole Star is seen to get higher in the sky.

Now, although the Pole Star is not exactly at the north celestial pole, it is a convenient guide to the eye as to the location of that very important mathematical point, and what we learn from its behaviour as our latitude is changed is that the altitude of the Pole above the horizon is equal to the latitude of the place of observation.

One of the methods employed for finding the latitude of a place is accordingly to determine the altitude of the Pole. This can be obtained by an instrumental measurement of the altitude of the Pole Star, from which, if the time of observation be known, the altitude of the true Pole, which occupies the centre of the small diurnal circle traversed by the star, can be computed. Tables which save an immense amount of labour in the calculations involved are given in the "Nautical Almanac," and in "Whitaker's Almanac."

Another method of finding the elevation of the Pole is to take advantage of the fact, that at intervals of twelve sidereal hours the Pole Star passes the meridan alternately above and below the Pole. If, then, one finds the altitudes at the upper and lower transits, and corrects them for refraction, the average of the readings is a measure of the altitude of the true Pole, and therefore of the latitude. Other stars which are circumpolar may be employed for the same purpose, and this method has the great advantage that a knowledge of the correct time, or of the exact position of the star observed, is superfluous. The disadvantage is that the correction for refraction, especially in low latitudes, cannot be made with the necessary degree of accuracy. It must be remembered that an error of only 1′ in latitude implies a mistake of a mile measured on the earth's surface.

Other methods, however, are available. As we go southwards, not only does the Pole Star become lower in the sky, other stars in the southern part of the sky become higher at the same rate that the Pole Star descends. Other stars can therefore be utilised, and in order that refraction may affect the observations as little as possible, stars of known declination near the zenith are observed. Suppose an observer, situated at O (Fig. 34) on the earth's surface, observing a star S on his meridian, O Z will represent his zenith, and O E, parallel to the Equator, will be the direction in which he will see the celestial equator where it crosses his meridian. The declination of the star, represented by the angle S O E, has been previously determined with great accuracy, and the angle S O Z, the zenith distance of the star, is the angle which he measures. In the case illustrated by the diagram, the difference between the declination and the zenith distance will give the angle Z O E, which is evidently equal to the latitude O C Q. To get rid of the ever troublesome refraction of our atmosphere, stars which pass as nearly as possible through the zenith are selected for observation, and stars both to north and south are observed.

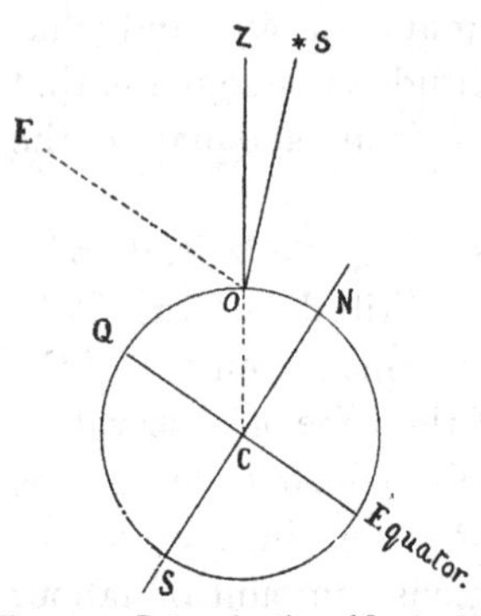

FIG. 34.—*Determination of Latitude.*

Another way of determining the latitude, which is very commonly employed, is known as Talcott's method. The observations are made with the aid of a zenith telescope. The latitude being approximately known, two stars are selected which transit nearly at the same time and nearly at the same distance from the zenith, one to the north and the other to the south. That which transits first is brought to the centre of the field of view, which is marked by a spider thread. The instrument is then reversed in its bearings so that it points at the same angle on the opposite side of the zenith. When the second star comes into the field, the telescope is kept fixed, and a moveable spider thread is made to coincide with the star passing through the field. The distance between the spider threads furnishes a measure of the difference in zenith distances. Half the sum of the declinations added to half the difference of zenith distances gives the latitude when this method is employed.

Various other methods have been devised for the precise determination of latitude, but the foregoing will sufficiently serve to illustrate the processes followed when the observations are made on land.

Before the invention of astronomical instruments, latitude was approximately measured by the lengths of shadows. At the summer solstice, at noon, the shadow of a vertical stick is at its shortest, while at the winter solstice it is longest. By measuring these lengths, a diagram can be made showing the altitude of the sun at noon on each occasion. Midway between these will be the altitude of the celestial equator where it crosses the meridian. Since the altitude of the Pole is equal to the latitude, the altitude of the Equator, subtracted from 90°, thus gives the latitude.

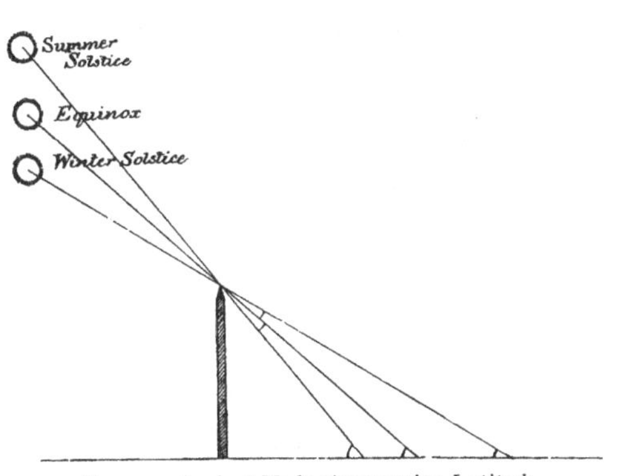

FIG. 35.—*Ancient Mode of measuring Latitude.*

It will be noted that this *gnomon* experiment also furnishes a measure of the obliquity of the ecliptic. The gnomon was in use by the ancient Chinese, and it is also believed that the Egyptian obelisks which are now embellishing various cities were originally erected for the same purpose.

DETERMINATION OF LONGITUDE.—As we have imagined an observer travelling in a north or south direction in connection with the measurement of latitude, let us consider what will happen to an observer who travels only in longitude—that is, east or west. At the starting-point, he will see the

Pole at a certain altitude, and the stars will perform their diurnal revolutions at a certain inclination to the horizon depending upon his latitude. If he travels towards the east, the Pole will remain at the same angle above the horizon, and he will detect no difference in the apparent movements of the stars. What then is there to indicate that he has changed his place at all? The answer is simple; he will find that the sun and stars cross the meridian earlier, and if he be 15° east of his first station they will transit an hour sooner, because it takes the earth an hour to turn through that angle. If he travel westward in the same way, the earth must turn through a greater angle to bring him back to the same star, so that the stars will appear to cross the meridian later.

The determination of longitude is accordingly based upon a measurement of the difference in the times of transit of sun or stars at the place of observation, and the place from which longitude is reckoned.

Let us take Greenwich as the start-point for our longitudes, and suppose we are in Dublin. The sun, or a star, will cross the meridian of Dublin at a certain interval after it has passed that of Greenwich, and if we measure this interval, the angle turned through by the earth in that time will determine the longitude. With a transit instrument one can readily tell the exact moment when the star crosses the meridian of Dublin, but how is one to know the exact moment at which the star crossed the meridian of Greenwich without going there?

Looking at the question in another way, let us remember that the clocks in Dublin register local time, that is time reckoned from the passage of the sun over the meridian of Dublin, while the Greenwich clock indicates times based on the transit of the sun over the Greenwich meridian. Evidently the difference of these times is the difference of longitude, and our question becomes, how to find the time at Greenwich when stationed at the observatory in Dublin.

In all modern work, the telegraph is employed whenever it is available, the two stations being directly connected. An observer at Greenwich is thus enabled to transmit a signal to the observer in Dublin at the exact moment a star passes through the centre of his transit instrument, and the latter observer then notes the interval which elapses before the same star passes the central line of his own instrument. If the signals were transmitted instantaneously, the interval elapsed from the reception of the signal to the observed transit of the same star would give the longitude as reckoned in time.

Practically, what is done is for each observer to determine his local sidereal time very accurately, with the aid of his transit instrument, and in this way to find the error of his clock. It is then only necessary to compare the two clocks, and this is done in the following way: the clock at Greenwich has an attachment by which an electrical contact is made every second, and this is switched in to the telegraphic circuit, so that the Dublin observer receives a signal every second so long as the clock is connected. These signals are automatically recorded by a chronograph, together with similar signals from the Dublin clock, and the times to which each of them corresponds is easily identified. Immediately afterwards the Dublin clock is switched into the circuit, and records its beats on the chronograph sheet at Greenwich, alongside those sent by the Greenwick clock. In this way the differences between the clocks can be very accurately measured, and the longitude can then be reckoned in degrees and minutes by allowing 15° for each hour. Before the invention of the telegraph, less accurate methods were of necessity employed. Among others the entrance of the moon into the earth's shadow during an eclipse was noted by an observer desiring to know his longitude. As we have already seen, this occurrence is independent of the observers' position on the earth, so that if he records the local time of the observation and compares with the calculated Greenwich time of the commencement of the eclipse, he can find his longitude. Similarly, the eclipses of the satellites of Jupiter may be utilised to signal Greenwich time to an observer situated elsewhere. Unfortunately, the shadows are too ill-defined at the edges to permit very accurate determinations in this way.

METHODS EMPLOYED AT SEA.—One of the most important applications of astronomy to the needs of everyday life is in enabling the navigator on the open ocean to determine the situation of his ship. Without the help supplied by astronomical predictions the sea would be truly trackless, and commerce by sea would be almost impossible.

A sextant and two or three good chronometers, together with a copy of the current "Nautical Almanac," furnish the means of ascertaining the geographical position of a ship. With the aid of the sextant, the sun's greatest angular distance above the sea horizon—that is, its meridian altitude—is measured, and from the known declination of the sun at the time, the latitude is deduced in exactly the same way as in the case of an observation of a star (p. 124).

The sextant also enables the observer, by measuring the sun's altitude in the early morning or evening, to determine the local time, as already explained (p. 83). Greenwich time is kept by the chronometers, and the difference between this and the local time is a measure of the longitude. More than one chronometer is carried by a ship, for fear that a single one might fail, through accident or other causes, to give correct readings. The rate of each has been previously very accurately gauged, and by taking the average indications, Greenwich time is known with considerable accuracy.

Should the chronometers fail, or any doubt be thrown upon their accuracy, there is another method by which the Greenwich time, and thence the longitude, can be ascertained. This is the *lunar method*, in which the heavens become the equivalent of the dial of a clock, while the moon, with its rapid easterly movement, plays the part of the hands.

In the words of Dr. Lardner, this is "a chronometer of unerring precision ; a chronometer which can never go down, nor fall into disrepair ; a chronometer which is exempt from the accidents of the deep ; which is undisturbed by the agitation of the vessel ; which will at all times be present and available to him wherever he may wander over the trackless and unexplored regions of the ocean."

From the known movements of the moon, its position with regard to the sun, planets, or conspicuous stars, at definite Greeenwich times, can be calculated in advance, and "lunar distances" are accordingly tabulated in our nautical almanacs. We find, for instance, that the apparent distances of the moon from the star Regulus, as they would appear from the earth's centre, were as follows on Jan. 1, 1896:—

6 P.M.	G.M.T.	35° 50′ 22″
9 P.M.	"	34° 3′ 23″
12 P.M.	"	32° 16′ 12″

To utilise these predictions for the purpose in hand, the observer would measure with the sextant the apparent distance of the moon from Regulus at a known local time, and he would then compute what the apparent distance would have been if his observation had been made from the earth's centre. From the tabulated distances, he would then be able to find the Greenwich time at which his observation was made; and, as we have seen, the difference between this and local time is a measure of the longitude.

CHAPTER X.

THE EXACT SIZE AND SHAPE OF THE EARTH.

GEODESY.—We have already seen that the earth is a sphere, or of some form which differs but little from a sphere, and a rough method of determining its size, on this supposition, has been indicated. Now we have to inquire more minutely into the size and shape of our planet, for, as we shall see presently, a knowledge of these facts is essential to the adequate explanation of the various movements of the heavenly bodies, besides forming the basis of all our knowledge of the distances which separate us from the other bodies which people space. As an illustration of the importance of an exact knowledge of the size of the earth, it may be remarked that Newton's grand law of gravitation was kept from the world for ten years, owing to an error in the generally accepted value of the earth's radius, which was afterwards rectified by the labours of a French astronomer, Picard.

A great amount of labour has been expended in the endeavour to arrive at the true size and shape of the earth, and the name *geodesy* is given to the science which deals with these operations. As a secondary object, geodesy is concerned with the measurement and description of tracts of country.

AN ARC OF MERIDIAN.—The measurement of the size of the earth is accomplished by first measuring relatively small parts of its surface, and then applying geometrical principles, in order to determine the whole circumference. If the earth were a true sphere, and we could measure the exact distance in miles between two places on the same meridian, a subsequent determination of the difference of latitudes of the two places would enable us to find the length of a degree, measured on the earth's circumference. As there are 360° in a circle, the circumference would be 360 times the length of a degree, and the diameter of the earth would be the length of the circumference divided by 3·14159, this number expressing the constant ratio which exists between the circumference and diameter of a circle of any size whatsoever.

The determination of the size and shape of the earth thus involves two distinct sets of operations; first, measures of distances; and second, astronomical observations to determine the angular measurements of the arcs on the earth's surface comprised between stations separated by known distances. When two such stations lie on the same meridian, the arc measured in this way is called an *arc of meridian.* We have already seen what means are available for finding the latitudes and longitudes of places on the earth, and it now remains for us to apply a yard measure, or its equivalent, to the precise measurement of the distance between places which are many miles apart.

THE BASE LINE.—In the first instance a line of unimpeachable straightness is measured with scrupulous accuracy. The measuring-rod which has been most successfully employed is one consisting of a combination of brass and steel bars, which automatically corrects itself for changes of temperature in very much the same way that the balance-wheel of a chronometer, or of a good watch, corrects itself so as to perform its swing in equal periods at all temperatures. Several of these compensated rods are used, and they are enclosed in wooden boxes which are provided with levels and sights. When in use the outer boxes rest on adjustable trestles, and instead of putting the rods end to end they are placed a certain definite distance apart by the use of microscopes, which are themselves mounted on compensating bars. The first rod is put in position and levelled, and the others are successively placed in line with it by means of the sights. As the ground ceases to be perfectly flat it becomes necessary to raise the level of succeeding bars, but they are kept in the same vertical plane. Six bars are frequently employed in laying out a base line, and in order to protect them from extremes of temperature they are usually kept covered with long tents. In this way a distance of several miles can be measured with no greater probable error than a couple of inches, and the ends of such a measured base line are marked on metal plugs built in columns of masonry. The chief base lines measured in connection with British map construction were on the sandy shores of Lough Foyle in Ireland, 41,614 feet in length, and on Salisbury Plain, 36,578 feet long.

TRIANGULATION.—When a base line has been accurately measured in this way, a distant object which is clearly visible from both ends is observed with the aid of an instrument called the *theodolite*, and the angles between the base line and the lines joining its ends with the object are very carefully determined. Thus if A B in Fig. 36 represent the base line, and C a conspicuous object several miles away, the angles C A B and C B A are measured, and then it becomes easy to determine the distances A C and B C by trigonometrical calculations. A check on the accuracy of the observations is obtained by transferring the theodolite to C and measuring the angle A C B. The sides of the triangle may then be employed as new base lines for the measurement of other distances. With the theodolite at C, another object, D, is sighted, and the angle D C A is measured; similarly, with the theodolite at A, the angle C A D is determined, and from these observations the distances of D from the points A and C are easily computed. These distances again become available for base lines, and so the triangulation can be extended indefinitely.

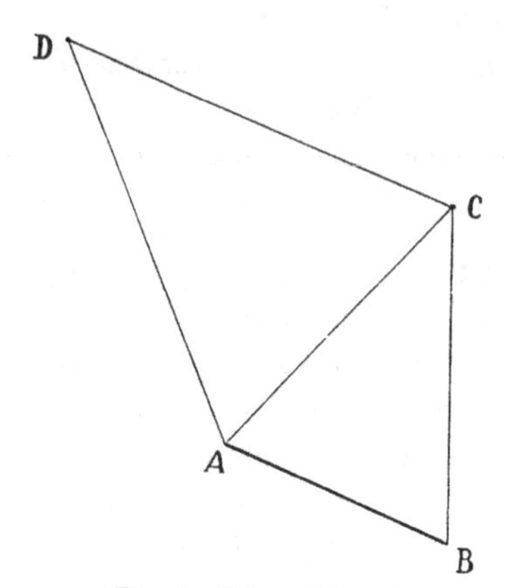

Fig. 36.—*Triangulation.*

In a mountainous country, the sides of the triangles are often as much as 100 miles in length. Signals on the Wicklow Mountains in Ireland have been observed from Ben Lomond in Scotland and from Scafell in Cumberland. The stations are chosen so that none of the angles to be measured are very small, and in this way the chances of error are greatly reduced. Hence the triangles in the immediate neighbourhood of the base line are comparatively small, but the sides are gradually extended as the survey proceeds.

The process of triangulation forms the basis of the construction of accurate *maps*, and for this purpose the great triangles are sub-divided by a secondary triangulation, so that the exact situations of a very great number of places are determined. These, again, serve for another set of still smaller triangles, with sides perhaps a mile in length; and finally the details are filled in by local chain surveys and draughtsmanship.

There is another point of some importance in connection with these triangulations when on a large scale. The larger triangles must be corrected for the curvature of the earth's surface. The construction of the theodolite is such that two adjacent sides of any triangle, measured from their intersection, are referred to the same horizon; but when the instrument is transferred to another corner of the triangle, the adjacent sides are referred to a new horizon. The sum of the three angles of a triangle in these geodetical surveys thus exceed two right angles, whereas in plane triangles they are always equal to two right angles; the difference is called the *spherical excess*, and in the computations the observed angles have to be corrected on this account.

Thus, after an extremely laborious survey, it becomes possible to determine with great accuracy the distance between any two places whatever, and so the number of miles between two places at the extremities of an arc of meridian is ascertained. An arc of meridian extending nearly 18° has been measured in India, and another over 25° long extends from Hammerfest in Norway to the mouth of the Danube.

EXACT SHAPE AND SIZE OF THE EARTH.—From the facts which have been gleaned by the measurements of arcs of meridian in different parts of the world, it is found that the length of a degree of latitude as measured on the earth's circumference increases towards the Poles. In latitude 66° N. a degree is about 3,000 feet longer than a degree near the Equator. This means that the curvature of a meridional arc is greatest at the Equator, whence it is concluded that the earth is flattened at the Poles. The figure which best accords with the observations is the ellipse, and thus it becomes possible to calculate the polar diameter, although no arcs have been measured in the immediate neighbourhood of the Poles.

Arcs of longitude, extending between two places which have the same latitude, have also been measured and applied to the determination of the figure of the earth, and, indeed, any arcs between two places of known latitude and longitude can be utilised.

When all the facts are brought together it is found that the earth's polar diameter is about 26 miles shorter than the average equatorial diameter, while an equatorial section of the earth is also elliptical, the diameter passing through longitude 14° E, being two miles longer than the one at right angles to it. According to the calculations of Colonel Clarke, R.E., we have the following principal dimensions :

Earth's mean equatorial semi-diameter	= 3,963·296 miles.
" " polar "	= 3,950·738 "
Polar compression	$\frac{1}{293 \cdot 46}$

A solid which has a shape like that of the earth, with three axes of unequal lengths, is called an *ellipsoid.*

A very important consequence of the ellipsoidal form of the earth is that lines which are vertical—that is, perpendicular to the surface of water—do not pass through the centre of the earth, unless they are at the Poles or at certain points on the Equator.

There is every reason to suppose that at one time the earth was in a molten condition, and in response to physical laws, such a mass of matter could not retain a spherical form when set in rotation, although the sphere would be its natural shape if at rest. This has been demonstrated by a variety of experiments.

Thus, taking it generally, the shape of the earth is very intimately associated with its rotation, and it will subsequently appear that the same holds good for the sun and planets. Those bodies which have the most rapid rotation show the greatest flattening in the direction of the polar diameter.

In addition to direct measurements of the earth, there are

other ways of studying the shape of our planet. One of these depends upon observations of the swing of a pendulum at different parts of the earth's surface; as the time of oscillation of a pendulum depends upon the force of gravity, which itself varies with the distance from the earth's centre, it is evident that this method is a practicable one. It is true that the matter is complicated in various ways, but after everything has been taken into account, these pendulum observations indicate, not only that the earth is flattened at the Poles, but they show further that the amount of polar compression deduced from geodetical work is in all probably very near the truth.

Again, the movement of the moon around the earth is found to be subject to certain irregularities which would not exist if the earth were a perfect sphere. These inequalities being deduced from observations of the moon's position, the amount of polar flattening necessary to produce them can be calculated, and this is found to agree very closely with the value derived from the measurements of arcs of meridian.

DIFFERENT KINDS OF LATITUDE.—If the earth were a smooth spherical body, the latitude of a place would be simply equal to the angle made by a line joining it to the earth's centre with the plane of the Equator. Owing to the bulging out of the earth in its equatorial part, however, it becomes necessary to distinguish between different kinds of latitude. If we adopt the definition given above, the name of *geocentric latitude* is given to the angular measurement. Taking the earth as a smooth geometrical spheroid, and assuming it to have certain dimensions, the angle which a line perpendicular to the surface makes with the plane of the Equator determines the *geographical latitude*. As the line perpendicular to the surface does not pass quite through the centre of the earth, the geographical and geocentric latitude differ by as much as 11′ in mid-latitudes, although nearly agreeing at the Poles and on the Equator.

As there are no direct means of finding the direction of a line passing through the earth's centre, or of one perpendicular to the imaginary standard spheroid, geocentric and geographical latitudes must be calculated from the *astronomical latitude*, which is determined by observations of the elevation of the Pole, or its equivalent. The astronomical latitude is the angle between the direction of gravity and the Equator, and is therefore to a small extent dependent upon local irregularities of the earth's surface.

A knowledge of geocentric latitude is chiefly of use in making corrections for parallax, in order that the data calculated for the earth's centre may be precisely corrected for the place of observation, or *vice versâ*, as in the case of a lunar distance measured for the determination of longitude, or in the calculation of a solar eclipse.

VARIATION OF LATITUDE.—For some years past a widespread interest has been taken in the question of a possible change in the position of the earth's axis with regard to its surface. The subject is by no means a new one, for as far back as two thousand years ago, such variations were suspected. Changes amounting to several degrees were then believed to have occurred, but it is now certain that the supposed variation was due solely to the imperfection of the observations. As astronomical science became more and more precise, even before the discovery of aberration, it became evident that if any changes of latitude were taking place at all, they must be very minute.

In its geological aspect, the possibility of great changes of latitude having occurred in the past history of our globe is evidently well worth serious investigation. Granted a sufficient change in the position of the earth's axis, the climate of London might become Arctic, or that of Greenland tropical. From this point of view the subject has been mathematically investigated by Professor G. H. Darwin, and it appears that if only the varying distribution of land and sea indicated by the geological records be taken into account, past changes of more than about three degrees are very improbable. Admitting that at any time during the life-history of our globe the earth was sufficiently plastic to be deformed by earthquakes or other disturbances, it is possible that changes amounting to 10° or 15° may have occurred.

Opinion is perhaps best reserved as to what has happened in the past. We are on surer ground when we consider the variations of latitude which are now going on.

Many competent observers have investigated the present movements of the Pole, and it has been conclusively demonstrated that changes in the position of the earth's axis do really occur. Dr. Küstner, of Berlin, commenced a series of observations for a different purpose in 1884, and found that some anomalous results could only be explained by supposing that the latitude of Berlin was from 0″·2 to 0″·3 greater from August to November, 1884, than from March to May in 1884 and 1885. Great interest was excited by this striking result, and steps were at once taken to test its truth. Old observations were re-discussed and compared, and new observations were made, with the final result that the movement of the earth's axis of rotation was placed beyond dispute. It was not until Dr. Chandler attacked the problem, however, in 1891, that the nature of the changes became clear. His masterly analysis indicated that the observed variations in latitude arise from two periodic fluctuations superposed upon each other; one of these has a period of 427 days, and a semi-amplitude of 0″·12, while the other is an annual change which has ranged between 0″·04 and 0″·20 during the last fifty years. The resultant of the two movements produces changes which are seemingly very irregular in amount and of varying period, but a cycle is completed about every seven years. When the two sources of difference are at their maximum at the same time, the total range reaches about two-thirds of a second of arc. In consequence of the inequality of the annual part of the change, the apparent average period between 1840 and 1855 approximated to 380 or 390 days; widely fluctuated from 1855 to 1865; from 1865 to about 1885 was very nearly 427 days, afterwards increased to near 440 days, and very recently fell to somewhat below 400 days.

At the present time the variation of latitude is being very carefully investigated by the International Geodetic Association, and the latest results obtained are illustrated diagrammatically in Fig. 37. The mean position of the Pole is at the centre of the diagram,[1] and the horizontal line to the right of this point is directed towards Greenwich. The re-

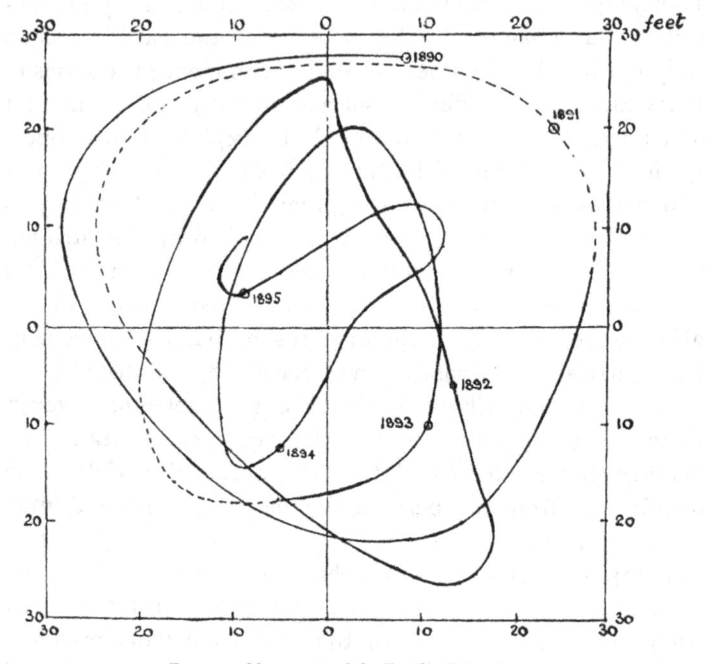

FIG. 37.—*Movements of the Earth's Pole*, 1890-95.

markable spiral curve shows the wanderings of the Pole about its mean position during five recent years. To simplify matters, the amount of deviation is represented in feet instead of in angular measure, and it will be seen that although the variation of latitude may be of considerable interest and importance in astronomical matters, it really does not amount

[1] The diagram is based upon one given by Prof. Albrech in the *Astronomische Nachrichten*, No. 3333. The dotted part of the curve could not be directly derived on account of insufficient observations.

to very much in matters terrestrial, the greatest change in the position of the Pole not amounting to more than 20 yards. Nevertheless, it is not inconceivable that it may yet have to be reckoned with in questions relating to boundary lines which depend upon latitude determinations.

CHAPTER XI.

THE DISTANCES AND DIMENSIONS OF THE HEAVENLY BODIES.

PARALLAX.—The problem of determining the distance of a heavenly body resolves itself into a measurement of its *parallax*, that is, of the apparent change of its position brought about by a change in the situation of an observer. If one be seated in a room, about a yard from a window, a very simple experiment may be made to illustrate the meaning of this term. Closing one eye, the observer will see a vertical line, such as the partition between two panes, projected upon some particular part of an opposite building; when the other eye is used the line will apparently be displaced, and the nearer one is to the window the greater will be the displacement or parallax. As the heavenly bodies are so far away, each of our eyes sees them in the same directions. Indeed, the stars are so distant that to *all* persons situated on our planet their apparent positions are identical. With the members of the solar system, however, the case is different; the earth has an appreciable size as seen from them, so that when viewed from different parts of the earth they will not appear in exactly the same part of the heavens.

The earth's rotation changes the relation of an observer's position with regard to a heavenly body in pretty much the same way as a change in his actual position on the globe.

When an object in the zenith is observed, it will appear in precisely the same part of the sky as if it were seen from the centre of the earth, but as it approaches the horizon it will be displaced. Hence the term *diurnal parallax*, meaning the displacement of a heavenly body depending upon the observer's position as affected by the earth's rotation. Taking it in its general astronomical sense, the parallax of a heavenly body is the angle between the two lines which join it to the observer and to the centre of the earth respectively. Thus, in Fig. 38, let O be an observer, Z his zenith, and C the centre of the earth; then the parallax of a body S is the angle O S C. As the observer's position is changed to O′ by the earth's rotation, the parallactic angle is increased to O′ S C. If S be on the horizon, that is, when O′ C is perpendicular to O′ S, the parallax is a maximum, and is then called the *horizontal parallax*. The horizontal parallax of a body is therefore the greatest angle subtended by the earth's radius as seen from the body. We have seen, however, that the earth's radius is not of the same length in all parts, and it is therefore necessary to specify more particularly which radius is in question. The standard adopted is the equatorial radius, and, when this is employed, our greatest parallactic angle is called the *equatorial horizontal parallax*.

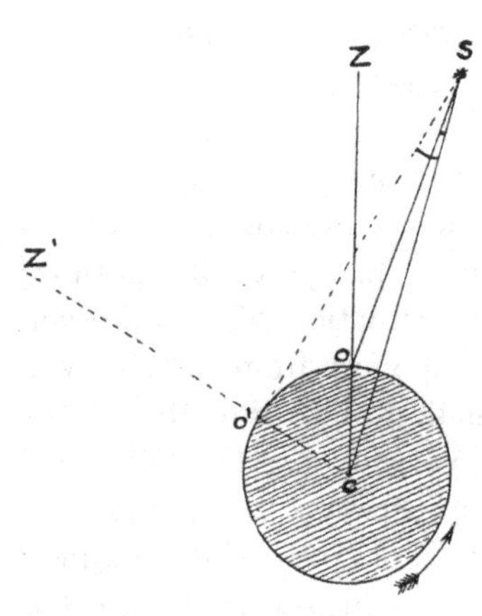

FIG. 38.—*Parallax of a Heavenly Body.*

In the case of all the heavenly bodies the parallaxes are very small; that of the moon averages about 57′, while that of the nearest planet does not exceed 40″. The parallax of a body evidently diminishes as the distance increases.

DISTANCE DEDUCED FROM PARALLAX.—When the

parallax of a heavenly body has been determined, it becomes a simple matter to calculate the corresponding distance; thus, in Fig. 38, the distance C O′ represents the earth's equatorial radius, O′ S C is the equatorial horizontal parallax, C O′ S is a right angle, and the required distance is C S. By a simple trigonometrical rule this distance is the earth's radius divided by the sine of the parallax. In the case of a small angle, the sine is very nearly equal to the angle itself divided by the angle corresponding to an arc of a circle equal in length to the radius. As there are 206,265 seconds in an arc equal to the radius, the sine of a small angle may be taken as the angle itself, expressed in seconds, divided by this number. Thus, if p be the equatorial horizontal parallax of an object reckoned in seconds of arc,

$$\text{Distance} = \frac{\text{earth's equatorial radius}}{\text{sine } p}$$

$$= \frac{206{,}265 \times \text{earth's equatorial radius}}{p}$$

We shall see presently that the average parallax of the sun is $8''{\cdot}80$, and its average distance, as derived from the application of this formula, is accordingly about 92,790,000 miles.

DIAMETERS.—It is a familiar fact that the further an object is removed from us the smaller it appears. The ascent of a balloon at once suggests itself as an excellent example. It is necessary, therefore, to distinguish very carefully between the apparent and the true size of an object. A halfpenny at a distance of nine feet from the eye will just cover the moon if the line of sight be directed towards that body, but we should not say the moon is the size of a halfpenny, because we know perfectly well that a disc twice the size would produce just the same appearance if removed to double the distance. Apparent size must, accordingly, be reckoned in angular measure, and we say, for example, that the moon has an apparent diameter of a little more than half a degree.

When the angular diameter and distance have both been

measured, the real diameter, in miles, can at once be deduced by a simple inversion of the process of determining the distance of an object from its known parallax. Thus, in Fig. 39

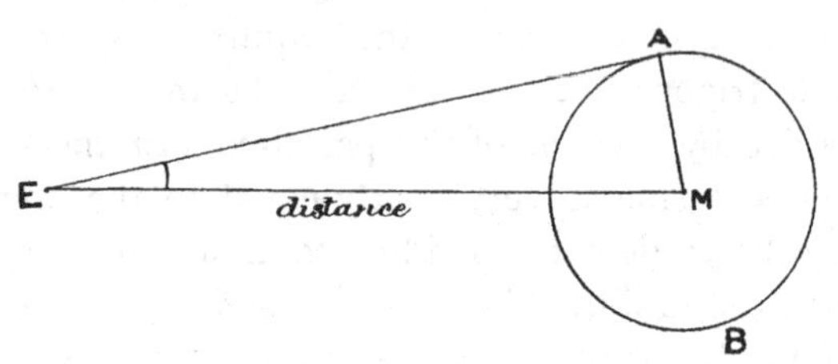

FIG. 39.—*Diameter of a Heavenly Body.*

let A B represent the moon or other heavenly body, and E the centre of the earth. The angle M E A is the angular semi-diameter, and E M the required distance; then, since the angle E A M is a right angle,

$$\text{A M} = \text{M E} \times \text{sine M E A}$$

That is,

Semi-diameter in miles = distance in miles × sine of angular semi-diameter.

Or,

Diameter = twice the distance × sine of angular semi-diameter.

Since the apparent diameters are always small, the sine may be taken as equal to the circular measure; that is, the number of seconds which the angle contains divided by 206,265.

DISTANCE AND SIZE OF THE MOON.—If the moon were a fixed body outside the earth, its parallax could be easily determined by a single observer, who, in that case, would note the apparent displacement produced by his rotation. It has, however, a very complex movement, and it is therefore difficult to separate the real change of position from the parallactic change. The best method is one in which two observers, far removed from each other, can observe the moon's position at nearly the same instant, so that the effect of its movement is very small and can be sufficiently allowed for. A necessary consequence of this condition is that the two observers should be placed as nearly as possible on the same meridian. Observations with the object of determining the lunar parallax have accordingly been made at Greenwich and the Cape of Good Hope. From the known positions of these places and the size of the earth, the distance between them is very accurately known, and this serves as a base line in a triangulation of the moon.

If G and C, in Fig. 40, represent Greenwich and the Cape respectively, the celestial equators at the two places will be in

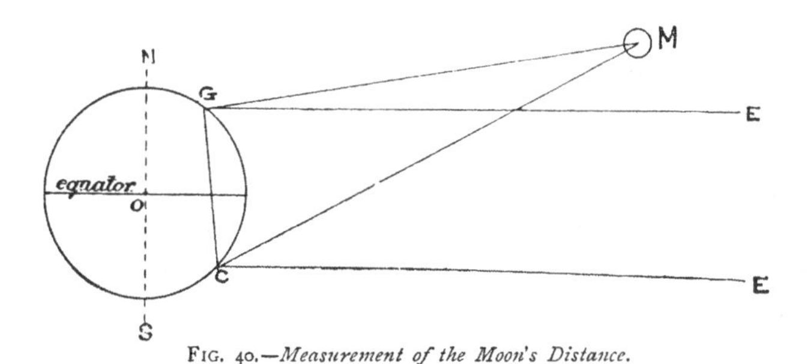

FIG. 40.—*Measurement of the Moon's Distance.*

the directions G E and C E. M being the moon, its declination, as measured at G, will be the angle M G E, and as measured at C it will be the angle M C E′. Since G E is parallel to C E′, the difference of these declinations (when both are north declinations, as in the diagram) will be the value of the parallactic angle G M C, which is about $1\frac{1}{2}°$. From these data it is easy to calculate the distance of the moon either from Greenwich, the Cape, or the earth's centre. In this way the distance of the moon is found at some particular moment, and the additional knowledge of the shape of its orbit enables us to determine the semi-major axis of the orbit, which is nothing more than the average or mean distance of the moon. The mean equatorial horizontal parallax of the moon is 3,422″·5, and the corresponding mean distance from the earth is 238,855 miles.

The average apparent diameter of the moon, as it would appear from the centre of the earth, is 31′7″, from which it results by the method already stated that the true diameter is 2,162 miles.

The apparent diameter of the moon is affected by the observer's position upon the earth, as well as by the situation of the moon in its orbit. An observer to whom the moon is directly overhead is nearly 4,000 miles nearer to it than another observer who has it on his horizon. Tables have accordingly been drawn up to indicate the *augmentation* of the moon's apparent diameter as it rises above the horizon. The greatest possible apparent diameter is about 36″.

Everyone must have noticed that when the moon is rising or setting, it looks much larger than when it is high up in the sky, an appearance which does not seem to accord with the fact that its measured angular diameter is least when on the horizon. It is evident, however, that the seeming increase of size is a subjective phenomenon, due to our incapacity to correctly judge distances.

RELATIVE DISTANCES OF PLANETS.—The relative distances of the planets from the sun were found long before any of the actual distances were known with any reasonable degree of accuracy. Kepler discovered the relation which exists between these distances, and expressed it in his third or harmonic law, which states that "the squares of the periodic times of the planets are proportional to the cubes of their mean distances from the sun."

In the case of the interior planets, the angles of greatest elongation furnish the means of finding their distances from the sun as compared with that of the earth. Thus, if V in Fig. 41 represents Venus, E the earth, and S the sun, the angle E V S is a right angle when Venus is at greatest elongation. The observed value of the angle S E V is 46°, and this definitely determines the shape, though not the size, of the triangle S E V. The distance of Venus from the sun, S V, is thus found to be 0·72 times the distance of the earth from the sun, S E. If Venus be at inferior conjunction, that is, at V′, its distance from the sun will be represented by 72, if the earth's distance from the sun be denoted by 100.

This method can also be applied in the case of Mercury, but as the orbit is so eccentric, it is necessary to take the average of a large number of greatest elongation angles.

The process of determining the relative distance of an exterior planet, such as Jupiter, is a little more complex, but involves no considerable difficulties.

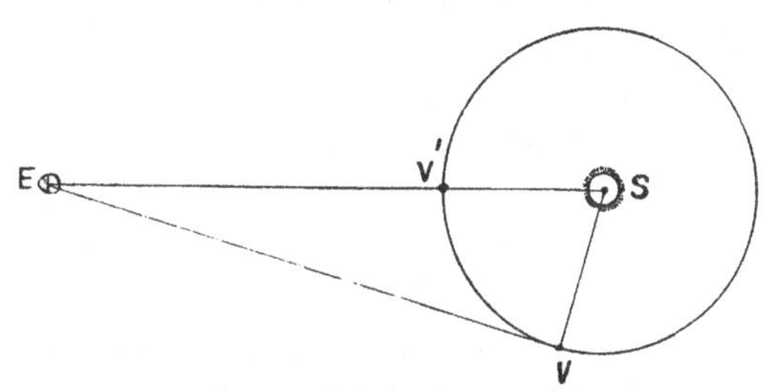

FIG. 41.—*Relative Distance of Venus.*

There is a curious relationship between the relative distances of the planets, which is commonly known as *Bode's law.* A series of figures, 0, 3, 6, 12, 24, 48, 96, 192, 384, each, with the exception of the second, being double the preceding one, is written down, and the number 4 added to each. Then the resulting numbers approximately represent the relative distances of the planets from the sun. Thus:—

4	7	10	16	28	52
Mercury	Venus	Earth	Mars	Asteroids	Jupiter

100	196	388
Saturn	Uranus	Neptune

It is interesting to note that this law was announced in 1772, when the asteroids and the planets Uranus and Neptune were still unknown, so that there was a break in the series corresponding to the number 28. The discovery of Uranus in 1781, and the fact that its distance agreed roughly with Bode's law, strengthened the conviction that an unknown planet revolved round the sun in an orbit between those of Mars and Jupiter. An association of astronomers was then formed to search systematically for the missing planet; but the actual discovery was made in 1801 by Piazzi, the Sicilian astronomer, who had not joined the association. The new planet was a very small one, and its discovery was rapidly followed by the detection of several others. At the present time, more than 400 of these asteroids, or minor planets, are known, and their average distance fits in very well with Bode's law.

THE SUN'S DISTANCE.—One of the grandest problems which astronomical science requires us to solve is the determination of the sun's distance. Starting with a knowledge of the earth's dimensions, the subsequent measurement of the sun's distance enables us to get a clear idea of the scale, not only of the solar family to which we ourselves belong, but of the whole sidereal universe. No wonder then that a vast amount of astronomical energy has been expended on this investigation.

The problem, however, is beset with many practical difficulties, and the greatest possible skill is required to cope with it. In the first place, the parallax of the sun is so small that the method employed for the moon fails, and it can only be determined by indirect means.

We have already seen that the constant of aberration gives us a means of determining the size of the earth's orbit, and consequently the distance of the sun. When proper allowance is made for the eccentricity of the orbit, this method is a very valuable one.

Other methods which have been employed depend upon the measurement of the parallax of one of the nearer planets, from which the distances of all the planets, including the earth, from the sun, can be found from our previous knowledge of the relative distances. Mars and some of the asteroids have been thus utilised at their oppositions, and Venus when at inferior conjunction.

The parallax of Mars can be determined in the same way as that of the moon, either by concerted observations at two distant places, or by a single observer who utilises the earth's rotation to provide him with a base line. The actual measurements do not consist of direct estimations of the right ascension and declination of the planet, but of its angular distances from stars among which it appears, the measurements being made with micrometers or heliometers. In this way certain errors due to refraction, etc., are minimised. To

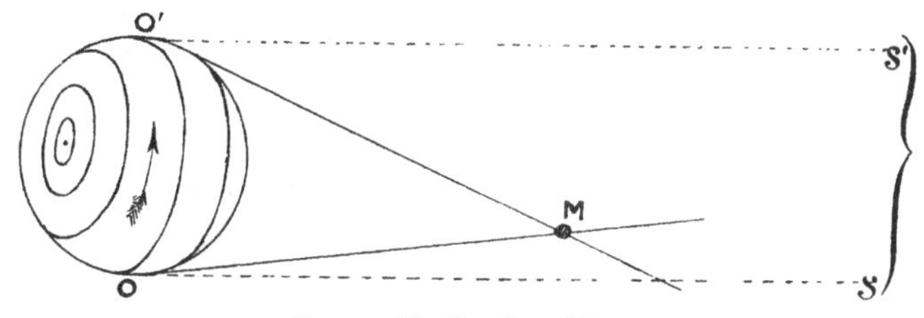

FIG. 42.—*The Parallax of Mars.*

take an extreme case, let the planet M (Fig. 42) be rising to an observer at O ; it will then be seen in the direction O M, while a neighbouring star will be seen along the line O S. After twelve hours the rotation of the earth will have carried our observer to O′, and he will now see the planet in the direction O′ M, while the star will remain in the same direction, O′ S′ In each case he would measure the angle separating the planet from the star, and would thus obtain the values of the angles S O M and S O M, which, in the case shown in the diagram, would be together equal to the angle O M O When corrected for the observer's latitude, and for the planet's change of place in the interval, the equatorial horizontal parallax of Mars would be determined. Then the distance of Mars from the earth would be known, and at opposition this is the difference between the distances of the earth and of Mars from the sun ; the ratio between the latter is already known, and their actual distances at once follow.

TRANSIT OF VENUS.—The planet Venus at inferior conjunction is near enough to the earth to have a considerable parallax, but the method employed in the case of Mars cannot be used, as the planet is not visible when between us and the sun, except on the very rare occasions when it transits across the sun's disc. When a transit occurs, the distance of the planet from the earth can be measured in essentially the same way as that of Mars at opposition, when two observers work together. The difference is that the apparent place of the planet is referred to the sun's disc instead of to neighbouring stars. Suppose the conditions to be as represented in Fig. 43, E being the earth, V the planet, and S the sun. Two observers on the earth, at *a* and *b*, will see the planet projected on different parts of the sun's disc. If we at first regard them as being at rest, the observer at *b* would see the planet cross the sun along the line C D, while to the one at *a* it would appear to cross the line F G. The times of crossing would, under the assumed conditions, depend upon the orbital velocity of Venus, and a measure of these times at the two stations would determine the relative lengths of the chords C D and F G. We already know that the distance of Venus from the sun is to its distance from the earth at inferior conjunction in the proportion 72 to 28. (See p. 145.) The rectilinear distance between the two places is also known, and the distance *x y* between the chords is $\frac{72}{28}$ of that from *a* to *b*, whatever the actual distance of the sun may be. We thus know the ratio of the lengths of two parallel chords, and the distance between them in miles, from which it is a simple matter to find the diameter

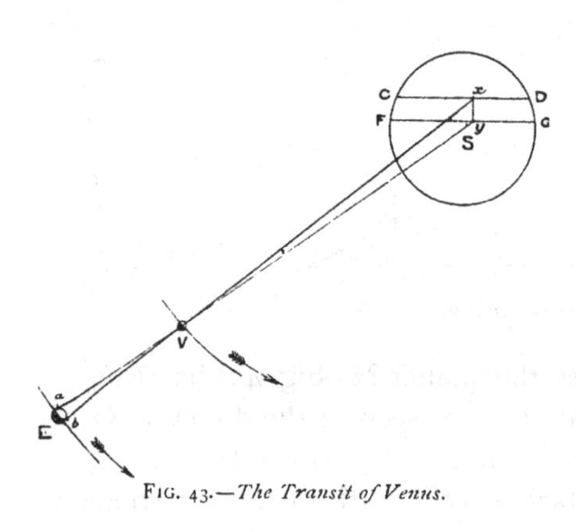

FIG. 43.—*The Transit of Venus.*

of the sun's disc in miles. The angular diameter of the sun is measured with a transit instrument, and to find the sun's distance we have simply to calculate the distance at which a body of known size subtends a known angle.

We have supposed the observers at rest, but they are in reality carried forward by the earth's orbital motion, and are turned about the earth's axis. The first of these movements will affect both observers in the same degree, and will simply lengthen the duration of the transit. The effect of rotation, however, depends upon the position of the sun and planet, with regard to the observer's meridian. At sunrise, an observer is carried by the rotation of the earth almost directly towards the sun, while at sunset he is carried away from it. The rate at which the planet traverses the sun's disc would, therefore, be little affected by the earth's rotation at sunrise or sunset. About mid-day, however, the effect of the earth's rotation is to accelerate the apparent motion of the planet, and to shorten the time of transit. If the beginning of the transit be observed at sunset, and the end soon after sunrise, as it may well be in high latitudes, the duration of the transit is retarded by the earth's rotation. Corrections for rotation, however, are not difficult to apply.

In this method of observing a transit of Venus, which was suggested by Halley, when it was impossible that he would live to see it carried out, the places of observation must be widely separated in latitude, and the beginning and end of the transit must both be observed.

Another method of utilising a transit of Venus is known as Delisle's method. In this case the two stations are near the Equator, and each observer notes the Greenwich time of internal contact, when the planet fully enters upon the sun's disc.

Owing to various causes, chief among which is the so-called " black drop," the time of ingress and egress cannot be actually recorded with the desired degree of accuracy, and the transit of Venus is no longer looked upon as the best method of determining the distance which separates us from the sun.

Some of the results which have been obtained for the solar parallax are as follows:—

Transit of Venus, 1874, contact observations,	.	8″·859
" " " photographs,	. .	8·859
" " 1882, contact observations,	.	8·824
" " " photographs,		8·842
Gill's observation of Mars, 1877,	. . .	8·780
Galle's " Flora, 1873,	. . .	8·873
Gill's " Juno, 1874,	. . .	8·765
" " minor planets, 1896,	. .	8·80

From a discussion of all the available data, Professor Harkness considers the most probable value of the solar parallax to be 8″·80905, with a probable error of 0·00567″. Turning this into miles, we find the distance of the sun to be 92,796,950 miles, and this is in all probability not more than 60,000 miles in error. This agrees very closely with Dr. Gill's latest value, which has been accepted by the superintendents of the British and American nautical almanacs.

THE SUN'S DIAMETER.—The real diameter of the sun is found from the parallax, and its mean angular diameter in the manner already explained (p. 142). Taking the distance as 92,780,000 miles, and the mean apparent semi-diameter as 962″, we have

$$\text{Sun's diameter} = \frac{2 \times 92{,}780{,}000 \times 962}{206{,}265}$$
$$= 865{,}400 \text{ miles.}$$

The sun's diameter is the same in all directions, so far as our measurements give any information on the point, so that there is no appreciable polar flattening corresponding to that of the earth and some of the other planets. This result is what we should expect from the relatively slow rate at which the sun turns upon its axis.

DISTANCES AND DIAMETERS OF PLANETS.—It has already been pointed out that our knowledge of the relative distances of the planets from the sun enables us to determine their absolute distances when the distance of one of them has been ascertained. In this way the determination of the earth's distance leads us to those of the other planets.

Our additional knowledge of the planetary orbits further permits the calculation of the distance of any planet from the earth at a stated time. If, then, the angular diameter of a planet be measured with a micrometer attached to a telescope, the absolute diameter in miles can be determined in the same way as that of the sun or moon.

To take an actual example, the equatorial angular diameter of the globe of Saturn, as measured by Prof. Barnard with the great telescope of the Lick Observatory on April 14, 1895, was 19″·4. It was then computed that if the observation had been made from the sun this would have been reduced to 17″·9. The distance of Saturn from the sun being 9·5388 times the earth's distance, it results from this measurement that the true equatorial diameter of the ball of Saturn is 76,500 miles. A number of independent measures made at intervals from March to July gave an average value of 76,470 miles for the diameter.

CHAPTER XII.

THE MASSES OF CELESTIAL BODIES.

MASS AND WEIGHT.—As a matter of daily experience, we know that a certain effort is required to prevent a body from falling to the ground, and the larger the bulk of any particular kind of matter, the greater is the effort demanded. Again, equal bulks of different kinds of matter require unequal efforts to sustain them in the hand. From facts such as these we get the idea of *weight*, and we say that one body is heavier than another when it has the greater tendency to fall to the ground. For the purposes of everyday life, the weight of a body is used as a measure of the quantity of matter which it contains, and the standard of weight in our own country is that of a certain piece of platinum kept at the Exchequer Office, in London, which is call a *pound*. The weight of the same piece of matter varies at different parts of the earth's surface, and also at different distances from the ground, and it is evident, therefore, that weight is not a very scientific measure of the quantity of matter which a body contains. The standard of comparison must be one which is invariable not only in all parts of the earth, but, if we wish to investigate the quantity of matter in the celestial bodies, it must be unalterable through all parts of the universe.

One's first idea is that the bulk, or space which a body occupies, will furnish a means of measuring the quantity of matter which it contains, but here again we find that the volume of a body can be varied without either adding to or subtracting from it, its weight remaining constant. A piece of ice, for example, occupies a greater space than an equal weight of water.

It is evident then that some other property of matter must be used as a measure of quantities. Now, there is every reason to believe that the same piece of matter, in whatever part of space it may be situated, requires the same force to set it moving with the same speed in a given time. By the continued application of a force, a body will first be set in motion, and at the end of a second it will have a certain speed; in the next second the velocity will have increased by an amount equal to that acquired at the end of the first second, and so on for subsequent intervals. For example, if at the end of a second the velocity were 3 feet per second, at the end of the next second it would be 6 feet per second, and after other equal intervals it would be successively 9, 12, 15, and so on. In this way the velocity is increased uniformly, and is said to be uniformly accelerated, while the gain per second is called the *acceleration*. The greater the force applied, the greater will be the acceleration it produces, and the acceleration can be used as a measure of the force at work.

If the same force be applied to different quantities of the same substance, the acceleration produced will be in inverse proportion to the quantities. We thus arrive at the important result that two bodies, whatever their nature, contain equal quantities of matter, or have equal *masses,* when equal forces give them the same acceleration. The mass of a body can thus be ascertained by observing the acceleration due to the action of a known force.

As a matter of observation, it is found that all bodies, whatever their composition or size, fall to the ground from the same height in the same time if the observations be made at one place. This means that the forces corresponding to weights produce equal accelerations in all bodies at the same place, and it follows, therefore, that the weights of bodies at the earth's surface, are proportional to their masses. Hence, it is that weight can be practically employed in comparing masses, or quantities of matter, for the purposes of everyday life. It must be clearly understood, however, that a *mass* of a pound is in reality quite distinct from a *weight* of a pound, the former specifying a certain quantity of matter, and the latter its tendency to fall towards the earth.

THE LAW OF GRAVITATION —The idea that weight is due to the attraction of the earth for all bodies in its neighbourhood was first suggested by Newton, and an extension of this idea led him to formulate the great law which underlies the whole science of astronomy. All bodies near the earth's surface are acted upon by forces proportional to their masses, and the same acceleration is produced in all of them if they are allowed to fall to the ground. Falling freely for a second, all bodies whatsoever, when the resistance of the air is eliminated, pass through a little over 16 feet, and acquire a velocity of just over 32 feet per second. The acceleration due to gravity is thus $32\frac{1}{6}$ feet per second for bodies near the earth's surface. If the experiment be made at the top of a high mountain, the distance fallen through and the acceleration acquired in a second is found to be less.

If we could ascend still higher, the acceleration produced in falling bodies would be again reduced, and, in the light of what has gone before, it is evident that the force with which bodies tend to fall to the earth is diminished as the distance from the earth's surface is increased. It was such considerations as these which led Sir Isaac Newton to formulate the law that *the force with which a body is attracted towards the earth diminishes in inverse proportion to the square of the distance from the earth's centre.* Terrestrial means of testing the truth of this statement are obviously very limited, and hence it was that Newton looked to the moon for its verification. If the law holds good at the distance of the moon, an object so far removed and not acted upon by other forces, should fall towards the earth, and as its distance is about sixty times that of a body at the surface from the centre of the earth, the acceleration produced should be only $\frac{1}{3600}$th part of that imparted to bodies near the surface. In other words, since a body near the surface falls through 16 feet in the first second, one at the moon's distance should only fall through about $\frac{1}{20}$th of an inch. If, then, the moon be subject to the earth's attraction, this fall towards the earth must be exhibited in some form or other, although the fact that the moon does not fall down upon the earth shows that there is some counteracting tendency.

Observations have shown us that the moon moves in a curved path. It has been put in motion somehow, and since there is no reason why it should turn to one side or the other, or come to rest, unless some forces are acting upon it, it would tend to go on uniformly in a straight line for ever. That its movement is curvilinear is at once an indication of the action of a force besides that which originally set it in motion. This force is directed towards the earth, and the moon is drawn out of its rectilinear path just as far in any specified time as it would fall towards the earth if at rest.

Let E and M in Fig. 44 represent the earth and moon respectively. Then, if the moon were not hindered in any way, it would move in the direction M *b*, and would reach the point *b*, let us say, at the end of a second. It is, however, found to be at the point *a*, and it has therefore fallen towards the earth through the distance *b a*. The size of the moon's orbit and the angle through which it moves in a second being known, it is easy to calculate the distance *a b*, which is found to be about $\frac{1}{20}$th of an inch, as demanded by Newton's law.

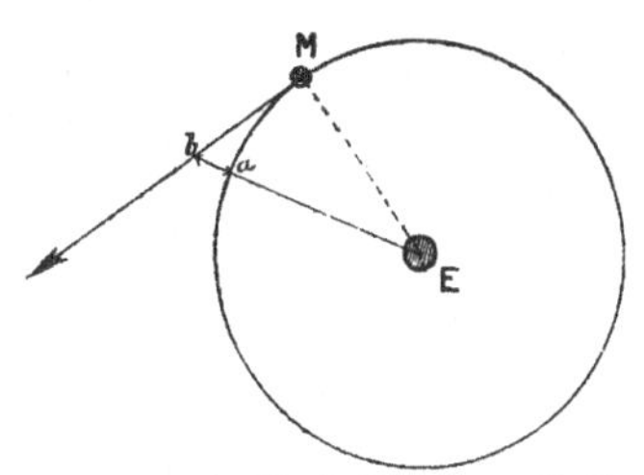

FIG. 44.—*The Moon's Curvilinear Path.*

In his first attempt to thus verify the law of gravitation, Newton failed for the want of a sufficiently accurate knowledge of the earth's diameter, but a few years later a new arc of meridian was measured, and he had the untold satisfaction of demonstrating its truth.

The curved path of the moon is, indeed, similar to that of a projectile. A cannon ball thrown out horizontally will reach the ground after describing a curved path ; but if it could be projected from a great elevation, with sufficient velocity, its forward movement would prevent its ever reaching the earth's surface at all, and a new satellite of the earth would have been manufactured.

The same kind of reasoning can be applied to the paths of the earth and planets around the sun, and Newton demonstrated that the laws of Kepler were a necessary consequence of the law of gravitation extended beyond the system of the earth and moon. By mathematical reasoning it was proved that if one body describes an elliptic orbit around another, and the line joining them describes equal areas in equal times, the attractive force must be directed to the central body, and, moreover, must vary inversely as the square of the distance between the two bodies. In this way the movements of the planets round the sun are perfectly explained by supposing that an attractive force, similar to that which causes bodies to fall to the earth's surface, is exerted between all masses of matter, and hence the origin of the term *Universal Gravitation.* In its complete form, the law of gravitation states that "any particle of matter attracts any other particle with a force which varies directly as the product of the masses, and inversely as the square of the distance between them."

Confirmation of this grand law, which controls the movements of all the vast array of heavenly bodies, is furnished by many other phenomena. We see one of its effects in the tides, and another in the disturbances of the movements of planets brought about by their mutual attractions. Even in the depths of stellar space the same law holds good for those systems of stars which are sufficiently close together for their attractions to produce effects which we can study at our immense distance from them.

The cause of gravity is still one of the greatest mysteries of physical science, although many ingenious attempts have been made to furnish an explanation of its mode of action.

MASS OF THE SUN.—When we know the distance of the sun, and the time in which the earth travels completely round it, it is easy to calculate the fall of the earth towards the sun in the same way that the moon's fall towards the earth is determined.

The distance which a body 93,000,000 miles distant falls towards the sun in a second is thus found to be 0·116 of an inch. A body at the earth's surface is about 4,000 miles from the centre, and it falls $16\frac{1}{12}$ feet in a second; if removed to a distance of 93,000,000 miles, its fall towards the earth would be reduced inversely as the squares of 4,000 and 93,000,000, and would amount to ·000,000,349 of an inch. This is only $\frac{1}{332,000}$th part of the fall due to the sun's attraction, and hence it is concluded that the mass of the sun is 332,000 times that of the earth.

Strictly speaking, the accelerations produced by the sun and earth should be compared, but the fall during the first second is proportional to the acceleration due to gravity, and the same result is therefore obtained. It may be observed also that the fall of the earth towards the sun would not be appreciably effected if it were twice the size. All bodies fall towards the earth at the same rate, whatever their weights, and so in the case of a planet, the distance fallen towards the central sun is independent of the planet's mass; the greater the mass the greater the attractive force.

The sun occupies about 1,300,000 times the space occupied by the earth, and as its mass is only 332,000 times that of the earth, it follows that the sun's density is only about a quarter that of the earth.

MASSES OF PLANETS.—The process employed for the determination of the sun's mass can be utilised for finding the masses of those planets which are accompanied by satellites. From the known distance of the planet, the size of the orbit of a satellite can be calculated in miles, and knowing the period of revolution of the satellite, its fall towards the planet can be determined. This fall is then compared with that of the planet's fall towards the sun, and the mass of the planet in terms of the sun's mass is thus arrived at.

A convenient way of employing this method is to make use of a modification of Kepler's third law. If m be the mass of a planet in terms of the sun's mass, M, a and T respectively denote the semi-axis major of the orbit of the planet and its time of revolution round the sun; a' and T similar quantities pertaining to the satellites' revolution round the planet: The following formula gives the relation of the masses:—

$$\frac{m}{M} = \left(\frac{a'}{a}\right)^3 \left(\frac{T}{T'}\right)^2$$

This formula can be applied in the case of Mars, Jupiter, Saturn, Uranus, and Neptune, but fails in the case of Mercury, Venus, and the asteroids, which, so far as we know, have no satellites.

The mass of Jupiter obtained in this way can be further checked by the influence of this giant planet upon other bodies in its neighbourhood. This planet has such an enormous mass that it produces very notable effects on the motions of Saturn, the asteroids, and of comets which travel in its neighbourhood, and, by measuring the amounts of these *perturbations*, the mass of the planet can be deduced.

This method of perturbations is at present the only one by which we can obtain a knowledge of the masses of those planets which have no satellites. The motion of Mercury is disturbed by its nearest neighbours, Venus and the earth; that of Venus by the earth and Mercury. The differences between the observed positions of the planets and those calculated on the supposition that the others did not affect them, give the necessary data for the computation of the masses. The process, however, is one requiring profound mathematical knowledge, and even yet the mass of Mercury is not very certainly known.

The asteroids, again, present no little difficulty. Their feeble light and small size point to small masses, and their mutual perturbations are almost insensible, except when two of them come into line with the sun. They produce no appreciable effects upon the movements of comets, so that it is almost impossible to determine their individual masses. Each asteroid, however, tends to produce a revolution of the major axis of the orbit of the nearest planet, Mars, and all tend to give it a motion in the same direction. If the total mass of all the asteroids put together were a quarter of the earth's mass, a measurable displacement of the position of Mars would be produced. Professor Newcomb has recently shown that such a displacement actually occurs, but cannot amount to more than $5''\cdot 5$ per century. From this it has been recently calculated that the total mass of the asteroids is probably about $\frac{1}{115}$th that of the earth's mass.

MASS OF THE MOON.—As the moon has no satellite, we must again have recourse to indirect methods if we wish to know anything as to its mass. Various processes are open to us; but although the moon is so near to us, it is more difficult to determine its mass than that of the most remote planet in our system.

It has already been explained (p. 77) that as the earth is accompanied by the moon, it is really the centre of gravity of the two bodies which obeys the laws of planetary movement. As this point lies between the centres of the two bodies, at distances which are in inverse proportion to the masses, the centre of the earth describes a small monthly orbit, which, as we have already seen, produces a small monthly inequality in the sun's apparent movement.

By a careful investigation of this monthly oscillation of the sun, it has been found that the centre of gravity of the earth and moon must lie within the earth at a distance of about 2,900 miles from the centre. This is about $\frac{1}{81}$th of the moon's distance, whence it follows that the mass of the moon is about $\frac{1}{81}$th that of the earth.

Other methods of ascertaining the moon's mass are also available. Among these are the investigation of the parts played by the moon in the production of the tides which swell our shores, and in the displacement of the earth's axis which causes "nutation."

MASSES OF SATELLITES.—The earth's satellite is of exceptional magnitude in comparison with its primary, and the method of finding its mass from the situation of the centre of gravity cannot be applied to the satellites attending other planets. In the case of the satellites of Jupiter and Saturn, the masses have been approximately determined by their mutual perturbations, these generally resulting in a revolution of the major axes of the orbits. Even this method fails for the satellites of Mars, Uranus, and Neptune, so that practically nothing is known with regard to their masses.

MASS AND DENSITY OF THE EARTH.—So far we have been concerned entirely with relative masses, referring the masses of the various orders of the heavenly bodies either to the earth or sun. Although this is usually all that is required for astronomical purposes, it is of great interest to determine the absolute mass of the earth, and from this the absolute masses of the heavenly bodies can at once be deduced.

We already know the dimensions of the earth, and therefore the number of cubic miles or feet which it occupies. We know also the weight or mass of a cubic foot of water or lead, and if the earth were of uniform specific gravity throughout its bulk, and composed of water or lead, we could at once calculate its total mass. It is, however, neither water nor lead; but if we can compare the mass of the earth with what it would be if composed of either of these substances, we can deduce either its mass or its specific gravity.

A very simple method of "weighing" the earth has been employed with much success by Professor Poynting. The experiment was carried out at the Mason Science College, Birmingham, with a large bullion balance in which the beam was 123 centimetres long. Two spheres of lead and antimony, each weighing about 21 kilograms, were suspended from the arms of the balance. Another sphere of lead and antimony, weighing 153 kilograms, was successively brought by means of a turn-table under each of the two smaller weights. The alteration in the weights of the attracted balls were measured by observing the deflection of the beam, this being immensely magnified by a simple optical arrangement in which a mirror reflecting a pencil of light was made to turn through 150 times the angle moved through by the beam itself. The weight corresponding to a given deflection of the beam was determined by observing the disturbance produced by the addition of "riders" of known weights. In order to reduce the chances of error, the large weight was balanced on the turn-table by another mass of half the weight and at twice the distance from the centre, this being necessary in order that the attracting weight should rotate horizontally. The effect of this additional mass was calculated and allowed for, and the weighings were also repeated with the weights in various positions. The principle of the subsequent calculation is briefly as follows:—A mass A of lead and antimony of known bulk attracts another mass B with the force measured; if A were of the same size as the earth, the attraction would be increased by as many times as the earth is larger than A. If the average specific gravity of the earth were the same as that of the mass A, this calculated attraction would be equal to the weight of B. The ratio of this calculated weight of B to the actual weight accordingly gives the proportion between the specific gravity of the experimental ball and the average specific gravity of the whole earth. From this experiment it was estimated that the mean density of the earth is 5·4934 times that of water.

The same principle is applied in the case of the famous Cavendish experiment, and its subsequent modifications by Baily, Cornu, and Boys.

Another method of finding the earth's density, and therefore its mass, is chiefly of historical interest. This is known as the "mountain method," and was carried out in 1774 by Maskelyne, Hutton and Playfair on the Schiehallion Mountain, in Perthshire. A plumb-line suspended at the north side of the mountain is drawn towards the mountain, and so will not hang quite vertically. If removed to the opposite side of the mountain it will be deflected in the reverse direction. The amount of this deflection can be measured by reference to the stars, the positions of which are in no wise influenced by the attraction of the mountain. A survey of the mountain was next made in order to determine its bulk, and then the average specific gravity of the rocks composing it was determined with the greatest possible accuracy.

The volume of the earth is 9,933 times that of the mountain, and its attraction would be this number of times greater if it were composed of the same materials as the mountain throughout. It was found to be in reality 17,781 times as great as the attraction of the mountain, and as this is 1·79 times 9,933, it follows that the average specific gravity of the matter composing the earth would be 1·79 times that of the rocks which build up Schiehallion. The mean specific gravity of the rocks being 2·8, the mean density of the earth was thus found to be 5·012 times that of water.

As a general result of all the observations which have been made, the value of the earth's density may with much probability be considered to be not far from 5·576, or a little over $5\frac{1}{2}$ times that of water.

Whatever may be the composition of the earth's interior, it is clear that the density must increase as the centre is approached.

This knowledge of the earth's density, in conjunction with the known number of cubic miles occupied by the earth, readily enables us to determine that the total mass of the earth is about 6,000,000,000,000,000,000,000 tons.

CHAPTER XIII.

GRAVITATIONAL EFFECTS OF SUN AND MOON UPON THE EARTH.

THE TIDES.—The familiar phenomena of the tides are of such importance to commerce in so many parts of the world that they have been carefully investigated from very early times. The necessities of coast navigation would soon lead to the recognition of a periodic character in the tides, as well as to their association with the age and position of the moon. With the march of science, an explanation of tidal phenomena was therefore sought in the motion of the moon. A great impetus was given to this inquiry by Newton's generalisation, and the tides were shown to be a necessary consequence of the gravitational attraction of the sun and moon. Regarding the earth merely as a cosmical particle, we have seen that its orbital motion is perfectly explained by the gravitational attraction of the sun, and some of its minor movements by the attractions of other members of the solar system. The law of gravitation, however, compels us, in a closer investigation of these mutual attractions, to regard each globe as an assemblage of particles, each of which individually influences and is influenced by other particles. If such a collection of particles be spherical and perfectly rigid, it will behave precisely as a simple particle in which the whole mass is concentrated.

When we cease to consider the earth as a mere particle, we must regard the waters of the oceans as being free to move over the more rigid crust of the globe. Imagine our globe to be a spherical mass completely surrounded by a liquid envelope. At any moment one half of this is presented towards the moon. The solid earth we may conceive to be attracted by the moon as a simple particle ; but the water on the side nearest to the moon is attracted with a greater force than the solid globe, because of its greater proximity to the attracting body, and it has therefore a tendency to heap itself up directly under the moon. Being free to move, the water thus remains heaped up under the moon, notwithstanding the earth's rotation, and if there were only one such elevation, there would only be one tide a day. Observation shows us that there are two high tides a day, and the water must therefore be heaped up on the side of the earth which is turned away from the moon. This is perfectly true, though seemingly at first sight inconsistent with the moon's attraction. The fact is that the solid earth is attracted by the moon with greater energy than the water on the side most remote from it, so that the heaping up of the water on the side away from the moon is to be regarded as due to the earth having left it behind.

There is thus a double tidal wave produced by a spheroid of water which, in the simple case we have considered, has its axis directed towards the moon, as in Fig. 45. The earth, rotating within this liquid shell, successively brings different parts of the solid earth to the points of high and low water. If the moon were fixed, we should then experience two high and

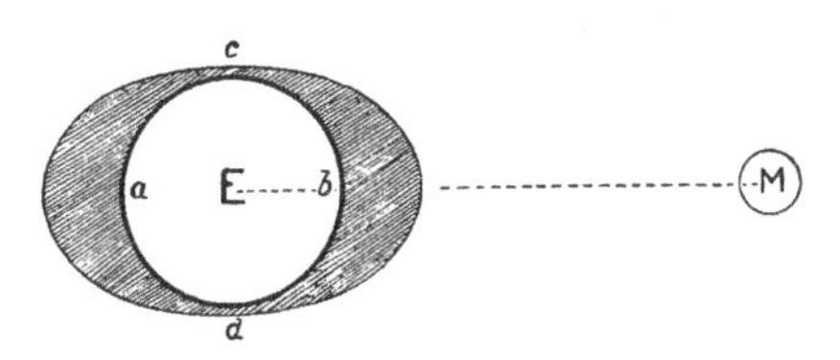

FIG. 45.—*The Tides.*

two low waters every day, but as it revolves in the same direction that the earth rotates, the average interval between two successive meridian passages is 24 hours 51 minutes. This, then, is the period in which alternate high waters or alternate low waters are experienced.

A similar train of reasoning applies to the attraction of the sun upon different parts of our planet, so that there are solar as well as lunar tides. Nevertheless, the moon is the dominating cause, for although the total attraction of the sun upon the earth is about 200 times that of the moon, its differential attraction upon the opposite sides of the earth, which is alone effective in producing tides, is only about $\frac{2}{5}$ths that of the moon.

A simple mathematical investigation shows that the tide-raising force of a body is proportional to its mass, and approximately in inverse proportion to the cube of its distance from the affected body. Thus, it appears that if the moon were removed to 1·36 times its present distance, solar and lunar tides would be equal.

At the times of new and full moon, the sun and moon will produce two tidal spheroids of water upon our imaginary earth, having their axes coincident, and an exceptionally high tide will occur. This is a *spring tide.* When the moon is at its quarters the two ellipsoids tend to neutralise each other, and an exceptionally low or *neap tide* results. Two spring tides and two neap tides thus occur in each synodic month of $29\frac{1}{2}$ days.

The height of the tide will also be affected by the variations in the distance of the moon. If the moon be at perigee the tide will be greater because of the smaller distance, and if this occur at new or full moon there will be a very high spring tide, while a less notable spring tide will occur when the new or full moon is at apogee.

The combination of the solar and lunar tides gives rise to what is called the *priming* and *lagging* of the tides. At new and full moons the combined tides will produce a spheroid of water with its axis directed towards the moon. When the moon is a few days old, however, the crest will take up a position intermediate between the direction of the moon and that of the sun, and high water will therefore be accelerated. The same thing will happen during three or four days after full moon. Three days before full or new moon the combination of the two tides will displace the crest towards the sun, and therefore in advance of the moon, so that high water will be retarded. The retardation and acceleration correspond to lagging and priming respectively.

At the quadratures the combined tides simply reduce the height of the crest, since there is no reason why the deviation should be to one side any more than to the other. On account of priming and lagging, the tides on successive days are accelerated or retarded by as much as 13 minutes when the effects are greatest.

Sufficient has been said to indicate that tidal phenomena are very complex even when we suppose the earth to be very simply constituted. When we take into account the actual configuration of the land and the consequent restrictions in the movements of the water, these complications are increased tenfold. Yet, by continued observations, the recurrence of tides at any port can be predicted with tolerable accuracy. It is observed that there is a certain pretty regular interval of time between the moon's meridian passage and the time of next high water; this is different at different ports, but is so nearly constant at a given place as to be called *the establishment of the port.* Observations being made at a great many places, the peculiar movements of the tidal wave can be investigated. For this purpose, it is convenient to draw on a map what are called *co-tidal lines*, that is, lines passing through places at which high water occurs at the same moment. It then appears that it is only in the Southern Pacific where the water is of sufficient extent to permit the formation of the tide crest. The effect of this wave, which commences twice a day, is gradually spread over different parts of the world, but before it reaches most places other waves have commenced a similar journey. The tide at London, for example, coming round the north of Scotland and down the North Sea, really started in the Southern Pacific 66 hours before, and in the same way the tide at New York is a little over 40 hours old.

The height of a tide is thus regulated by the conditions of the sun and moon with regard to the earth when the primary tide was formed, and not by their relation when a tide is actually observed.

In the Pacific Ocean the tides are very feeble, but near the coast they vary enormously, and sometimes reach great heights. At Bristol the difference between high and low water sometimes amounts to fifty feet, and in the Bay of Fundy, Nova Scotia, it has been as much as a hundred feet.

The peculiarities of the tides at many places are due to interference. The primary tidal wave striking the British Islands travels partly up the English Channel, and partly round to the North Sea by the north of Scotland. At some places on the east coast the two waves almost neutralise each other, while at others there are even four high tides in a day.

The circumstances under which tides occur at a given place can only be determined by actual observations, as theory is at present utterly inadequate to deal with the manifold complications brought about by the configuration of the land, and the varying depth of the water.

TIDAL FRICTION.—The regular influx of the tide supplies us with a source of mechanical energy, which in the future will no doubt become of immense importance to mankind. A great mass of water is raised to a higher level, and by suitable contrivances it can be made to do useful work during its subsequent flow to the ocean from which it came. Ordinarily, however, the water simply rushes back without its energy being utilised, and the potential power is merely transferred to another locality. It is manifest, however, that a certain amount of tidal energy is lost by friction as the water rolls to and from the rocky shores. This energy is converted into heat, and finally radiated into space, or dissipated. Now, the principle of the conservation of energy tells us that energy can neither be created nor destroyed, although its form may change from a useful to a useless one.

It follows, therefore, that the energy lost through the tides must be abstracted from one source or another, and it has been shown that this energy is really derived from the earth's rotation. As the earth steadily ploughs its way through its liquid envelope, the tides act as a break, and its rotational velocity is reduced; it is part of this lost energy of rotation which is dissipated by the tides.

One tendency of tidal friction is accordingly to lengthen the period of the earth's rotation, and, therefore, to increase the length of the day. There are, however, counteracting causes, so that there is no certain direct evidence that the day has actually lengthened in historical times.

All the energy of rotation which is lost by the earth is not, however, dissipated by the tides. Some of it is transferred to the moon, with the result that the velocity of our satellite, and consequently the size of its orbit, must be increasing. From this it is inferred that the moon was formerly very much closer than at present, and an elaborate investigation of the conditions of its retreat has led Professor G. H. Darwin to his interesting theory of "tidal evolution." (See p. 236.)

Professor Darwin has shown that if the term "tide" be extended to include distortions of the earth and moon at an earlier stage of their history, when both were fluid or viscous, a similar grinding down of the energies of rotation of both bodies must have taken place. The axial rotation of the moon, under these circumstances, would be retarded by the attraction of the earth on the tides raised in the moon, while that of the earth would also be slowed down, but in a less degree because of the moon's smaller mass.

Cause of Precession.—On account of the spheroidal form of the earth, we may regard it as a sphere which is surrounded by a ring of protuberant matter at the Equator. Now the attraction of the sun upon the spherical part will be quite independent of the position of its axis of rotation, and will, therefore, not affect the position of the Equator. It is different, however, with the ring; at the solstices the ring is inclined to the line joining its centre with the sun, and the near side is subject to a greater attraction than the side more remote from the sun. On account of this difference of pull, there is a tendency for the ring to move into the plane of the ecliptic, and this is what would happen if the ring were not in rotation. The practical outcome of this tendency, combined with the rotation, is to produce the twisting of the plane of the ring, and, therefore, of the plane of the Equator. At the equinoxes the plane of the ring passes through the sun, and although there is still a difference of attraction on opposite sides of the ring, the differential force is entirely directed to the sun, and therefore cannot produce any precessional effect.

The ultimate tendency to turn into the plane of the ecliptic thus depends upon the *difference* of the attractions on opposite sides of the ring, or rather that part of the difference which acts in a direction perpendicular to the Equator.

The terrestrial ring cannot change the position of its plane without taking the whole earth with it, and the rate of precession is thus very slow. The effect of solar precession alone would cause the equatorial plane to twist round with but little change of inclination; or the earth's axis would travel with a conical movement round a perpendicular to the ecliptic passing through the earth's centre.

It will be remarked that as the force-producing precession is identical with that which is effective in producing the tides, the moon must have a greater precessional effect than the sun. This is quite true, and on the average the precession-producing force of the moon is $2\frac{1}{2}$ times that of the sun. When the moon is on the celestial equator, as it is twice a month, the differential force acts in the plane of the ring, and no precessional effect results. On the other hand, the greatest effect is produced by the moon when the earth's Equator is most inclined to the line joining the earth and moon. The amount of this greatest inclination is different in different months according to the position of the moon's nodes. In consequence of the revolution of the moon's nodes, the moon's orbit is inclined to the Equator at all angles from 18° to 28°, and back again to 18° in a period of 19 years. The precessional effect of the moon thus has a principal period of 19 years, while that of the sun has a period of a year during which it has two maxima and two minima. The summation of the effects of the sun and moon gives us the *luni-solar precession*, which is very variable in its actual rate, but averages about 50″·2 per annum.

NUTATION.—If the precession-producing force were of constant amount, there would be no change in the inclination of the earth's axis to the ecliptic. When the force is increasing, the equatorial ring is slightly tilted towards the ecliptic, and when it is decreasing the converse takes place. As the moon has the preponderating effect, these changes in the inclination will evidently depend mainly upon the changing value of the moon's precessional force; that is, they will have a period of 19 years. Thus, if P, Fig. 46, represents the pole of the ecliptic, the north celestial pole would travel in a circle of $23\frac{1}{2}°$ radius about P if precession were uniform. Suppose, then, the celestial pole to be at a when the moon's node is on the Equator—that is, when the inclination of the moon's orbit to the Equator is greatest—from this time the integrated effects of the moon's precessional force will be decreasing, and the inclination of the Equator to the ecliptic will be increased; the celestial pole will consequently recede a little more than the average from the pole of the ecliptic, so that after $9\frac{1}{2}$ years it will be at b instead of c. During the next $9\frac{1}{2}$ years the inclination of the moon's orbit to the ecliptic will be gradually getting smaller, the precessional force will be proportionately reduced, and the obliquity of the ecliptic will be increased, so that the north celestial pole will have arrived at d after the lapse of 19 years. The prolongation of the earth's axis thus describes a wavy curve, each wave extending

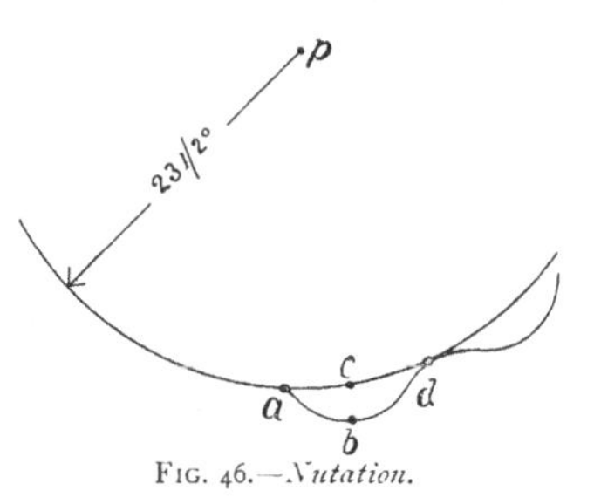

FIG. 46.—Nutation.

over 19 years, so that there are about 1,400 waves during the great precessional cycle. This approach and recession of the two poles is called *nutation*, or nodding of the earth's axis. The most recent investigation of its maximum amount, by Dr. Chandler, gives it as 9″·202. Besides the principal nutation there are others of very much smaller amount, due to the monthly changes of the moon's declination and to the annual change of the sun's declination.

The most obvious effect of nutation is that upon the inclination of the earth's axis to the ecliptic—the "nutation in obliquity." There is, however, a displacement of the equinoctial point, and corresponding nutations in longitude and right ascension.

As pointed out by Sir John Herschel, we have in nutation a splendid example of a periodical movement in one part of a system giving rise to a motion having the same precise period in another.

EFFECTS OF PRECESSION.—The effects of precession may be conveniently summarised here, although some of them have necessarily been mentioned elsewhere:

(1) The first point of Aries revolves completely round the ecliptic, so that it passes through all the constellations of the zodiac in a period of 25,800 years. The "signs" of the zodiac, accordingly, no longer correspond with the constellations after which they are named.

(2) The Pole Star is constantly changing, since the north celestial pole travels round the pole of the ecliptic at a distance of about 23½° in a period of 25,800 years. About 14,000 years ago the bright star Alpha Lyrae was the Pole Star.

(3) The position of the north celestial pole is in time changed by 47°, and there may accordingly be this change in the north polar distances or declinations of all stars whatsoever. As the position of the ecliptic is almost constant, the celestial latitudes of stars will be but little affected by precession.

(4) The right ascensions and longitudes of stars, being reckoned from the shifting first point of Aries, are them-

selves changeable, passing through all possible values in the precessional period.

(5) The tropical year is shorter than the sidereal year by the time taken for the earth to travel through 50″·2—that is, 20 minutes 23 seconds.

(6) Celestial globes and maps, as well as star catalogues, can only represent the right ascensions and declinations of stars at a specified epoch.

CHAPTER XIV.

INSTRUMENTAL MEASUREMENT OF ANGLES AND TIME.

GRADUATED CIRCLES.—Astronomy is essentially a science of precision, and the progress of our knowledge has to a large extent been dependent upon the increasing power of accurately measuring angles and time.

Let us see, first of all, how to measure angles.

A circle is divided into 360 degrees, each degree again into 60 minutes, and each minute into 60 seconds of arc; and yet, a second of arc is not a small enough quantity for many astronomical purposes. Now, unless a very large circle be employed, it is mechanically impossible to even mark the minutes of arc directly upon it, and if a very large circle were constructed, the distortion of its shape produced by its own weight would be sufficient to mar its accuracy.

What is actually done then is to get a circle of convenient size, and to graduate it, as well as the highest mechanical skill is capable of, into such parts as may leave distinct and equal spaces between the separate divisions. A competent instrument maker would, for instance, put 4,320 divisions on the *limb* of a circle 16 inches in diameter, two consecutive divisions thus being 5′ apart. For work of the highest precision it is necessary to strictly investigate the errors of the divisions and to correct for them in all observations.

For the further sub-division of these graduations, verniers or reading microscopes are introduced.

THE VERNIER.—A graduated circle being attached to an instrument, what one has to do is to take a *reading* with reference to some fixed mark. If the fixed mark is seen to fall precisely on one of the divisions of the circle when observed with a magnifying glass, the reading can be written down exactly. If there be no such coincidence, some means are required for accurately reckoning the fraction of a division. One method in general use on small instruments, and where extreme precision is unnecessary, is to employ a subsidiary scale which is called a *Vernier*, in honour of the Frenchman who invented it. This can be applied indifferently to a scale of degrees and parts of degrees on a graduated circle, or to a straight scale. With the aid of this device it becomes possible to measure angles with no greater probable error than a few seconds of arc.

THE READING MICROSCOPE.—If a greater degree of accuracy than 10″ be required, the vernier is superseded by a *reading microscope*. This is a compound microscope (Fig. 48) by which the scale can be observed, and at the focus of its eye-piece is a pair of spider threads which can be moved by a fine screw S. Looking into such a microscope, one sees a magnified picture of a very small part of the scale running through the field of view, as in Fig. 47. Running across the field, in the same direction as the marks on the scale, are the spider threads *a b*, which can be given a right and left movement by means of the screw. At the top of the field is the part called the "comb," having its edge cut with saw-like teeth; like the threads, this is at the focus of the eye-piece. The scale is divided

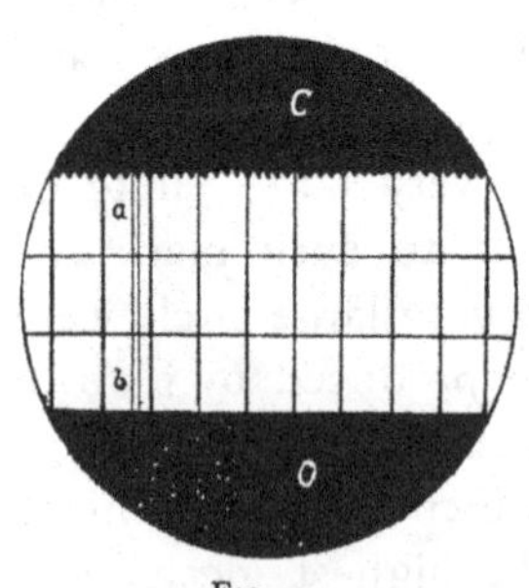

FIG. 47.

so that the smallest part is 5′, and in that case the teeth of the comb are arranged so that five of them equal a scale division. The reading microscope is a fixture, and the circle is brought into the position in which its reading is required by moving the instrument with which it is connected. The zero of the microscope is a point at the middle of the comb, and one has to determine what part of the scale corresponds with it. In order to do this, the threads or "wires" are moved until the next division lies between them, and the amount which the screw has been turned from the position of zero is read off on the graduated head of the screw. The dimensions of the parts, and the magnifying power of the microscope, are adjusted so that the screw must be turned five times to carry the wires through a space equal to a division on the scale. One division, therefore, will move the wires through 1′, and as the screw head is divided into 60 parts, a movement of $\frac{1}{60}$th of a revolution will shift the wires through a second of arc. Even fractions of a second can be thus measured.

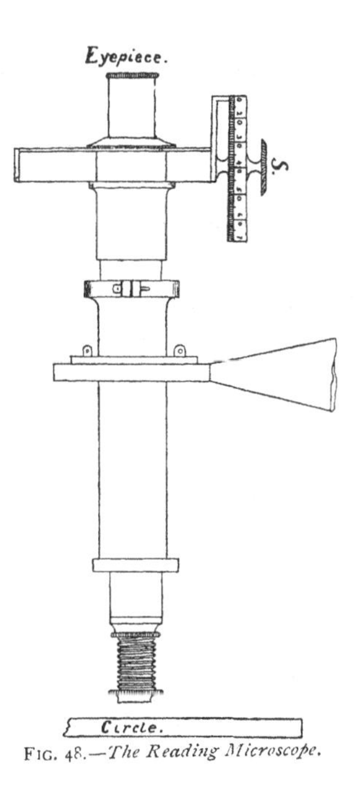

FIG. 48.—*The Reading Microscope.*

The introduction of this method of measuring minute angles is due to Ramsden, who first applied it at the end of the last century. The microscopes themselves are used for measuring fractional parts of the graduations of the circles, and usually four to six of them are applied to different parts of the same circle. In this way, errors arising from flexure of the circle, fluctuations of temperature, want of exact circularity, etc., are eliminated, so that finally, after taking every conceivable precaution, the astronomer can measure angles with the accuracy which is absolutely necessary in many branches of research.

ASTRONOMICAL CLOCKS.—Means for the exact estimation of time are of no less importance in an observatory than arrangements for the accurate measurement of angles. Astronomical clocks are constructed with extreme care, but in principle they do not differ from ordinary time-keepers. As sidereal time is of the greatest use in an observatory, the hour hand only makes one revolution a day, and the face is provided with a seconds hand, which is plainly visible. The pendulum is of such a length that it performs its swing in a second. One of the most important improvements in clocks was the introduction of the "compensation" principle, whereby the equivalent length of a pendulum remains constant in spite of fluctuations of temperature. The mercurial pendulum which one very frequently sees in a watchmaker's establishment has a glass or steel cylinder near the bottom partly filled with mercury; as the rod lengthens by increased temperature, the centre of gravity is raised by a corresponding amount, on account of the upward expansion of the mercury, and the rate of swing remains constant when the quantity of mercury is properly adjusted. The chief defect of this plan is that the mercury and the steel rod do not respond equally well to a change of temperature.

In the most approved clocks the pendulum rod is a compound one, consisting of rods, or concentric tubes, of zinc and steel. The pendulum bob is hung on a steel rod suspended from the top of a zinc tube, which in turn is fixed at the bottom end to a larger tube of steel; a rod attached directly to the latter is suspended by a flat spring in the usual manner. By this arrangement the unequal expansions or contractions of the different parts due to changes of temperature neutralise each other, so that a constant rate is the result. The tubes are pierced with numerous holes so that the inner and outer ones acquire the same temperature almost at the same time.

The rate of a clock is disturbed slightly by changes in the pressure of the atmosphere. When the air is densest there is a greater resistance to the swinging of the pendulum, and the clock will go more slowly. Although this only amounts to a small fraction of a second a day, it must necessarily be taken into account in such an establishment as that at Greenwich, to which all the country looks for the precise control of time-keepers. In the standard clock at Greenwich a magnet is raised or lowered by the changing height of a barometer, and its varying attraction upon a certain piece of iron attached to the pendulum compensates for the differences produced by change of pressure.

Pendulum clocks are obviously unsuitable for use at sea, so that *chronometers* are usually employed on ships. These are like large watches, very carefully constructed, with "compensation" balance wheels, and can generally be relied upon as good time-keepers.

After all precautions, however, no astronomer would put his faith in any clock for any length of time, as the best of them is liable to change its rate rather irregularly. The "error" of the clock is therefore very frequently determined by the observation of certain standard stars with the transit instrument. The stars can be relied upon to come to the meridian at the proper time, and any apparent departure from this time must be set down to the account of the clock.

THE CHRONOGRAPH.—A good clock, however, is not the only requirement of an observatory. It is necessary further to be able to record very precisely the moment at which an observation is made. If the clock be in the immediate vicinity of the observer, the time can be noted by counting the beats of the pendulum, and a practised observer will, by this "eye and ear" method, record times to the nearest tenth of a second. Mere estimation, however, is not very reliable, so that a mechanical method, which also permits greater subdivision of the second, is very generally adopted. The instrument is called a *chronograph*, and, although constructed in various forms, its function is to record on a sheet or strip of paper the regular beats of the clock, as well as the signals made by the observer. In one form of the instrument the recording sheet is fixed on a cylindrical drum which is made to revolve once a minute by a small clock. Beneath the drum is a pair of prickers worked by the armatures of electro-magnets. One of these magnets is in connection with the clock, and a simple arrangement sends an electric current through it every second, with the result that the seconds are marked by small punctures on the paper. As the cylinder revolves, the marker travels slowly lengthwise, so that the clock record runs spirally from one end to the other. To facilitate the identification of the punctures, one is omitted at the end of every minute. When an observation is made, the observer presses a button, and a current is sent through the second magnet, with the result that a puncture is made alongside those made by the clock. In this way the exact moment at which an observation is made can be easily registered, and read off at any convenient time.

At Greenwich a room is set apart for a number of chronographs, each in communication with an instrument in the various observatories.

CHAPTER XV.

TELESCOPES.

THE REFRACTING TELESCOPE.—The function of a telescope is two-fold. First, to magnify the heavenly bodies, or, what comes to the same thing, to make them look as if they were nearer to us, so that we can see them better. Second, to collect a much greater number of rays of light than the unassisted eye alone can grasp, so that objects too dim to be otherwise perceptible are brought within our range of vision.

There are two forms of telescope, distinguished as *Refractors* and *Reflectors*. The simplest form of refracting telescope is exemplified by the common opera-glass, and large refractors are not essentially different. Such instruments depend for their action upon the formation of an image by a lens. One can easily illustrate this by producing upon the wall of a room an inverted image of a candle or gas flame with a spectacle lens (one adapted for a long-sighted person), or with one of the larger lenses from an opera-glass. Having such an image, it may be magnified by means of another lens, just as one may magnify a photograph with an ordinary reading glass. Technically, the lens which forms the primary image is called the *object-glass* of the telescope, and that which is used to magnify this image is called the *eye-piece.* The object-glass is usually a large lens, which is placed at one end of a tube, while the eye-piece is a much smaller lens, placed at the other end. Means are provided for adjusting the distance between the two lenses so as to admit of distinct vision.

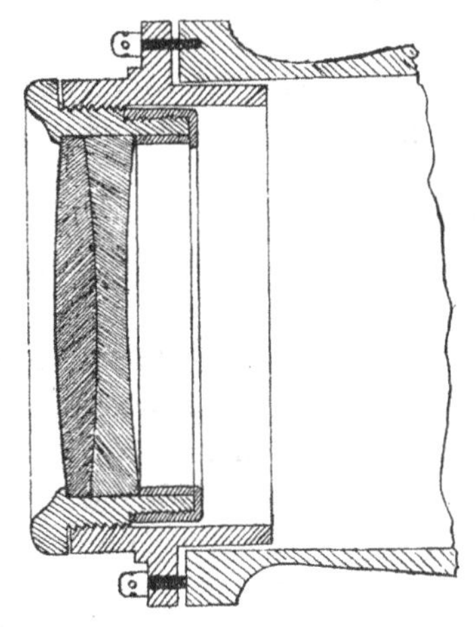

Fig. 49.—*The Achromatic Object-Glass.*

Matters are, however, not quite so simple as has been stated. There is a very great difficulty introduced by the fact that a lens made out of a single piece of glass gives an image which is surrounded by fringes of colour, so that some device has to be adopted in order to destroy, as far as possible, this enemy of good definition. In the early history of the telescope, this so-called *chromatic aberration* was considerably reduced by making small object-glasses of very great focal length.[1]

Lenses of 100-feet focus, however, are not easy to employ as object-glasses, and astronomy was, therefore, greatly benefited by Dollond's invention of the *achromatic lens*

[1] The focal length of a lens is the distance from its centre at which an image of a very distant object, such as the sun, is formed.

in 1760. This is a compound lens, usually consisting of a double convex crown-glass lens and a concavo-convex, or double concave, lens of flint glass. The curvatures of the lenses, and the optical properties of the two kinds of glass composing them, are such that the colour due to one of them is practically neutralised by that due to the other acting in opposition. A section of such an object-glass, with the "cell" in which it rests, is shown in Fig. 49.

In this way the focal length of the lens, and, therefore, the length of the telescope tube, can be kept within reasonable dimensions, while the definition is improved. There is, however, usually a little outstanding colour, due to the imperfect matching of the two lenses, and if one looks through a large refractor, even of a good quality, a purple fringe will be noticed round all very bright objects. This only affects a few of the brighter objects, while millions of others which are dimmer may be seen free from spurious colour.

It may be remarked that the curved surfaces of the lenses forming telescopic object-glasses must not be parts of spheres. If they are, the images will be rendered indistinct by *spherical aberration*, and the optician has to design his curves to get rid of this defect at the same time as chromatic aberration.

A new form of telescopic objective, consisting of three lenses, which has many important advantages, has recently been invented by Mr. Dennis Taylor, of the well-known firm of T. Cooke & Sons, York.

Such a lens as this illustrates the perfection which the optician's art has now attained. Six surfaces of glass have to be so accurately figured that every ray of light falling upon the surface of the lens shall pass through the finest pin-hole at a distance of eighteen times the diameter of the lens.

THE REFLECTOR.—In a reflecting telescope, the object-glass of the refractor is replaced by a concave mirror. In order that such a mirror may reflect all the rays from a star to a single point, its concave surface must be part of a paraboloid of revolution, that is, a surface produced by the revolution of a parabola on its axis. If a spherical surface be employed, all the rays will not be reflected to a single point, and the images which it gives will be ill-defined. Yet it is astonishing to find that the difference between a parabolic and spherical surface, even in the case of a large mirror, is exceedingly small. Sir John Herschel states that in the case of a mirror four feet in diameter, and forming an image at a distance of forty feet, the parabolic only departs from the spherical form at the edges by less than a twenty-one thousandth part of an inch.

An image being formed by a mirror, it is next to be viewed with an eye-piece just as in the case of a refracting telescope. Here there is a little difficulty, for if the eye-piece be applied in the direct line of the mirror, the interposition of the observer's head will block out the light. Several ways of overcoming this have been devised, but the plan most generally followed is that which Newton adopted in the first reflecting telescope which was ever constructed. With his own hands Newton made a small reflector, $6\frac{1}{4}$ inches long and having an aperture of $1\frac{1}{3}$ inches, with which he was able to study the phases of Venus, and the phenomena of Jupiter's satellites. This precious little instrument is now one of the greatest treasures in the collection of the Royal Society of London. The general design of this telescope is shown in

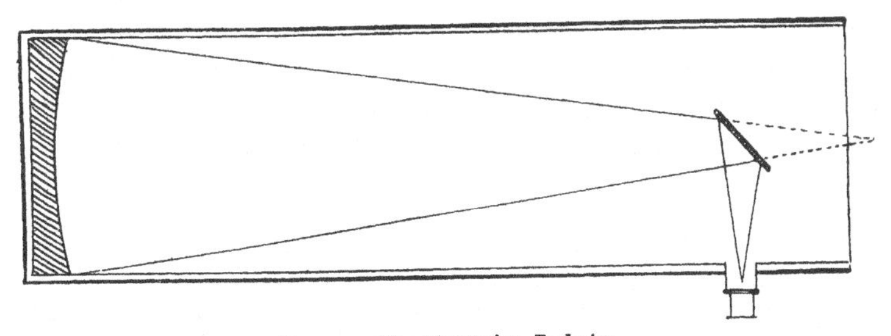

FIG. 50.—*The Newtonian Reflector.*

Fig. 50. The concave mirror is at the bottom of the telescope tube, and normally it would form an image of a star near the end of the tube. A plane mirror, however, of small size intercepts the rays and reflects them to the side, where they converge to a focus. This image is observed and magnified by an eye-piece, as in the refractor. It is true that in this arrangement the plane mirror, or *flat*, renders the central part of the principal mirror ineffective, but the loss of light is very much less than would be the case if the eye-piece were placed in position to view the image centrally.

In the hands of Sir William Herschel the reflecting telescope was greatly developed. The great telescope with which he enriched astronomical science had a mirror four feet in diameter, and its tube was 40 feet in length. With the view of utilising the whole surface of the mirror and dispensing with a second reflecting surface, the four foot mirror was placed at a small angle to the bottom of the tube, so that its principal focal point was no longer at the centre, but at the side of the tube.

In practice, however, it is found that the Herschellian form of reflector does not give the best definition, and it is now very seldom seen.

Among other forms, the "Cassegrain" is perhaps the most important. During the last year or two this form has received a great deal of attention, more especially in regard to its special adaptability for photographic purposes.

In the Cassegrain telescope, the plane mirror of the Newtonian form is replaced by a small convex mirror which is part of a hyperboloid of revolution, its axis and focal point being coincident with those of the primary mirror. The rays are in this way reflected back to the mirror at the bottom of the tube, and in order that the image may be seen, it is necessary to cut out the middle part of the mirror to admit the eye-piece.

Although the small mirror must theoretically be hyperbolic, tolerable definition is obtained even if it be spherical or ellipsoidal, and its actual departure from these forms is so slight as to be beyond detection by measurement, so that the figuring of such mirrors can only be tested in the telescope. For photographic purposes this telescope has the very important advantage that a short telescope is equivalent to a very long one of the Newtonian form, or refracting telescope, so that the image of sun, moon, or planets formed at the focus is very large in comparison with the size of the telescope. A modification of this form of telescope, in which the small mirror is out of the path of the rays falling upon the larger one, and no longer obstructing the central part, has been recently revived by Dr. Common, and has become generally known as the "Skew Cassegrain."

In reflecting telescopes the mirrors were formerly made of speculum metal (an alloy of copper and tin), and the word *speculum* is even now commonly employed to signify a telescopic mirror, although it is usual to make the mirror of glass, with the concave surface silvered and highly polished.

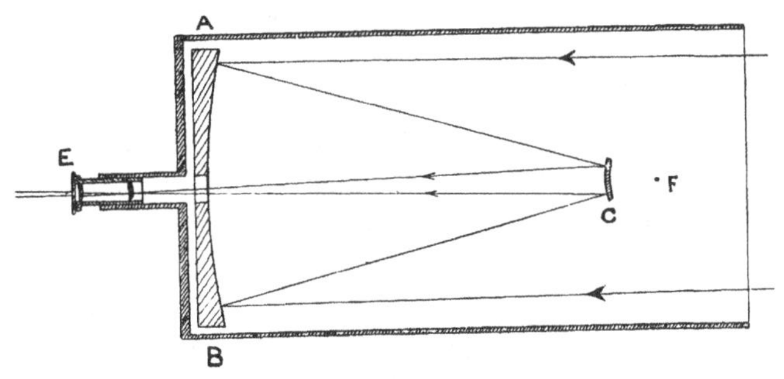

FIG. 51.—*The Cassegrain Reflector.*

One is frequently asked for an opinion as to which is the better form of telescope, the reflector or refractor, and it is a question that one finds some little difficulty in answering. On one point, however, all are agreed, namely, that the reflector has the advantage in regard to its achromatism; it is indeed perfectly achromatic, while the so-called "achromatic" refractor is at best only a compromise. For the rest, one cannot do better than quote the evidence of Dr. Isaac Roberts before the International Astro-photograpic Congress:—"The reflector requires the exercise of great care and patience, and a thorough personal interest on the part of the observer using it. In the hands of such a person it yields excellent results but in other hands it might be a bad instrument. The reflector gives results at least equal, if not superior, to those obtained with the refractor, if the observer be careful of the centering, and of the polish of the mirror, and keeps the instrument in the highest state of efficiency; but when entrusted to an ordinary assistant the conditions necessary for its best performance cannot be so well fulfilled as the same could be in the case of the refractor." One great practical advantage of the reflector is that there are fewer optical surfaces, so that a large reflector may be obtained for the price of a much smaller refractor.

EYE-PIECES.—So far we have regarded the eye-piece of a telescope as a simple lens, but it is evident that the spherical and chromatic aberration of such a lens will interfere with its performance. For occasional use, however, even a simple lens is very serviceable if the object observed is brought to the centre of the field of view.

Compound eye-pieces are of various forms, each having certain advantages, the desiderata being freedom from colour and "flatness of field"—that is, stars in different parts of the field are to be equally well in focus. Those most commonly employed are the Ramsden and Huyghenian eye-pieces. The former consists of two plano-convex lenses of equal focal lengths, having their curved faces towards each other, and being placed at a distance apart equal to two-thirds of the focal length of either lens. Such an eye-piece can be used as a magnifying-glass, and it is therefore placed outside the focal image formed by the telescope with which it is used; on this account it is called a *positive eye-piece.* This kind of eye-piece is not quite achromatic, but its flat field of view gives it a special value for many purposes.

In the Huyghenian eye-piece there are again two lenses, made of the same kind of glass. That which comes nearest to the eye has a focal length of only one-third that of the *field* lens, and the distance between the two lenses is half the sum of the focal lengths. This form of eye-piece cannot be used as a magnifying-glass in the ordinary sense, and as the field lens must be placed on the object-glass or mirror side of the focus, it is called a *negative eye-piece.* The Huyghenian eye-piece is more achromatic than the Ramsden, and is more widely used when it is only required to view the heavenly bodies. In instruments employed for purposes of measurement, a positive eye-piece is essential in order that the spider threads may be placed at the focus of the telescope. The images formed by an astronomical telescope are upside down, and neither of the eye-pieces described reinverts them.

A special form of eye-piece is therefore used when a telescope is employed for terrestrial sight-seeing. The desired result is obtained by the introduction of additional lenses, but there is a corresponding reduction of brightness.

For viewing the sun some device is necessary to reduce the quantity of light entering the eye. To look at the sun directly, even with a small instrument, is very dangerous. The arrangement usually adopted is a *solar diagonal,* in which the light is reflected from a piece of plane glass before entering the eye-piece ; the piece of glass is wedge-shaped, so that the reflection from one surface only is effective ; if the glass had parallel sides, the solar image would be double.

MAGNIFYING POWER.—The magnifying power of a telescope depends upon the focal length of the object-glass, or speculum, and that of the eye-piece. Optically, it is equal to the former divided by the latter, so that the greater the focal length of an object-glass, or the smaller the focal length of the eye-piece, the greater will be the magnifying power. In a given telescope, the object-glass, or speculum, is a constant factor, and the magnifying power can only be varied by changing the eye-piece. The focal length of the Lick telescope, for example, is about 600 inches; with an eye-piece which is equivalent to a lens of one-inch focus, the magnifying power would be 600; with a lens of half an inch focus, it would be 1,200, and so on.

The magnifying power which can be effectively employed, however, depends upon a great variety of circumstances. First, the clearness and steadiness of the air ; then there is the quality of the object-glass, or speculum, to be considered ;

and also the brightness of the object to be observed, for when the object is very dim, its light will be spread out into invisibility if too high a power be used.

In practice, good refractors perform well with powers ranging up to 80 or 100 for each inch in the diameter of the object-glass. Thus, on sufficiently bright objects, a six-inch telescope will work well with a power of about 500, while a 30-inch may be effectively employed with powers between 2,000 and 3,000.

ILLUMINATING POWER.—It has already been pointed out that magnification is not the only function of a telescope. As a matter of fact, the most powerful telescopes in the world fail to produce the slightest increase in the apparent size of a star, for even if these objects be brought to apparently a 3,000th part of their real distances, they are still too far away to have any visible size. But although a star cannot be magnified, it can be rendered more visible by the telescope, for the reason that the object-glass collects a greater number of rays than the naked eye. The pupil of the eye may be taken to have a diameter of one-fifth of an inch; a lens one inch in diameter will have 25 times the *area* of the pupil, and will therefore collect 25 times the amount of light from a star; a two-inch lens will grasp 100 times, and a 36-inch 32,400 times as much light as the pupil alone. Practically all these rays collected by the object-glass, or speculum, of a telescope cannot be brought into the eye; some are lost through the imperfect transparency of the glass, or the imperfect reflecting power of the speculum. Still, allowing a considerable percentage for loss, there is an enormous concentration of light when a large telescope is employed.

THE ALTAZIMUTH MOUNTING.—Having got a telescope, we have next to see how it can be best supported, for unless it be a very small instrument indeed, it will be impossible to hold it in the hand like a spy-glass. However a telescope be mounted, provision must be made for turning it to any part of the sky whatsoever. Very frequently one of the axes on which the instrument turns is vertical, while the other is horizontal. Such a stand for a telescope is called an *altazimuth mounting*, for the reason that it permits the instrument to be moved in altitude and in azimuth.

As a rule, one finds only small telescopes mounted in this manner. The objection to it is that, as one continues to observe a heavenly body, two independent movements must be given to the telescope in order to follow the body in its diurnal movement across the heavens. If we commence observing a star newly risen, for example, the telescope must trace a stair-like path in order to follow it, as it ascends into the heavens.

THE EQUATORIAL TELESCOPE.—A much more convenient method of setting up a telescope is to mount it as an *equatorial*. The essential feature of this instrument is that one of the axes of movement, instead of being vertical, is placed parallel to the axis of the earth. This is called the *polar axis*, and, when the telescope is turned around such an axis, it traces out curves in the sky which are identical with those described by the stars in their diurnal motions. If, then, the telescope be directed to a star or other heavenly body, it can be made to follow the object and keep it in view by a single movement. The axis at right angles to the polar axis is called the declination axis, and is necessary in order that the telescope may be moved towards and from the Poles so that all the heavenly bodies above the horizon may be included in its sweep.

One very important advantage of the equatorial is that as only one motion is required to keep a star in view, so long as it is above the horizon, the necessary movement may be furnished by clockwork. A good equatorial is accordingly provided with a driving clock, which is regulated so that it would drive the telescope through a whole revolution once a day. Unlike an ordinary clock, the driving clock of a telescope is regulated by a governor, in order that the instrument may have a continuous and not a jerky movement.

The telescope is also provided with clamps and fine adjustments, one each in R. A. and declination, in order that it may be under the control of the observer. It is evident that the telescope must be capable of moving independently of the driving gear, so that it may first be placed in the desired direction ; when this is accomplished, the R. A. clamp is used to put the telescope in gear with the clock. The declination clamp is them made to fix the telescope firmly to the declination axis. Fine adjustments in both directions are necessary, because it is impossible to sight a large instrument with such precision as to bring an object exactly to the centre of the field of view.

Some of the driving clocks fitted to equatorials are very elaborate. As clocks regulated by governors are not such reliable timekeepers as those regulated by pendulums, arrangements are made by which the accuracy of a pendulum can be electrically communicated to a governor clock. One of the best forms of electrically-controlled clocks is that devised by Sir Howard Grubb.

Another important feature of an equatorial is that it can be provided with circles which enable the telescope to be pointed to any desired object of known right ascension and declination. One of these is the declination circle, attached to the declination axis and read by a vernier fixed to the sleeve in which the axis turns ; this is adjusted so as to read 0° when the telescope points to any part of the celestial equator, and 90° when it is directed to the Pole. The other circle is attached to the polar axis, and determines the position of the telescope with regard to the meridian ; this is called the *hour circle*, and is divided into 24 hours. When the telescope is on the meridian, the hour circle reads zero, so that its reading in any other position gives the hour angle of the telescope. Having given the right ascension and declination of a heavenly body which it is desired to observe, the telescope is turned until the declination circle reads the proper angle, and the hour circle indicates the hour angle which is calculated for the particular moment of pointing the telescope. [The hour angle is the difference between the right ascension of the object and the sidereal time of observation.] In this way it is easy to find objects of known position which are invisible to the naked eye, and one can even pick up the planets and brighter stars in full sunshine. Conversely one can determine from the circles the right ascension and declination of any object under observation, but for various reasons only approximate results can be obtained in this way. The chief use of the circles on an equatorial is therefore to provide a means of pointing the telescope.

Telescopes of 4 inches aperture and upwards are usually provided with a smaller companion called a *finder*. This has a larger field of view than the main telescope, so that objects which are of sufficient brightness can readily be picked up and brought to the centre of the finder, the adjustments being such that the object is then also at the centre of the field of the large telescope.

There are, of course, many practical details connected with the working of an equatorial with which space does not permit us to deal. It may be remarked, however, that the adjustment of the polar axis is very simply performed by first inclining it at an angle approximately equal to the latitude of the place where it is set up, and setting it as nearly as possible in the meridian by means of a compass or by observations of the sun at noon. The final adjustment is then made by a series of observations of stars of known position.

SOME OF THE WORLD'S GREAT TELESCOPES.—Thanks to the wide public interest taken in astronomical matters, a large number of powerful telescopes has been set up in various parts of the world. To the British Islands belongs the honour of possessing the largest telescope in the world. This is the giant reflector erected by Lord Rosse, in 1842, at Parsonstown, the mirror being 6 feet in diameter, and the focal length 60 feet. Many very valuable observations were made with this instrument in its early days, but of late years it seems to have fallen into disuse. One reason may be that the mounting is not of the most convenient form, and makes the telescope unsuitable for photographic work.

Coming next in point of size to the Rosse telescope is the

reflector erected at Ealing, by Dr. A. A. Common. The glass mirror of this telescope is 5 feet in diameter, 5 inches thick, and weighs more than half a ton. Dr. Common aimed specially at constructing the largest possible telescope which could be equatorially mounted and provided with a driving clock, and he was only limited to an aperture of 5 feet by the impossibility of obtaining a glass disc of larger size. He has attained such great skill in this work that he was able to produce a perfect mirror 5 feet in diameter in three months time, although no less than 410,000 strokes of the polishing machine were required.

The telescope is of the Newtonian form, and the mounting is quite unique. The polar axis consists of an iron cylinder, made up of boiler plates, 7 feet 8 inches in diameter, and about 15 feet long. From the top of the cylinder, near its outer edge, two horns, each 6 feet long, project outwards, and the tube of the telescope swings on trunnions attached to the ends of the horns. The main part of the telescope tube is square, built up of steel angle iron, and carries the mirror at its lower end; the upper part of the tube, which carries the "flat" and eye-piece, is round, and of tinned steel strengthened by a skeleton frame-work.

It is evident that such an enormous instrument as this cannot be made to travel by clockwork with the necessary uniformity without some very efficient arrangement for reducing friction. Dr. Common's plan—and it is here that his instrument is unlike others—is to make the hollow polar axis water-tight, and to fix it in a tank of water. At the bottom of the polar axis is a ball and socket joint to keep it in position, and at the top is another bearing, which can be adjusted so that the polar axis lies truly in the meridian. It was found necessary to introduce 9 tons of iron into the bottom of the hollow polar axis in order to sink it to the proper angle, and to put sufficient weight on the bearings to give stability to the instrument. In this way the great mass is brought into the region of manageability, and the driving clock, which is driven by a weight of 1½ tons, is able to do its work efficiently. Such, in general outline, is this wonderful telescope, which, although not so large as Lord Rosse's famous instrument, is undoubtedly its superior in light-grasping power and general utility, and more especially in its adaptability for photographing the heavens.

Among other large reflecting telescopes now in use are the four-foot reflectors at Melbourne and Paris, and the three-foot reflectors at South Kensington and the Lick Observatory, California.

The largest refracting telescope yet constructed is one of 40 inches aperture for the new Yerkes Observatory of the University of Chicago. It is interesting to note here that Professor Keeler, in his report as an expert upon the preformance of the object-glass, considers that there is "evidence for the first time that we are approaching the limit of size in the construction of great objectives." Unlike a mirror, a lens can be supported only upon its circumference, and it is the bending by its own weight that proves detrimental to its defining power. If the lens be made thicker with a view of overcoming this defect, the absorption of light by the glass increases, so that there is in the end no special gain by increasing the size.

The length of the Yerkes telescope is 62 feet, and it will be provided with all accessories pertaining to astrophysical research. The Yerkes telescope, however, is not yet in actual use, and meanwhile the world-renowned Lick telescope, of 36 inches aperture, keeps the lead among active big refractors. The story of the foundation of this monster instrument is not much less wonderful than the telescope itself. Brought up in poor circumstances, with few opportunities for intellectual development, James Lick, nevertheless, amassed a fortune in business, and having few relations, he was anxious to dispose of his wealth in such a way as to bring him that fame which he had failed to achieve in other directions. Although it is very probable that he had never looked through a telescope in his life, the idea of a large telescope had taken a very firm hold upon his mind, and, thanks to the influence of his advisers, it was definitely announced in 1873 that Mr. Lick's bid for immortality was to take this form. Several sites were examined by experts, and finally Mount Hamilton, California, 4,200 feet above sea-level, was selected. An excellent road, 26 miles in length, made at the cost of the county authorities, connects the observatory with the nearest town, San José, 13 miles distant.

Owing to various delays, operations were not commenced until 1880, and five years were consumed in clearing away 72,000 tons of rocks and in erecting the buildings.

Mr. Lick had stipulated for the erection of "a telescope superior to and more powerful than any telescope yet made," and Messrs. Alvan, Clark & Co. contracted to supply a lens of 36 inches aperture for the sum of 50,000 dollars. It turned out, however, that it was much easier to make such a contract than to fulfil it. To produce large discs of optically perfect glass, even in the rough, requires the greatest possible skill and patience, and this part of the work was undertaken by Feil & Co. of Paris. The flint glass disc was safely delivered in America in 1882, but the crown disc was cracked in packing. The elder Feil having retired from business, the duty of providing a new block of crown glass devolved upon his sons, who, after two years spent in vain attempts, ended in bankruptcy, and it was only through the elder Feil again resuming business that the much-required disc was finally completed in 1885. After the lapse of another year, the rough discs were fashioned, in the workshops of the Clarks, into the most marvellous of telescopic lenses.

The mounting of the object-glass is worthy of the occasion, as will be seen from our illustration (see page 40). The tube is no less than 57 feet long, and 4 feet in diameter in the middle part. An iron pier, 38 feet high, beneath which lie the remains of Mr. Lick, supports the equatorial head, and a winding staircase enables the observer to reach the setting circles. Inside the hollow pier is the powerful driving clock which turns the telescope to follow the heavenly bodies in their apparent movements. Finders of 6, 4, and 3 inches diameter, rods for the manipulation of the instrument, and all necessary accessories, complete what must long remain one of the most perfect instruments at the service of astronomical science. The 200,000 dollars expended upon it have already been amply justified by the work accomplished, while Mr. Lick's dream of immortality has become a reality.

The following list indicates some of the large refractors now (Feb., 1897) doing active service:—

Aperture.	*Observatory.*
36 inch	[Lick] California.
30 "	Pulkowa, Russia.
30 "	[Bischoffeim] Nice.
28 "	Greenwich.
27 "	Vienna.
26 "	Washington.
25 "	[Newall] Cambridge.
24 "	[Lowell] Mexico.
23 "	Princeton, New Jersey.

It is right to add, however, that opinion is still greatly divided as to whether these telescopes of large aperture really repay the expense and labour involved in their erection and use. On the very rare occasion when the "seeing" is practically perfect—which occurs perhaps only a few hours in a year—it is probable that the superiority of a large telescope is very marked, but under average conditions there seems to be little advantage over instruments of moderate size for many classes of observations.

Certain it is that a great deal of valuable work is done with comparatively small telescopes, ranging from six to fifteen inches aperture, and this in all departments of astronomical research. Hence, some of the most active observatories do not figure in the above list; among them may be mentioned the observatories of Harvard College (U.S.A.), Potsdam, Paris, Heidelberg, Cape of Good Hope, Edinburgh, South Kensington, Stonyhurst College, and the observatory of Dr. Isaac Roberts at Crowborough, Sussex.

HOUSING OF EQUATORIALS.—The building which accomodates an equatorial telescope must evidently be designed to admit of giving a clear opening to any part of the sky. Usually this is accomplished by making the roof, or *dome*, with a circular base, provided with wheels, which run on rails. It is then only necessary to open a narrow portion of the dome, extending from top to base, and to turn the dome until this aperture is in the required direction. One of the most elaborate domes now in existence is that built by M. Eiffel for the great refractor of the Nice Observatory. The lower part of the building is in the form of a square (see Frontispiece), having a side of about 87 feet, and a height of about 30 feet. The dome itself is 74 feet in diameter, and the moving parts alone weigh 95 tons.

As will be seen from the illustration, there are two shutters, each a little wider than half the possible opening; these run on short rails, and are moved simultaneously by means of an endless rope. The whole of the dome is built up of steel angle iron, covered with very thin sheet steel. In order to facilitate the manipulation of the dome, its great weight is buoyed up by means of a float attached to its base and immersed in a circular tank of water of a little greater size than the base of the dome. If any mishap occurs with this gigantic tank, the dome rests on wheels which run on a circular rail, so that the work need not be interrupted. The whole arrangement is very easily turned with the aid of a winch by one man when the dome is floating, but when resting on the wheels several men are required at the winch.

This brief description will serve to illustrate some of the problems which confront the possessor of a very large telescope. For smaller instruments, the observatories follow pretty nearly the same plan, except that it is unnecessary to provide an arrangement for floating the dome.

The observatory which shelters a reflecting telescope need not differ very greatly from one which contains a refractor. If the instrument be a Newtonian, it is generally convenient to sink the polar axis below the level of the floor in order that the observer may not be at too great a height from the ground, and in that case, the dome, or its equivalent, is all that is necessary. For his five-foot reflector, Dr. Common designed an observatory which is not of the ordinary form, but gives the necessary opening partly by means of large shutters, and partly by a revolution of the whole house. It is not everyone who is able to lay out £8,000 on such a dome as that erected at Nice by M. Bischoffeim.

The varying position of the eye end of a telescope, when it is turned to different parts of the sky, makes it necessary to provide comfortable and safe seating accommodation for the observer, more especially when the telescope is a very large one. In the case of the Yerkes telescope, the eye-piece will be 30 feet higher when observing near the horizon than when observing near the zenith, and the observer must necessarily follow the telescope. The most convenient arrangement in such a case is to raise or lower the floor of the observatory as occasion demands. The floor of the Yerkes Observatory is 75 feet in diameter, and by means of electric motors it can be given a vertical motion of 22 feet. A similar arrangement was provided for the Lick telescope from the designs of Sir Howard Grubb. With smaller instruments, observing ladders and adjustable chairs of various forms are employed.

THE EQUATORIAL COUDÉ.—A form of equatorial telescope which has possibly a great future before it is one introduced at Paris under the name of the *equatorial coudé*, or elbowed telescope. Its practical advantage is that the observer remains in a constant and comfortable position, so that revolving domes and elevating floors, or other arrangements serving similar purposes, are no longer necessary. The telescope tube is of two parts of nearly equal length, and what is ordinarily the lower half of the tube forms part of the polar axis, while the other half is attached to it at right angles. At the point of intersection of the two halves of the tube is a plane mirror, and there is another mirror in front of the object-glass. If the latter mirror were removed, such a telescope would only enable the observer to see objects lying along the celestial equator, but by its means objects in all parts of the heavens can be brought within range to an observer gazing down the hollow polar axis. The largest instrument is that at the Paris Observatory, which has an object-glass 23½ inches in diameter for visual observations, and another of the same size for photographic purposes.

FIXED TELESCOPES.—There is still another method of using a telescope. The telescope itself may be fixed, and the light of the heavenly bodies may be reflected into it by means of a mirror which is made to revolve so as to keep pace with their movements. Foucault devised an instrument called the *siderostat* for this purpose, and although it is not largely employed for telescopic observations, it is very widely utilised for spectroscopic work, where the spectroscope is of a kind not readily attached to a telescope.

Another instrument used for the same purpose has recently been brought forward under the name of the *coelostat*. This is simply a mirror which is made to turn on a polar axis in its own plane, and since a reflected ray of light moves through twice the angle that the reflecting surface turns through, the mirror is made to revolve at the rate of one revolution in two days. As the name indicates, the whole heavens appear stationary in such an instrument, whereas in a siderostat, only one star at a time appears at rest, while its neighbours slowly revolve round it.

PHOTOGRAPHIC TELESCOPES.—The application of photography to the study of the heavenly bodies marks one of the greatest advances of the present century. The instruments which are employed for this purpose range from the ordinary tourist camera to the largest telescope. Unlike a person sitting for a portrait, the heavenly bodies cannot be made to stand still for the purpose, and as instantaneous photographs can only be obtained in the case of the sun and moon, it is usually necessary to make the camera follow the stars very exactly during the time of exposure, in order that the images may fall on precisely the same parts of the photographic plate.

Some guiding arrangement is, therefore, essential, and

generally the photographic camera or telescope is attached to an ordinary equatorial which is driven by clock-work, or very carefully by hand if the camera be a small one. In the guiding telescope are two spider threads at right angles to each other, and it is by constantly keeping the image of a star at the intersection of these "wires" that the operator ensures the images remaining in a constant position upon the sensitive plate.

An ordinary portrait camera, in the hands of a skilled observer, yields very beautiful pictures, but they are naturally on a small scale. The field of view of such an instrument is so large that a whole constellation may be photographed with a single exposure.

Portrait lenses of 6 inches aperture in the hands of Dr. Max Wolf and Professor Barnard have given magnificent delineations of the Milky Way, and of the extremely faint nebulosities which are to be found in many parts of the heavens

For many purposes, however, telescopes of greater power are required, and here it may be remarked that the distance between the images of any two adjacent stars will vary in direct proportion to the focal length of the telescope. In the same way the size of the image of a planet, the moon, or a comet, increases as the focal length of the objective is increased.

Refracting telescopes which are employed for photography require object-glasses which are specially "corrected" for the photographic rays. White light is compounded of light of all colours, but it is the blue and violet constituents which are effective in producing photographic action on an ordinary sensitive plate. Now, an object-glass which is intended for visual puposes is made to focus at the same point as many as possible of the rays which are most effective to the human eye, that is the green, yellow, and red, and usually there is a blue or purple halo round the images of the brighter objects, which is, however, too feeble as a rule to interfere with visual observations. This blue halo, will evidently result in defective definition if the lens be employed for photography. By putting the plate at the point where the blue rays are most nearly focused, a better image is obtained; but for really good work a photographic object-glass must be so designed that all the blue and violet rays are brought to one and the same focus. Such a lens will consequently be a very poor one for visual observations. At the present time, 18 photographic telescopes, each of 13 inches aperture, and

FIG. 52.—*The Photographic Telescope employed by Dr. Isaac Roberts.*

corrected in this way, are at work in various parts of the world for the international star chart.

The new "photo-telescopic" object-glass now manufactured by Messrs. Cooke appears to be full of promise. In this lens all the colours of the spectrum are brought to almost exactly the same focal point, so that it serves equally well for photographic or visual purposes.

This difficulty in regard to achromatism does not exist in the case of the reflecting telescope, since rays of light of every colour are reflected at precisely the same angles. For this reason reflectors, when properly managed, give the best photographic results. Dr. Isaac Roberts and Dr. Common are especially identified with the application of the reflecting telescope for celestial photography. The instrument employed by the former consists of a 20-inch reflector, and a 7-inch guiding telescope of the refracting form. The two telescopes are mounted on the extreme ends of the declination axis of an equatorial, a photograph of which we owe to the kindness of Dr. Roberts.

Dr. Common does not employ a guiding telescope at all. The photographic plate which he places at the focus of the reflector is smaller than the field of view, so that by means of an eye-piece fitted with a cross wire at the side of the dark slide, he is able to watch a star near the edge of the field. Both eye-piece and dark slide are attached to a frame which can be controlled by two screws at right angles to each other. If the guiding star leaves the cross wire through errors in driving, or other causes, the eye-piece and dark slide are bodily moved after it by means of the adjusting screws. This method not only has the advantage of saving the cost of a guiding telescope, but reduces the effects of vibration consequent upon the correction of errors by moving the whole telescope.

For photographing the sun a special instrument called a *photoheliograph* is usually employed. This differs only from an ordinary photographic telescope in being provided with a secondary magnifier, by which means the focal image formed by the object-glass is amplified before falling upon the photographic plate. On a bright clear day, pictures of the sun 8 inches in diameter can be taken with an exposure of about $\frac{1}{500}$th of a second, and such a photograph will frequently record more facts as to the state of the solar surface than a whole day's observation. Lenses or mirrors of very long focus are also occasionally employed in solar photography, and in this way a large image is obtained without the use of a secondary magnifier.

Photographs of the moon and planets may be taken either

with or without a secondary magnifier, but in either case the exposures are longer than for the sun.

Finally, it may be added that the sensitive plates and processes used in astronomical photography do not differ from those employed by ordinary photographers.

CHAPTER XVI.

INSTRUMENTS OF PRECISION.

THE MERIDIAN CIRCLE.—The accurate registration of the positions of the heavenly bodies is one of the most important functions of an astronomical observatory. When the apparent places of an object at a sufficient number of different times have been duly recorded, it becomes possible to investigate the laws upon which its changes of position depend, and to predict its positions at subsequent times for the benefit of navigators and others to whom such predictions are of practical utility. For this purpose various instruments have been devised, but in all cases where it can be employed, the *transit circle*, or *meridian circle*, as it is indifferently called, is generally conceded to give the most trustworthy results.

With this instrument the observations are made when the celestial body under observation is crossing the meridian of the place where the instrument is set up, that is, when it "transits," or "souths." At this time the accuracy of the observations is least impaired by the ever-varying effects of atmospheric refraction.

The meridian circle consists of a refracting telescope—seldom exceeding 6 inches in aperture—which is fixed to a hollow axis at right angles to itself, and this axis is supported horizontally in an east and west direction, so that the telescope is only free to move in the plane of the meridian. A large graduated circle—or frequently two such circles—attached perpendicularly to the hollow axis, and read by microscopes fixed to the walls or iron pillars which support the axis, completes the essential parts of the instrument.

As the field of view of the telescope covers a considerable area, it becomes necessary to provide some means of marking the exact point within it which represents the meridional axis of the instrument. This is accomplished by placing at the common focus of the object-glass and the positive eye-piece a system of "cross wires," consisting of tightly-stretched spider threads, two of which are fixed horizontally and nearly in contact, and five or seven vertically at equal distances apart. What the observer has actually to do is to incline the telescope at such an angle that the star is seen to traverse the space between the two horizontal threads, and then to record the exact times, by means of a chronograph and sidereal clock, at which the star appears to cross each of the equidistant vertical threads. By thus making five or seven observations and taking the average, greater accuracy is attained.

The time observations, as we have already seen, determine the right ascension of the star under observation, while the declination is indicated by the readings of the graduated circle, if the latter is so placed as to read 90° when the telescope is directed to the Pole.

The ideal meridian circle is thus simplicity itself, but the mechanical difficulties encountered in making such an instrument are insuperable. Perfect right angles and perfect circles exist only in our minds, so that after all the undoubted skill and care bestowed on its construction, the actual meridian circle is only an approximation to the ideal. Still, when the instrument is provided with levels and other means for estimating its deviation from the meridian plane in which it ought to move, the actual observations are capable of correction by mathematical processes, so that the final statements of positions sensibly represent those which would follow from the use of a perfect instrument.

The greatest possible care is taken to secure rigidity in all parts of the meridian circle. The hollow horizontal axis is supported on bearings which rest either on heavy piers of iron or walls of masonry, and the axis and telescope tube are firmly joined together at their intersection. The bearings for the axis are turned with extreme precision, and, to reduce the friction upon them, the pressure of the instrument is counterpoised by an arrangement of balancing weights.

Adjustments are provided for every needful purpose. The cross wires are fitted in a small frame which can by suitable fittings be given a small movement in the field of view until the right place for them is found, while the horizontality of the axis and its correct direction can be secured by other adjusting screws.

Since most of the observations have to be made at night, the field of view will generally be dark, and the exceedingly delicate spider lines will be invisible unless some means of illuminating them be provided. Usually a very tiny mirror is fixed diagonally at the intersection of the axis and the telescope, where it is held in position by a stiff wire. A light shining through the hollow axis is thus reflected into the field of view, and the threads are rendered visible. The intensity of this illumination of the field can be regulated in accordance with the brightness of the star under observation.

The instrument having been erected, one of the first tests applied to it is to see that it is correctly *collimated*, or, in other words, that the optical axis of the telescope is perpendicular to the axis of movement. For this purpose the telescope is directed to some distant object, such as a building, and some mark which falls on the intersection of the central spider threads is noted. The axis is then reversed end for end by a mechanical arrangement, and the telescope again pointed at the same object. If the mark again falls on the intersection of the cross wires, the collimation is correct ;

if not, the wires are moved with the frame containing them until the error is corrected.

To test the horizontality of the axis, a spirit-level long enough to stretch across the bearings, and called the "striding level," is provided.

Various methods are employed for adjusting the instrument so that the telescope moves as truly as possible in the plane of the meridian. Collimation and level being correct, the telescope will move in a vertical plane, whatever may be the error in the direction of the horizontal axis, and therefore any star passing through the zenith will cross the centre of the instrument at the same moment that it crosses the meridian. A star away from the zenith, however, will not be seen on the cross wires when it crosses the meridian, unless the axis be truly east and west. Hence, by taking the difference of time between the observed transits of a star near the zenith and one a long way from the zenith, and turning the whole instrument in azimuth until this difference is equal to the difference of right ascensions of the two stars, the instrument is readily placed in the meridian.

Another useful method of adjustment is to observe the upper and lower transits of a circumpolar star. If the instrument moves truly in the meridian, the interval between the two transits will evidently be twelve sidereal hours.

Next, the declination circle has to be adjusted so that it reads 90° when the telescope is directed to the celestial pole, or zero when an equatorial star is under observation. An obvious way of doing this is to take the readings when Polaris, or other circumpolar star, is at upper and lower transits; the celestial pole lying midway between these positions, the average of the two readings, when corrected for refraction, should be 90°, and the circle would be shifted round in its fittings until this was the case.

Such, in mere outline, are the processes by which the meridian circle is set up. In actual practice, the greatest possible refinement is brought to bear on the adjustments, and every precaution taken to estimate the various errors so that due allowance may be made for them in the reduction of the observations. It has even been shown that the heat of the observer's body, by affecting the lower side of the telescope tube more than the upper, introduces sensible errors in the measures of declination. Hence it is important to use metals of high conductivity in the construction of meridian instruments, so that errors due to the varying temperatures of different parts may be reduced to a minimum.

As an illustration of a modern meridian circle, we select that of the Lick Observatory. (Fig. 53.) This instrument has an aperture of six inches, and embodies all the improvements which have been introduced by the Berlin firm of Repsöld & Co.

The observatory containing a meridian circle is usually a very simple structure, as it is only necessary to provide an opening to the sky along a north and south line. This is sufficiently provided for by a series of narrow shutters in a building of ordinary construction.

To prevent confusion it may be pointed out that the term "transit instrument" is frequently restricted to a meridian instrument which is not supplied with large circles for the accurate measurement of declinations, although it may have a small circle to assist in directing the telescope. The use of such an instrument is evidently limited to the determination of time and right ascension.

THE ALTAZIMUTH.—Although the meridian circle furnishes us with the most accurate method of determining celestial positions, its use is somewhat restricted by the fact that it can only be employed for the observation of objects on the meridian. It sometimes happens, however, that bodies cannot conveniently be so observed, and other methods become necessary. This is especially the case with the moon during the first and fourth quarters, when it crosses the meridian in daylight, and it is then that an instrument called the *altazimuth* is of special value. This is something like a transit circle in which the base supporting the piers is made to

FIG. 53.—*The Meridian Circle of the Paris Observatory.*

turn on a vertical axis, so that the telescope can be directed to any part of the heavens whatsoever. A fixed horizontal graduated circle, read by verniers or microscopes attached to the revolving part, gives the azimuth of the telescope when an observation is made, and the altitude is furnished by the vertical circles. The azimuth circle is adjusted to read zero when the telescope is pointed due north, and the altitude circle to zero when the telescope is horizontal. To secure the first adjustment, after correcting level and collimation, a star may be observed before it crosses the meridian, and again when it has exactly the same altitude after passing to the west; midway between the two positions would be due south, and the circle should read 180°. In adjusting the vertical circle, the telescope is made to point downwards to a trough of mercury, and it is known that the telescope is truly vertical when the reflected image of the cross wires is coincident with the wires themselves; the circle should then read 90°.

From a knowledge of the sidereal time at which a celestial body has an observed altitude and azimuth, the more useful co-ordinates of right ascension and declination can be calculated by spherical trigonometry.

One of the largest instruments of this class has recently been erected at Greenwich Observatory. The aperture of the telescope is 6 inches, and the rigidity of the various parts may be gathered from the fact that the instrument weighs something like six tons.

A *theodolite* is a small portable form of altazimuth specially adapted for the needs of surveyors, but occasionally employed in astronomical work.

THE WIRE MICROMETER.—Notwithstanding that an equatorial telescope is usually furnished with circles for estimating the positions of objects observed, or to serve as a guide in directing the telescope to objects of known position, it is not entitled to be called an instrument of precision in the sense we are now considering. The provision for driving by clock-work and other causes are antagonistic to constancy of adjustment, and hence determinations of positions by the circles alone might be many seconds in error. Most large telescopes, however, are provided with some form of micrometer which not only serves for the measurement of planets, lunar craters, and the like, but may also be used to measure the angular separation of adjacent stars. In this way, by making a "triangulation" of stars visible in the field of view, and including at least two which have had their precise positions determined by the meridian circle, the positions of objects can be measured with great accuracy.

This method is especially valuable in the case of comets, which may cross the meridian in daylight, and are often too dim to be seen with the altazimuth.

Several forms of micrometers are in use, but the so-called *wire* or *filar micrometer* is most commonly seen in our observatories. The essential parts are very similar to those of the reading microscope (p. 172). Two parallel spider threads are so arranged on sliding frames that they may be brought into coincidence, or separated, by means of very finely-cut screws. Perpendicular to these are two fixed threads almost close together. The system of "wires" is viewed by a positive eye-piece, and the whole is attached to a draw tube so that it may be placed in position at the eye end of the telescope. In order that the wires and telescopic images may be sharply defined at the same time, the plane of the wires must be at the principal focus of the object-glass. The screws are provided with large heads which are graduated so as to show the hundredth of a revolution, and counting wheels register the numbers of complete turns.

Matters are so arranged that when both counting-wheels indicate zero, the spider threads are coincident. Then, supposing one of the screws be turned through a revolution, the threads will be separated by a definite amount; an equal and opposite movement of the other screw will double the separation, and in all cases the distance between the threads will be registered in turns, and fractions of turns of the screws.

The next proceeding is to ascertain what is called the "value," in angular measure, of the micrometer screw. This value will evidently depend upon the pitch of the screw and the focal length of the telescope to which the micrometer is applied, so that measurements merely stated in terms of revolutions of the screw would serve no useful purpose. It can easily be calculated that the images of two stars which are 28′ 39″ apart will be separated by an inch at the focus of a telescope of 10 feet focal length; then, if the screws have 100 threads to the inch, the angular separation of the wires corresponding to a single revolution will be one-hundredth part of 28′ 39″, that is, 17″·15, and the latter would be the value of that particular micrometer when used with the telescope in question. If the focal length of the telescopic object-glass were 20 feet, the linear separation of the images of two such stars as we have considered would be 2 inches, and the value would therefore be halved, so that measures of twice the accuracy would be possible. Since the stellar images and the cross wires are equally magnified by the eye-piece, the value of the screw is in no way affected by using eye-pieces of different powers.

In practice it is necessary to determine the value of the micrometer screw by actual measurement. For this purpose, the wires are separated by a known number of revolutions, say twenty, and the micrometer is adjusted so that a star of known declination travels exactly between the two fixed wires when the telescope remains at rest. With the telescope still fixed, the number of seconds required by the image of the star to traverse the distance between the separated wires is noted, and knowing the angle through which the star must have moved in that interval, the angular value of one turn of the screw is at once deduced. For work of extreme precision each individual turn of the screw must be separately evaluated, and allowances must also be made for changes of temperature.

When measuring the apparent diameter of a planet, the two threads are separated until the image just lies between them, and the sum of the readings of the two screws multiplied by the angular value of one turn gives the diameter in seconds of arc. The distance having been formed by other observations, the diameter of the planet in miles can be determined in the manner to which reference has already been made (p. 142).

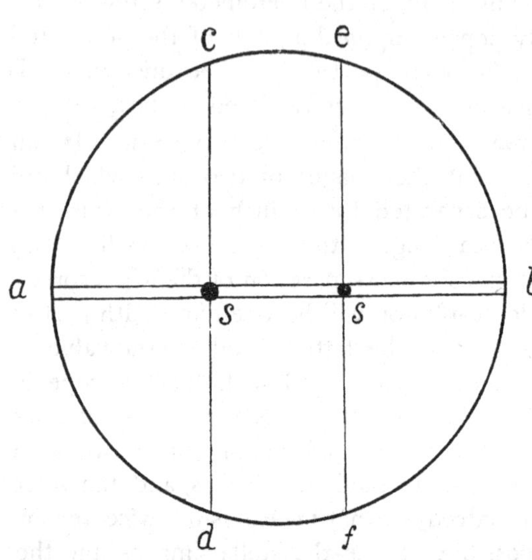

FIG. 54.—*The Micrometer applied to a Binary Star: a b, Fixed Threads; c d, e f, Movable Threads; s s, Components of Binary Star.*

One of the most important applications of the micrometer is in the measurement of double and binary stars. In this case the fixed threads are made to enclose the two stars, and the movable threads are made to bisect the star images. (Fig. 54.)

THE POSITION CIRCLE.—It is frequently necessary to be able to specify a direction, as in the case of a planet's equator, or the line joining the components of a double star. Such directions are expressed by "position angle," which may be defined as the angle from the north point, reckoned from 0° to 360° through east, south, and west. For these observations, a *position circle* is usually attached to the micrometer. This is a circle graduated from 0° to 360°, which can remain fixed in position as regards the telescope, while the part containing the wires and micrometer screws can be rotated by means of a rack and pinion. A vernier attached to the movable frame indicates the required angles.

To adjust the position circle the vernier is set to zero, and the telescope directed to a star; the circle and micrometer are then together turned round until the diurnal movement of the star, which is east and west, makes its image to traverse the space between the fixed wires. The movable threads will then lie in a north and south direction. The circle remains in this position during subsequent observations, while the micrometer is rotated until the movable threads are in the required direction, the position angle then being read off on the circle.

THE HELIOMETER.—Another means of measuring small angles for astronomical purposes is afforded by the instrument called the *heliometer*, which, as the name will at once suggest, was invented for measurements of the sun. This instrument is a telescope mounted equatorially, but differs from the ordinary telescope, inasmuch as the object-glass is cut across the centre, and means are provided for separating the two halves by moving one or both parts in the direction of the line of bisection, and also for measuring the amount of displacement. The cell containing this somewhat peculiar object-glass can be rotated so that the line of division of the lens may be placed in the same direction as the line representing the distance to be measured.

The action of the instrument depends upon the fact that any small part of a lens is competent to form a complete image of a celestial body, so that when an object-glass is bisected, and the two halves separated laterally, two distinct images will be produced, each differing only from the image formed by the complete lens in being less bright.

To measure the distance from a star to a planet, let us say, as in observations of the parallax of Mars, the lenses are separated to such an extent that the image of the star formed by one half, coincides with that of the planet formed by the other half, and the amount of separation noted. As a check, the measurement is repeated with the lenses separated in the opposite direction. The angular value corresponding to a known separation of the semi-lenses being determined, just as in the case of the micrometer screw, the angle between star and planet at once follows. Angles ranging from a few minutes to about two degrees can be measured in this way with great accuracy.

In the hands of Dr. Gill, of the Cape Observatory, the heliometer has yielded very valuable results in connection with the distances of the sun and stars.

OTHER INSTRUMENTS.—There are other instruments which may fairly be classed as instruments of precision, but space permits little more than a mention of their names.

The *zenith telescope* is a telescope specially designed for the measurement of the angular distances of stars from the zenith, for precise determinations of latitude by Talcott's method.

The *prime vertical instrument* is nothing more than a transit instrument, so arranged that the observing telescope swings in a vertical plane which is perpendicular to the plane of the meridian. From the observed times at which a star passes the prime vertical on the eastern and western sides, the latitude of the place of observation can be ascertained with great accuracy.

It is perhaps at sea that the labours of astronomers are of most direct value in everyday affairs, and it is precisely here that the instruments of high precision cannot be employed, in consequence of the absence of firm supports. Nevertheless, there is one instrument—*the sextant*—which yields results that satisfy all requirements when carefully constructed and placed in good hands. A graduated arc extending over about 60° (from which the name is derived) is supported by a light framework, and pivoted truly on the centre of the arc is the radius bar, or index arm, which carries a vernier for reading off the angles to be measured. A plane mirror is fixed to the index arm, over the centre of movement, and another, of which only half is silvered, is fixed to the frame near its outer edge. A small telescope parallel to the surface of the frame is directed towards the fixed mirror, so that the continuation of its axis is in line with the boundary between the silvered and clear part of the glass. Thus, while one object may be seen by direct observation through the clear glass, another, in quite a different direction, may be seen after reflection from the surfaces of the two mirrors.

The sextant is chiefly used for measuring the altitude of the sun, about noon for the determination of latitude, and in the morning or evening for the correction of chronometers. In such observations, the sextant is held in the right hand, with its plane vertical, and the sea horizon is sighted directly with the telescope; the index arm is then moved until the reflected image of the sun is brought into coincidence with the horizon. The reading is then taken, and if the adjustment is such that zero is indicated when the reflected and direct images of the same object are observed, it will give the altitude. The actual angle recorded by the sextant is only half that between the objects observed, but by numbering half degrees as whole ones, the true angles are read off directly. For observations of the sun the instrument is provided with coloured glasses of different shades, attached so that they can readily be interposed to reduce the intensity of the light.

CHAPTER XVII.

ASTROPHYSICAL INSTRUMENTS.

So far we have been concerned with instruments which enable us to ascertain the positions, dimensions, and appearances of the various orders of heavenly bodies; but we can go further than this, and learn something of the physical and chemical constitutions of the glittering orbs by which we are surrounded. We can, for instance, bring instrumental aid to bear upon the determination of the brightnesses of the heavenly bodies, and by means of that powerful appliance of modern astronomy—the spectroscope—we can study the chemistry of all those bodies which shine by light of their own, and which are not so feebly luminous as to be out of our range.

PHOTOMETRY.—The naked eye was alone employed in observations of stellar brightness until quite recently. Each step in the advance of astronomical research, as in most other branches of science, however, depends upon the greater precision of observation which can be introduced, and so we now find the eye to be assisted in these inquiries by a *photometer* of some kind or other. The general purpose of photometry will be familiar to all in connection with such practical matters as the determination of the illuminating power of coal gas. The methods here employed, however, are not directly applicable to the comparatively feeble light-sources which have usually to be dealt with in astronomical photometry.

As will be more fully explained in another part of this work, the stars visible to the naked eye are divided into six grades of magnitude. The brightest of them are classed as first magnitude, while those only just visible to the naked eye are of the sixth magnitude. Now that telescopes are used, this division of stars into magnitudes must be continued in some form or other, so as to include telescopic stars. From photometric comparisons it has been ascertained that the average star of the first magnitude may conveniently be reckoned 100 times as bright as a sixth magnitude star. Hence, the light-ratio corresponding to a difference of a single magnitude is 2·5. Thus, a star which is $2\frac{1}{2}$ times less bright than one of the sixth magnitude ranks as seventh magnitude, and so on. Fractions of magnitudes are also necessary to express the results which can now be obtained.

LIMITING APERTURES.—For the reason that a large telescope enables us to see stars which are too dim to be visible in a smaller one, the brightnesses of stars may be compared with more or less satisfactory results by reducing the aperture of a telescope until the star in question ceases to be visible. This is called the method of *limiting apertures*, and in practice a telescope intended for this work would be provided with a series of diaphragms, or other arrangement for conveniently reducing the effective area of its object-glass. A telescope which has an object-glass 10 inches in diameter should just show stars of the fourteenth magnitude under favourable conditions; a star which could just be seen when this aperture was reduced to an inch would be of the ninth magnitude, and so on.

There are numerous reasons why this method fails to give satisfactory results, but one of the most important is that the image of a star becomes more diffuse with each reduction in the aperture of the telescope. At best it must evidently fail for a comparison of stars which are visible to the naked eye.

WEDGE PHOTOMETER.—One of the simplest and best methods of estimating star magnitudes is afforded by the *wedge photometer*. This is a strip of neutral-tinted glass about six inches in length, and a quarter to half an inch deep, tapering from one end to the other, so as to present a gradual reduction in depth of tint from the thick to the thin end. A similar wedge of clear glass, tapering the opposite way, is cemented to this, in order to get rid of prismatic action. Compensated in this way, and mounted in a suitable frame, the wedge is placed in front of the eye-piece of a telescope, and is pushed along until the star under examination is just extinguished. A scale is then read off, and from the results of a previous evaluation of the wedge in the laboratory, the corresponding star magnitude is easily deduced.

In order to eliminate the effects of differences in the state of the sky, the position of the wedge at which a standard star, such as Polaris, ceases to be visible, is determined, and then it is the difference of wedge readings upon which the final calculation is based.

The great value of the wedge in stellar photometry was demonstrated by the labours of the late Prof. Pritchard, to whom we owe the catalogue of the magnitudes of naked eye stars in the northern hemisphere known to the astronomical world as the "Uranometria Nova Oxoniensis."

OTHER PHOTOMETERS.—Some photometers depend for their action upon comparisons with terrestrial sources of light.

In some cases, an artificial star, consisting of a pinhole illuminated by a standard lamp, is brought into the same field of view as the star to be compared, and then, by polarising apparatus, the brightnesses of the two images are equalised. The amount of reduction of either of the stars is determined by a scale which measures the rotation of the polariscope, and in this way all the stars are compared with an artificial star of known brightness.

One of the most notable achievements in this field of astronomical work is that of Professor Pickering of the Harvard College Observatory, who invented and made splendid use of the so-called *meridian photometer.* Here the telescope has two object-glasses of equal aperture side by side, and in front of each is a silvered flat mirror inclined at an angle of 45° to the optic axes. The telescope is supported in an east and west direction, so that one mirror reflects the Pole Star into its object-glass, while the other can be rotated so as to reflect any other star which is on the meridian into the second object-glass. Again, by a polariscope at the eye end of the telescope the images of the two stars are made of equal brightness, and the readings give the data for calculating the required magnitude.

Photographs of the stars are also largely employed for the estimation of magnitudes, stars of different magnitudes being represented on the photographs by spurious discs of different sizes. If all stars gave out light of the same quality, the photographic method would be very trustworthy, but as the colours of the stars vary, the photographic and visual magnitudes are not invariably in agreement. A bright, reddish star, such as Betelgeuse, would photographically be only equivalent to a white star which was much less bright to the naked eye.

THE PRISMATIC SPECTROSCOPE.—Reference has already been made in these pages to the wonderful field of astronomical research which has been opened up by the discovery of the action of a triangular glass prism upon rays of light, and the subsequent improvements in the method of utilising this effect.

A prismatic spectroscope may be regarded as an arrangement which will enable us to get a pure spectrum, and to observe it to the best advantage. The light to be analysed is admitted through a narrow aperture called the *slit*, which is placed at the focus of a double convex lens. Emerging from this *collimator*, as a parallel beam, the rays pass through the prism, and after deviation and dispersion they fall upon another double convex lens, which brings them to a focus in the form of a spectrum. An eye-piece may then be employed to view the spectrum, or a sensitive plate may be placed at the focus to photograph it.

In a simple form of spectroscope the prism is supported at the centre of a graduated circular plate, to which the collimator is firmly fixed, while the observing telescope is attached to an arm pivoted at the centre of the plate. A vernier moving with the telescope indicates the position, on a scale of degrees, of any colour brought to the centre of the field of view.

The best results are obtained when the rays of light emerge from the prism at the same angle at which they enter it, in which case the prism is said to be at *minimum deviation*, for the reason that the deflection of the rays from their original path is then the least possible. As lights of different colours are refracted unequally, it is clear that the prism can only be at minimum deviation for rays of one particular colour at any instant. Frequently, however, there is an automatic arrangement by which, as the observing telescope is moved so as to bring different colours into the field of view, the prism is turned so as to be at minimum deviation for the colour actually under observation.

The appearances observed in the spectroscope are a series of images of the aperture through which the light is admitted. If the source of light be yellow, such as that of a spirit lamp flame when common salt is introduced, a yellow image of the aperture will be seen, and so on for other monochromatic radiations. When a white light is observed, images of every gradation of colour are formed, and in such a "continuous spectrum" the separate images cannot be recognised. The form of aperture most widely adopted is a narrow straight slit with parallel sides. In this case there is the least possible confusion, because the several images of the slit appear as so many spectrum "lines."

For observations of the sun, where the light is so intense, a great number of prisms, each drawing out the spectrum into a longer band, may be employed, so that the lines of the spectrum may be widely separated, and the peculiarities of each more closely investigated. For the fainter bodies, however, the instrument must generally be one of comparatively small dispersion, so that the light may not be spread out into invisibility. It will be evident that the longer the spectrum the greater will be the chances of accurate measurements.

Another way of obtaining great dispersion is to use prisms of the new dense Jena glass, one of which is equal to three or four of the flint glass prisms in general use.

There are various forms of the prismatic spectroscope. In some of them reflecting prisms are introduced to turn the rays back through the dispersive train, so as to get increased dispersion without increasing the number of prisms. In the so-called *direct vision spectroscope*, prisms of different kinds of glass are combined so that the rays of light leave them in nearly the same direction that they enter. Here the collimator and observing telescope are in the same straight line, and this is a great convenience in certain classes of observation.

THE GRATING SPECTROSCOPE.—Sometimes, especially in instruments designed for solar observations, the prisms are replaced by what is called a diffraction grating. Usually this consists of a piece of highly polished speculum metal, upon which is ruled a great number of equidistant parallel scratches or lines. A portion of the light falling upon the grating is simply reflected, while the remainder is spread out into two series of beautiful spectra, one on each side of the directly reflected beam. The two nearest to the directly reflected beam are called spectra of the first order, while following these are spectra of the second, third, and fourth orders; the length of spectrum increasing in each case, and all being available for observation if the light dealt with be sufficiently bright. The production of these spectra is due to the interference of light waves.

All gratings produce exactly similar spectra, so that the distances between identical lines as seen with one grating are always strictly proportional to their distances as seen with any other. With prisms, the relative separation of colours is by no means constant; a prism made of one kind of glass may, for example, separate the green and yellow more than another prism made from different material, while the separation of yellow and red might be the same in both cases. The grating spectrum accordingly affords a constant standard of reference, and what is called the "normal solar spectrum" is the spectrum of the sun mapped with the various dark lines in the relative positions shown by a grating spectroscope.

Prof. Rowland, of John Hopkins University, has introduced a form of grating spectroscope, in which the grating is ruled on a concave spherical surface of speculum metal. After passing through the slit the rays of light fall directly upon this concave surface, and are brought to a focus after reflection, so that no lens except the eye-piece used for visual observations is required. Several of these gratings, having mostly a radius of curvature of about 21 feet, and a ruled surface of about 5½ inches × 2 inches, with 20,000 lines to the inch, are in use at the present time. Some idea of the difficulties to be faced in making these magnificent aids to research may be gathered from the following remarks of Mr. J. S. Ames:—"It takes months to make a perfect screw for the ruling engine, but a year may easily be spent in search of a suitable diamond point. . . . When all goes well it takes five days and nights to rule a 6 inch grating having 20,000 lines to the inch. Comparatively no difficulty is found in ruling 14,000 lines to the inch."

With the aid of these wonderful gratings, the solar spectrum can be photographed with perfect definition, and extending over a total length of several yards. Thousands of the tell-tale Fraunhofer lines are rendered visible in this way.

MEASUREMENT OF SPECTRA.—The spectra of many substances, including hydrogen and iron, are so characteristic as to be recognisable at a glance by an experienced observer, but one must as a rule resort to measurement for the identification of lines, or for the purpose of locating unknown lines for future reference. One of the simplest methods of measurement is that of reading the position of the observing telescope upon a graduated circle, when the line is seen at the centre of the field. If supplemented by a micrometer eye-piece, for differential measures with regard to known spectra, this method is extremely convenient. As recorded on arbitrary scales of this character, the position of the same line would be represented by a number which would be different for every instrument, and it is therefore necessary to reduce all measurements to a common scale; that now universally adopted is the natural one of wave-lengths. The position of a line in the spectrum depends upon the length of the waves constituting the rays of light which produce it, so that a measure of wave-length completely specifies the situation of a line whatever spectroscope may be employed. Light waves are excessively minute, but by the use of the diffraction grating they can be measured with great accuracy. So small are they, that the most convenient unit of wave-length is the ten-millionth part of a millimetre[1]—or tenth metre, as it is technically named. Expressed in this way, the wave-length of the glorious red line seen in the spectrum of hydrogen is 6563·07, while that of the blue line characteristic of the same gas is 4861·51.

When the positions of a certain number of lines of known wave-length have been read off on the scale of any spectroscope, the required wave-lengths of other lines are ascertained by a graphical interpolation, or by calculation. Elaborate tables of the wave-lengths of the lines in the spectra of the sun and chemical elements have been prepared by various

[1] In a British inch there are 25·4 millimetres.

investigators, and these are in constant demand by all workers in the field of astrophysics.

THE TELESPECTROSCOPE.—For the examination of the spectra of the heavenly bodies, a spectroscope is attached to the eye end of a telescope from which the eye-piece has been removed, such a combination forming a *telespectroscope.* The slit is placed at the principal focus of the object-glass of the main telescope, and an image of the object to be observed is thus produced upon it. If the sun be under observation, any special part of it, such as a sun-spot or the chromosphere, may be separately investigated by bringing the corresponding part of the image upon the slit.

In the case of the sun, moon, comets, planets, or nebulæ, the image is one of sensible size and the spectrum lines have a perceptible length. With a star, however, the image is only an illuminated dot upon the slit, and the spectrum would have no appreciable breadth, so that all but the strongest lines would in general fail to show themselves. Accordingly, when observing star spectra, a cylindrical lens is placed in front of the slit, so that the stellar image is drawn out into a bright line, and the necessary breadth of spectrum and length of the spectrum lines are secured.

For photographing the spectra of the heavenly bodies it is simply necessary to replace the eye-piece by a small camera, and to expose a sensitive plate for a length of time dependent on the brightness of the spectrum. The spectrum of a terrestrial substance, such as hydrogen or iron, photographed in juxtaposition, is always a great convenience, and is essential for the investigation of stellar movements by the displacement of spectrum lines.

THE LICK STAR SPECTROSCOPE.—Among the most complete and perfect spectroscopes adapted for use with the telescope is that designed by Prof. Keeler for the great refractor of the Lick Observatory. It is illustrated in Fig. 55, and it will be at once evident that there are ample means for keeping the instrument under control. Towards the upper part of the diagram, on the left, is the eye end of the telescope, without the eye-piece. Two stout brass rods 3 inches in diameter and 6 feet long are attached by clamps to a revolving jacket which surrounds the end of the telescope tube, and on these the spectroscope is supported by clamps which allow of it being moved inwards or outwards from the focus of the telescope. The collimator of the spectroscope lies midway between the rods, and in order to facilitate the focussing of the image upon the slit, it has a small longitudinal movement independently of that of the whole spectroscope. The observing telescope is seen on the left of the diagram, while the grating rests on the circular graduated plate over which the observing telescope can be moved. The grating has 14,438 lines to the inch.

Three prisms can also be used with the spectroscope, two of them being single prisms of 30° and 60° refracting angles respectively, and the third a compound prism giving a very high dispersion. Two observing telescopes are provided, one being of extra power for use with the grating in solar spectroscopy.

The instrument is generously supplied with the small refinements which contribute so largely to easy and successful manipulation. Among these are a diagonal eye-piece for viewing the image of the object on the slit plate, electrical illumination of the graduated scale and wires of the micrometer eye-piece, and an automatic arrangement for keeping the prisms at minimum deviation.

There is a small totally-reflecting prism covering half of the slit, by which the light from an electric spark, or other source of luminosity, can be made to pass through the spectroscope so as to produce a series of known reference lines which serve as so many mile-posts for the measurement of the spectrum of the celestial body under observation. The induction coil, seen to the right of the diagram, is for the purpose of producing these electrical sparks.

In mounting the spectroscope, which weighs no less than 200 pounds, the eye end of the great telescope tube is first supported by a prop, and the long rods are inserted. The

FIG. 55.—*The Spectroscope adapted to the Eye End of the Lick Telescope.*

spectroscope is then placed on the rods, and balancing weights equivalent to the weight of the spectroscope are removed from the lower part of the telescope tube.

THE OBJECTIVE PRISM.—It is a very remarkable fact that many of the recent advances in our knowledge of the spectra of stars have followed from the revival of a method first employed by Fraunhofer in 1814, in which the slit and collimating lens, forming part of an ordinary spectroscope, are dispensed with. The rays coming from a star being already parallel, and the star itself being a virtual slit without length, a large prism placed in front of the object-glass of a telescope makes a complete stellar spectroscope. A prism employed in this way is known as an *objective prism.*

In place of the image of a star, which would be seen in the absence of the prism, a spectrum without appreciable width appears at the focus of the telescope, and the spectrum lines will be represented by mere dots. To turn these dots into lines so that they may be better visible, a cylindrical lens must be employed in conjunction with the eye-piece.

It is to the application of photography, however, that we owe so much, and in this case the cylindrical lens is removed, while a small camera replaces the eye-piece of the telescope. In this form the instrument is often called a *prismatic camera.*

The prism is so arranged that the spectrum lies along the meridian passing through the star, and it is then only necessary to allow the driving clock to be slightly in error in order that the spectrum may trail a short distance perpendicular to its own length, and in this way broaden the photographed spectrum. On the proper regulation of the clock rate, and consequent "trail" of the spectrum across the plate parallel to itself, depends very largely the success of the photograph obtained. The spectrum of a bright star must obviously be made to travel more quickly than that of a fainter one, and a short exposure suffices. For the same clock rate, and in the same time, a star near the Pole will give a shorter trail than one nearer the Equator, and declination must therefore be taken into account in adjusting the clock error for this method of photography.

One great advantage of the objective prism in the photography of stellar spectra depends upon the fact that all the light passing through the object-glass is utilised in the production of the spectrum, whereas in an ordinary telespectroscope a large percentage of the light is lost in the jaws of the slit. The large focal length of the telescope also enables a long spectrum to be obtained even with a single prism of small angle.

When the dispersion is only small, the spectra of stars as faint as the tenth or eleventh magnitude can be photographed by this method, so that sometimes as many as 200 spectra are registered with a single exposure. Here, again, the objective prism has an immense advantage over the telespectroscope.

Professor Pickering, of Harvard College, was among the first to recognise the value of the objective prism for the photography of stellar spectra, and the munificent endowment of this research, by Mrs. Draper, as a memorial to Dr. Henry Draper, has enabled him to produce the Draper catalogue of stellar spectra, giving the chief characteristics of the spectra of over 10,000 stars.

Professor Norman Lockyer, at South Kensington, has also been conspicuously successful in this department of astrophysical research. The chief instrument he employs is a photographic telescope of only six inches aperture, with an objective prism of 45° refracting angle. The spectra thus obtained show hundreds of lines in such stars as Arcturus, with very fine definition, so that they bear almost unlimited enlargement.

An objective prism of twenty-four inches aperture will form one of the accessories of the fine telescope which is now being erected at the expense of Dr. Frank McClean, for the Cape Observatory, and there can be no doubt that the use of this gigantic prism will add greatly to our knowledge of the chemistry of the fainter stars.

As yet there is no very practicable method of employing

the objective prism for determining the velocities of stars in the line of sight from the displacement of spectrum lines, and herein lies its one great disadvantage as compared with the telespectroscope. The difficulty is to ensure that the spectrum always falls absolutely in the same position with respect to the terrestrial spectrum, which must be photographed alongside for purposes of measurements. It is true that the spectrum of an approaching star is somewhat shorter, and of a receding star slightly longer than that of one at rest relatively to the observer, but these changes are so small as to little more than indicate the direction of movement even when a large instrument is employed.

Under the direction of Professor Norman Lockyer, the objective prism was very successfully used for photographing the spectra of the solar surroundings during the total eclipses of 1893 and 1896. In place of the picture of the solar corona, which would appear in the absence of the prism, the prismatic camera shows a spectrum consisting of bright rings. If, for instance, the corona were wholly composed of hydrogen, there would be a picture of it in red, blue-green, blue, and violet, corresponding to the lines ordinarily seen in the spectrum of that gas. These rings thus indicate the chemical nature of the corona, and at the same time show, by their differing forms, the distribution of different gases throughout its extent. The spectra of the solar prominences and chromosphere are also depicted during the brief time of their visibility, during an eclipse, with such distinctness that a series of " snap shots " is all that is required to give a lasting record.

THE SPECTROHELIOGRAPH.—A special form of spectroscope—called the *spectroheliograph*—has been devised by Prof. Hale, of Chicago, for photographing the sun in monochromatic light. It consists of a spectroscope, arranged for photography, in which the slit can be made to travel by clock-work across the sun's image, which is projected upon it by the telescope to which the instrument is attached. In front of the photographic plate there is a secondary slit, so that only a very restricted part of the spectrum reaches the sensitive film. The secondary slit is connected by mechanism with the primary one, so that as the latter traverses the sun's image, the former exposes different parts of the photographic plate to the light which passes through it, and in this way builds up an image of the sun in monochromatic light, matters being so arranged that light of the same wave-length always falls upon the secondary slit. By utilising the brightest lines which appear in the spectrum of the solar prominences, monochromatic images of those interesting appendages to our luminary have been successfully photographed without waiting for a total solar eclipse.

THE BOLOMETER.—Besides the luminous effects of the spectrum, there are heating effects which can be measured by the *bolometer*, an instrument invented by Prof. Langley. A very thin strip of metal is connected with a delicate galvanometer, and is arranged so that it can be passed along the whole spectrum. The electrical resistance of the strip varies according to its temperature, and the galvanometer at once signals any fluctuations which may occur. If, for instance, the strip comes to the place occupied by a dark line, there will be a notable fall of temperature. In this way, the bolometer is used to map the solar spectrum in the "infra-red" region—a part of the spectrum invisible to the eye, and of which we should otherwise have remained in ignorance.

ASTRONOMY

SECTION III.—THE SOLAR SYSTEM

By AGNES M. CLERKE

DONATI'S COMET, OCTOBER 9, 1858. (FROM LANGLEY'S "NEW ASTRONOMY.")

SECTION III.—THE SOLAR SYSTEM.

BY AGNES M. CLERKE.

CHAPTER I.

THE SOLAR SYSTEM AS A WHOLE.

THE solar system consists of one supereminent body, with a train of miscellaneous attendants. By its immense gravitative power, their movements are so governed that they not only revolve round it as a common centre, but accompany its march through space; they are, in various degrees, warmed and enlightened by its copious emissions of heat and light; they are linked with it by origin and destiny. Some, indeed, much more closely than others. Planets, satellites, and asteroids belong to the immediate family of the sun; periodical comets and revolving meteoric rings have been adopted into it. The planets are eight in number; the six nearest the sun—Mercury, Venus, the Earth, Mars, Jupiter, and Saturn—have been known immemorially; Uranus and Neptune were discovered respectively in 1781 and 1846. Mercury, Venus, and Mars form, with the Earth, a group of "terrestrial planets," so-called because they differ not very greatly in scale from our globe, and are constructed on nearly the same lines. The outer quartette of planets are giants by comparison, and show obvious symptoms of being in a very different physical condition. And it is noteworthy that the zone of asteroids, lying between Mars and Jupiter, divides the planetary classes.

The asteroids are sometimes designated minor planets; but the former term is preferable, as accentuating their distinctive character. For they are not simply diminutive planets. A planet revolves in solitary state within its own broad domain. The asteroids traverse intercrossing and entangled paths, indefinitely numerous, ranging widely in celestial latitude, and covering with their net-work nearly the entire chasm of space between Mars and Jupiter. The small bodies moving in them have doubtless been formed in a manner totally different from that by which the single body they seem to replace would have taken shape.

Satellites bear in many respects the same relation to planets that planets bear to the sun. They are united with them into secondary systems, one of which is particularly well known to us, since it is constituted by the earth and the moon. The existence of twenty-one satellites has been ascertained, and many more possibly remain to be detected. Their apportionment is singularly unequal. Only three of the twenty-one belong to the four small interior planets, while eighteen are attached to the four exterior giants. Moreover, both Mercury and Venus are solitary; so that the solar neighbourhood appears to be a region unpropitious to the development of subordinate systems.

Seventeen comets certainly, and many more probably, are domiciled in the solar kingdom. And even these preserve traces of an alien origin. They revolve round the sun in closed orbits, and are hence periodical in their apparitions; but their periodicity has to be qualified by a saving clause. They come up to time *barring accidents*. For their orbits, not being adjusted to stability, are liable to violent changes through the influence of the powerful masses, the tracks of which they intersect. In running up to, or back from perihelion, comets have to cross many railroads, so to speak, and do not always escape disturbing or destructive encounters with passing trains. Thus, many are entered in our astronomical visitor's book as lost or strayed. Halley's is the only well-secured cometary prisoner of the sun of imposing magnitude; the rest are of little spectacular, although of very high theoretic, interest. Comets are the only self-luminous members of the solar system.

Meteorites, besides being intrinsically obscure, reflect, owing to their minuteness, so little sunlight that they remain invisible until ignited in our atmosphere. They travel round the sun in annular systems, each mote-like component of which pursues its way, independently of the others, under the strict regimen of gravitational law. The number of these meteoric rings must be prodigious. Some hundreds have been brought to our acquaintance, which can only include such as cut the earth's orbit; and these must be an insignificant fraction of the whole. The innumerable closely-related orbits grouped into each ring are ill-regulated for the safety of the bodies moving in them, since they conform in no way to the rules of planetary circulation. Hence the numerous encounters with the earth announced by the luminous trails of shooting-stars.

Our system, as at present known, is 5,585 millions of miles in diameter. It is limited by the orbit of Neptune. But no less than three trans-Neptunian planets have been, on some show of evidence, alleged to exist. One of them, held by Professor Todd of Amherst College, U.S., to be responsible for some outstanding perturbations of Uranus, was placed by him in 1877 at a distance from the sun fifty-two times that of the earth (the radius of Neptune's orbit being measured by thirty of the same units); the two others, called into existence by Professor Forbes of Edinburgh in 1880, to account for the formation of two groups of comets with aphelia respectively at one hundred, and three hundred astronomical units, were believed to occupy those enormously remote positions. Although none of the three, in spite of telescopic and photographic search, has yet been found, the possibility is not excluded that the appearance on a long-exposed sensitive plate of a line in lieu of a dot as the representative of a seeming star, may in the future announce the annexation by the sun of a further immense slice of territory out in the depths of space. The boundaries of our system are thus only provisionally fixed.

Intra-Mercurian planets have proved equally recalcitrant to prediction; and it may safely be said that no globe of the superficial dimensions of an English county lies concealed in the comparatively narrow space available for its circulation. The necessity for the presence of "Vulcan" was deduced by Leverrier from an unexplained displacement of Mercury's perihelion, and a transit of the required body, supposed to have been observed March 26, 1859, was thereupon, in all good faith, brought forward by Dr. Lescarbault of Orgères. Another pseudo-discovery—this time of a pair of Vulcans—was made during the total eclipse of July 29, 1878; but neither on nor off the sun has the body needed to satisfy the French mathematician's theory been genuinely seen, and few believe that it will ever be forthcoming.

Professor Titius of Wittenberg pointed out in 1772 that the relative distances of the planets from the sun could be expressed by adding 4 to the series 0, 3, 6, 12, 24, 48, etc. Thus, if the distance of Mercury were called 4, those of Venus, the Earth, Mars, and so on, would severally be 7, 10, 16. The validity of this relation—known as "Bode's Law"—was strengthened by the conformity to it of Uranus and Ceres, neither of which had been discovered when it was enunciated; Neptune, however, proved to be much nearer to the sun than he should have been, and the formula hence ranks as an empirical one, not grounded in the nature of things.

Yet the grand outlines of the solar system are traced on a visibly symmetrical plan. The larger bodies composing it move nearly in the same plane, in orbits nearly circular, and at regulated intervals, augmenting rapidly outward. All revolve from west to east, or "counter clockwise," and this fundamental current of motion carries with it, besides the asteroids, all the periodical comets, save Halley's. Among secondary systems only the Uranian and Neptunian escape from its sway; there being a visible tendency towards deviations from rule towards the confines of the solar domain. These deviations, however, are not of a subversive character.

The planetary machine may continue working forever without a hitch. Such irregularities as would be likely to throw it out of gear are found only in parts of almost evanescent mass and negligeable influence. Two modes of action which should, in the long run, bring about a collapse, are non-existent or insensible. These destructive agencies are a resisting medium, and the progressive transmission of gravity. The presence of either should prove fatal in the same ultimate fashion. Along slowly narrowing tracks, the planets would descend, one after the other, into the ample lap of the sun. Their circulation is, however, to the best of our present knowledge, unimpeded and undeflected; the disturbances affecting it are self-compensatory.

But while the mechanical stability of the system is assured, its physical state is continually changing. And the change is always in the same direction. A degradation of energy steadily progresses. The sun is, in fact, spending his capital, and even with a millionaire of his stamp this cannot last. The time must come, if science is to be believed, when his radiative powers will have become exhausted. Five millions of years hence they will, in all probability, be much less efficacious than they are now. Within twice or thrice that interval they may have become almost extinct.

Planetary globes, too, grow old through the wasting of their internal heat. The moon seems in a measure to prefigure the future condition of all, should their decay not be arrested. Possibly the lunar stage is not the last. Death may, in the long ages to come, be succeeded by disintegration, when a ring of rubbish will be substituted for our "wan-faced" companion. To what purpose, then, our readers will ask, the mechanical perfections of a system destined eventually to be involved in darkness and destruction? To what purpose its exquisite balance, the nicely-adjusted relations of its members, its self-righting faculty, its compensatory springs? We can reply only by recalling that the extreme conclusions of science are invariably pessimistic, because they are reached without taking any account of the intelligent control perpetually, though insensibly, overruling the workings of blind forces. If, in one sense, heaven and earth pass away, we still know that, in good time, "a new heaven and a new earth" shall inscrutably arise. Not "faintly," then, but boldly and ardently, we "trust the larger hope" that renovation will succeed, or anticipate subversion.

Whatever *can* have an end *must* have had a beginning, and the origins of things have an especial fascination for our minds. As regards the history of the planetary world, we are not altogether in the dark. The problem of the maintenance of the sun's heat was satisfactorily solved by Helmholtz in 1854. Its radiative supplies, as he showed all but conclusively, are derived from gravitative power. As they are diffused into space, the cooled particles from which they proceed, clash together, and their arrested motion is converted into a fresh thermal stock. This implies a steady diminution, although to a surprisingly slight extent, in the bulk of the solar globe. It has been computed that a shortening of the sun's diameter by 380 feet yearly would suffice to keep this grand heat-producing machine in full working order; and at least ten thousand years should elapse before the contraction became measurable by any instrumental means at our command. Its progress should, nevertheless, eventually reduce our glowing luminary to an obscure, inert mass.

Now, evidently, its shining in the past was sustained in the same way as at present. The globe that blazes in our summer skies is, accordingly, but the shrunken remnant of what it once was. It is shrunken in proportion to the vast quantity of its former emissions. Hence, the farther we look back into the ages, the more voluminous its dimensions. And, sounding the utmost profundities of time, we arrive at an epoch when all the planets were swallowed up in a sphere girdled by the present orbit of Neptune.

The tenuity of this distended body was unimaginable. At ninety miles of altitude, our air is one hundred million times rarer than it is at sea level ; yet the primitive solar "nebula" was considerably more attenuated still. This aerial mass had, doubtless, been in some way impressed with a slow movement of rotation, which, by mechanical necessity, quickened as condensation progressed. The planets represent a few fragments detached during the process ; nearly the whole of its substance being compacted into the sun. How the fragments came to be detached is the crux of cosmogonists. According to Laplace's famous hypothesis, equatorial rings of matter separated successively from the parent nebula at certain critical epochs when gravity was overcome by the gaining centrifugal tendency due to accelerating rotation. These rings drew together into planets, from which satellites were generated by a repetition of their own birth-process. Many incongruities are, however, involved in this *modus operandi.* Only two need here be mentioned. Reason and experience teach us that globes of small interior consistence easily break up into rings, while cosmic rings show not the slightest tendency to collect into globes. Again, Laplace supposed that the production of each planet relieved a long antecedent strain. But nebulous stuff is almost absolutely incoherent. Hence it *cannot be stretched or strained.* As the nebula condensed and whirled, it would, accordingly, have left behind innumerable disaggregated particles, but no massive rings.

M. Faye of the French Academy has attempted to remedy these defects. The planets, he considers, were not abandoned, but formed at centres of condensation within the nebular matrix. The order of their formation would thus have been quite different from that assigned by Laplace, in whose theory the exterior globes were necessarily the earliest to take shape. M. Faye, on the contrary, argues Uranus and Neptune, from their retrograde rotation, to be the *youngest* instead of the *oldest* members of the solar system, while the terrestrial group belong to the first era of planetary development.

Astronomers are now virtually agreed that "The world was once a fluid haze of light," but by what precise means, in what succession, under what compulsion, its constituent bodies were set wheeling in the void, they are less ready to pronounce than were their predecessors, who, dazzled with the analytical triumphs of the eighteenth century, accepted unquestioningly the plan of creation it complacently transmitted to them. The complexities of world-making have, besides, been instructively illustrated by Professor G. H. Darwin's discovery that tidal friction was essentially concerned in the process. By an able mathematical investigation, he showed, in 1879, that it was particularly effective in modelling the earth-moon system, owing to the fact that our satellite, comparatively to its primary, is by far the largest in the solar system.

Tidal friction may be regarded under a two-fold aspect. Its effect in grinding down the speed of rotation has been explained in Section II. (page 166). The energy, however, thus apparently destroyed is only transformed. The rotational momentum subtracted from the earth is added to the orbital momentum of the moon, which thus travels (setting aside other causes of change) along continually widening spires. This retreat from the earth is even now going on, although with elusive slowness, amid the rise and fall of secular change. Its effects in past ages, nevertheless, coupled with those due to the slackening of rotation by the friction of the tidal wave —the two forming, as it were, the obverse and reverse of one medal—must have been of overruling importance. Laying hold of the clue they offer, Professor Darwin succeeded in tracing back the history of the moon through a "corridor of time" nearly a hundred million years long. It was then spinning at a vertiginous rate, round, and nearly in contact with the earth, which must have been fluid or plastic, while of about its present size. The *month* of that epoch was three or four hours in duration; the *day* was shorter still. The actual existence of the moon convinces us of this latter fact. Otherwise, the huge tidal wave raised by the moon upon the earth should have lagged, however slightly. Its attraction would have pulled the moon backwards at the decisive moment of its emergence into separate being, and led infallibly to its re-engulfment.

The origin of the moon has been, by Professor Darwin's analysis, made clearer than that of any other heavenly body. Certainty regarding such remote events is unattainable; but it is highly probable that our globe, at a late stage of its development, gave birth, amid the throes of disruption, to its solitary offspring. But the case is unique. The terrestrial system presents conditions not repeated elsewhere. Generalisations founded upon them are sure to be misleading. We have indeed gained, from all recent inquiries into cosmogony, the profound conviction that no single scheme will account for everything; that the utmost variety prevailed in the circumstances under which the heavenly bodies attained their present status; and that a rigidly constructed hypothesis can only misrepresent the boundless diversity of nature.

CHAPTER II.

THE SUN.

THE sun is an immense reservoir of radiant energy. For our daily uses we have no other store worth mentioning to draw upon, our fuel being the embalmed sun-heat of former ages; and all the physical and vital operations carried on over the whole globe derive their motive power from the same copious source. Yet only $\frac{1}{2,128,000,000}$th part of the sum total of solar radiations strike its comparatively diminutive surface; while all the planets combined intercept no more than $\frac{1}{234,000,000}$th of that inconceivable effluence.

The sun gives as much light as 600,000 full moons, or two and a half billions of the most powerful electric lights, or as 1,575 billions of billions of standard candles. And since his disc is the projection of a hemisphere, and is thus equivalent only to one-fourth the globular surface, these vast numbers must be quadrupled to represent the whole luminous emissions of this surpassing body. Their amazing profusion is the combined result of immensity of shining area, and vivid intrinsic brilliancy. Each square inch of the sun's surface has been estimated to integrate the lustre of twenty-five electric arcs,[1] and Professor Langley, by direct experiment, proved it to be 5,300 times brighter, and 87 times hotter, area for area, than the white-hot "pour" from a Bessemer converter; notwithstanding that the circumstances of the comparison were exceedingly "unfair to the sun."[2]

Radiant heat and light do not indeed differ in themselves, but only in their effects. The sun sends out into space ethereal waves of various lengths, but all of the same kind, subject to the same laws, and travelling with the same velocity of 186,000 miles a second. They appear, however, under diverse forms of energy according to the qualities of the substances upon which they impinge. Thus a small section of this long range of undulations affects our eyes as light, the human retina being so fashioned as to be able to *see* with their help. There is nothing in the nature of the rays themselves to make them visible, and it is in fact more than probable that other living creatures perceive vibrations to which we are blind. Our eyes are sensitive over nearly two octaves; from waves measuring about 760 millionths of a millimetre, to those of less than 400 millionths. In the solar spectrum the limits are roughly marked at one end by a great dark band in the deep red—Fraunhofer's "A,"—and at the other by "H," in the extreme violet. Beyond H extend undulations so short as to be visually imperceptible, while photographically active. This means that certain salts of silver are capable of taking up the energy they bring from the sun, and of using it to break their chemical bonds; while on differently prepared plates similar effects can be produced by rays in all parts of the spectrum,

[1] Proctor: "Old and New Astronomy," p. 327.
[2] Langley: "The New Astronomy," p. 108.

even in the ultra-red, where the undulations, too long to be sensible as light, are mainly felt as heat. Here, as Professor Langley has shown by "bolometric"[1] explorations, reside three-fourths of the energy distributed throughout the solar spectrum; nor is it impossible that this great stretch of heat waves may merge, without interruption, into electrical *rollers*, measured, not by millionths of a millimetre, but by metres, or even by kilometres. The important point to be borne in mind, however, is that the solar energy is diffused abroad by means of ethereal vibrations of a single type, but immensely varied size and frequency, and hence susceptible of dispersion into a spectrum.

The "solar constant" expresses the quantity of heat received by the earth from the sun. Its value, according to the most trustworthy determinations, is three calories per square centimetre per minute. This means that a vertical sun pours down upon each square centimetre of the globe heat enough (supposing the atmosphere out of the way) to raise the temperature of three grams of water by one degree centigrade in a minute. Putting it otherwise, the energy imparted would suffice to keep an engine of three-horse power continually at work on every square yard of the terrestrial surface. Or, if the heat were distributed uniformily in all latitudes, it would annually melt a complete ice-jacket one hundred and seventy feet thick.

The temperature of the body lavishing heat at this tremendous rate must obviously be very high; but enquiries on the point are necessarily limited to the actual emitting shell, or "photosphere." Their success is testified to by a noteworthy reduction of late in the range of uncertainty. The difficulty attending them consists mainly in our ignorance of any systematic relation between temperature and radiation. Excessively hot bodies lose heat much more rapidly, under the same conditions, than moderately hot ones; and empirical "laws of radiation" have been, over and over

[1] The "bolometer," invented by Langley, measures heat with exquisite refinement by means of its electrical effects.

again, arrived at as the upshot of long series of laboratory experiments. But such laws are only too apt to turn traitors if trusted without control; and since the thermal power of the sun vastly exceeds that of any terrestrial source, they are precarious guides in this particular research. Nevertheless, as the outcome of various improvements and refinements, it has, within the last few years, been prosecuted with excellent results. That obtained in 1894 by Messrs. Wilson and Gray deserves particular confidence. The *effective* temperature of the sun was by them fixed at 8,000°, or allowing for absorption in the solar atmosphere (measured by Wilson and Rambaud), at 8,800° centigrade. This estimate, which makes the sun's surface more than twice as hot as the carbons of the electric arc, is unlikely to be widely erroneous. The word "effective" signifies the condition that the photosphere is equivalent in radiative power to a stratum of lampblack; if it fall short of this standard, as appears probable, then the temperature must be raised by a corresponding amount.

The solar atmosphere, of which the absorptive effects have just been alluded to, is a shallow envelope, stopping predominantly the shorter wave-lengths of the light transmitted through it. Hence, if it were removed, the sun would appear, not only much brighter, but also much *bluer* than it does at present. The general darkening of the limb due to its action is apparent to visual, and conspicuous in photographic, observations. By its aid, "faculæ"—brilliant and elevated portions of the photosphere—were early detected. Invisible on or near the middle of the disc, they stand out in relief against its dusky edges as they are brought round, and carried off again by the sun's rotation.

The magnitude of this astonishing luminary fairly baffles our conceptions. Its mass is 745 times that of all the planets taken together. Its volume is such, that if Jupiter were located centrally within it, two of his Galilean moons, besides the lately discovered inner satellite, would have "ample room and verge enough" to revolve round him, keeping well inside the photosphere. The entire Uranian system could be easily accommodated in the same way; while Neptune and his satellite, and the earth and moon, could very nearly perform their evolutions side by side in the sun's excavated interior.

The sun is 865,000 miles in diameter, and in figure is sensibly spherical. Its surface is 12,000 times, its volume 1,300,000 times that of the earth. In mass it is equal to 332,000 earths. Its mean density, then, is only one-quarter that of the earth, or 1·4 times that of water. In other words, the terrestrial globe, if equally bulky, would contain four times the quantity of matter contained in the solar globe. Yet we know that it is largely made up of iron and still heavier metals; while gravity at its surface is 27·6 more powerful than it is here. Thus, the sun's materials are weighed down by an inconceivable pressure, and would be of a density utterly transcending our experience but for the counteracting agency of heat. The comparative insubstantiality of such a globe gives us some faint notion of the violent molecular agitation affecting every particle of its mass. Contrasted with the fires raging within, the surface temperature of 8,000° or 9,000° might perhaps be deemed moderate or cool. There is much evidence that it is throughout gaseous, although of a consistence approaching more nearly that of pitch or treacle than can easily be reconciled with established ideas as to the qualities proper to an aerial substance. Yet the laws governing the gaseous state are plainly those obeyed in the sun.

Its function, as a great thermal engine, is to produce and diffuse heat. For these purposes it is essential that the interior stores should be brought rapidly to the surface; and this is accomplished, not, as in solids, by conduction, but by actual transport, or "convection." Only the enormous elasticity of highly compressed gases could render this process swift enough to sustain the incessant outpourings of heat from the photosphere. It may be accompanied by an actual rise in temperature. If the sun be truly gaseous throughout, it *must* be so accompanied. The reason of this seeming anomaly is that a sphere of radiating and contracting gas developes by shrinkage more heat than it can dispose of by radiation. Whether or no the sun comes within the scope of this principle, known as "Lane's Law," cannot at present be decided. It is, in other words, an open question whether the sun is growing hotter or colder. Help towards answering it might have been expected from the study of geological climates; but their variations have evidently been due to a complexity of causes. At any rate, the sun's decline, if the inevitable turning-point has already been reached, is going on with extreme slowness.

The visible structure of the photosphere, or lustrous envelope of the solar globe, is, in itself, suggestive of the vertical circulation by which the indispensable communications between its interior and exterior are kept up. It is composed of brilliant granules and dusky interstices, the former representing, it is supposed, the vividly incandescent summits of uprushing currents, the latter the cooled, descending return-flows. It may be safely described as the limiting surface of thermal interchange, and is often spoken of as a cloud-sphere, or level of condensation, where the ascending vapours, like mounting volumes of water-gas in our atmosphere, are chilled into liquid droplets. To the brilliant luminosity of these incandescent droplets, the blaze of the solar emissions is ascribed. Or the droplets might equally well be solid particles on the model of the ice-spicules collected to form the delicate fields of cirrus in our upper air. The cloud theory of the photosphere is, however, hampered by the difficulty of finding a substance capable of liquefying or solidifying at a temperature of 8000° C. Carbon has generally been selected as the material of the solar "granules," but carbon evaporates at about 4,000°, and although its boiling-point might be raised by enormous pressure, there are no signs that the requisite conditions exist in the sun. Hence, some speculators turn towards electricity as the exciting agent of the photospheric radiance; but it would be waste of time to attempt, at present, to discuss the vague possibilities connected with an hypothesis which offers no holding ground for distinct reasoning.

The photospheric texture is often rent and perforated. This ragged condition (well exemplified in Fig. 1 from a photograph taken by Dr. Janssen at Meudon) is accompanied or caused by a violent disturbance of the sun's bodily circulation. A typical sun-spot consists of a dark opening, or "umbra,"

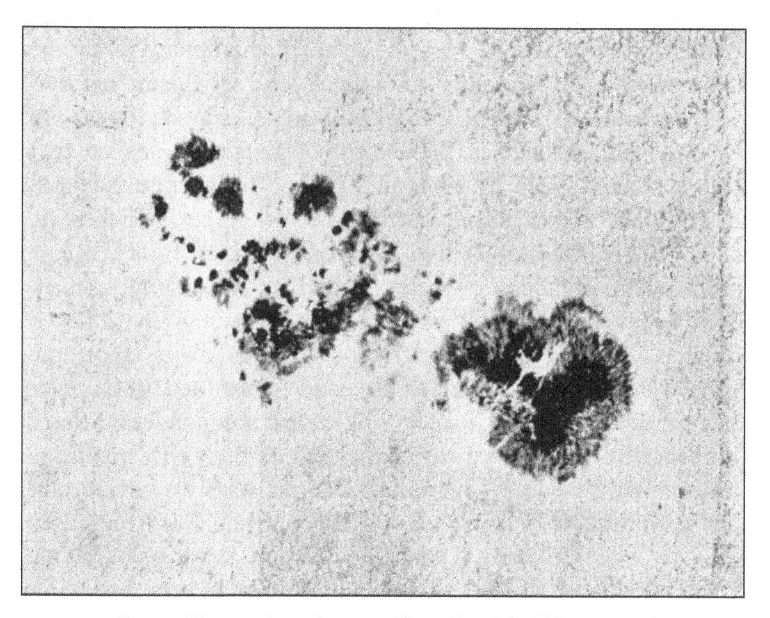

FIG. 1.—*Photograph of a Sun-spot.* (From *Knowledge*, February, 1890.)

within which a still darker "nucleus" can often be discerned. The umbra is garnished all round with a semi-luminous "penumbra," composed of elongated shining bodies placed side by side, and all, when undisturbed, pointing radially inwards towards the centre of the spot. The effect has been compared to that of "straw-thatching," although the solar "straws" are, at times, thrown somewhat wildly about. Where they hang over the *eaves* of the spot they are always brightest, because set most closely together. The penumbra may be called a modified extension of the ordinary mottled surface of the photosphere, the lustrous grains being drawn out into filaments, the "pores" into obscure interspaces.

Spots commonly occur in groups (as in our Figure) belonging to a single area of disturbance marked by the brightening, and probably by an elevation of the photosphere. The members of such families show curious and unexplained mutual relations. The size of these extraordinary formations is on the gigantic scale of all solar phenomena. They are often visible, individually or collectively, to the naked eye, and attracted notice accordingly in pre-telescopic times. In 1858, a spot opened to the extent of 144,000 miles, so that sixteen earths, side by side, might have been engulfed in it. A still more remarkable outbreak took place in February, 1892. Three thousand three hundred and sixty million square miles of the photosphere were riddled as if by some tremendous bombardment, the extreme dimensions of the affected district being 150,000 by 75,000 miles. This spot, the largest ever photographed at Greenwich, attained its acme on February 13th, when a magnetic storm and widely diffused auroral display attested the sympathy of the earth with commotions in the sun. Five times brought back to view by the sun's rotation, its history was followed from November until March; but this duration is not an extreme case, a spot having been known to survive throughout eighteen rotations. Although the group of February, 1892, covered $\frac{1}{700}$th of the sun's entire surface, its proportions were outdone by those of a spot and its immediate attendants, without counting outliers, measured by Sir John Herschel at the Cape, March 29th, 1837.

Spots are always associated with faculæ. The two are correlated phenomena. There is no certainty as to their order of precedence, if any fixed order there be, but faculæ both survive spots and develop apart from them. Not infrequently the faculæ garlanding a spot throw a "bridge" right across it (see Fig. 1). In stereoscopic views these brilliant projections show as veritable *suspension bridges.* They float almost palpably at a high altitude above the black gulf they span.

The distribution of spots is easily perceived to depend immediately upon the sun's rotation. Two zones of its surface, parallel to the solar equator, are alone infested by them. These may be defined as lying between 6° and 35° of north and south latitude; but the prohibition of spot-development is much more absolute in the polar than in the equatorial direction. One solitary macula has been observed in 50° north latitude.

The periodicity of sun spots was first recognised by Schwabe at Dessau in 1851. Since abundantly confirmed, it constitutes one of the fundamental data of solar physics. Once in about eleven years a "maximum" is attained; for months together the photosphere is never calm and unbroken; its agitated condition betrays the turmoil of the interior. The superabundance of spots is succeeded, after some years, by a scarcity, or "minimum," when the perturbing agencies appear to have sunk into repose, preparatory to another outburst of activity. In this highly irregular, although well-marked, cycle, the ascent is almost always much more rapid than the descent; the upspringing of the disturbance occupies, as a rule, not much more than half the time allotted to its quieting down. Nor is its intensity by any means uniform. High and low maxima alternate with, or succeed each other, with no obvious regularity. Sometimes we have a divided or double maximum, as in 1882-4, followed by an unusually swift ebb of agitation. The minimum of 1889 was premature and brief; for spots were again numerous in 1891, and developed prodigiously throughout the years 1892 and 1893. Only in January, 1894, a slight falling off became apparent, and the tranquillity which set in with 1895 may very probably reign with only temporary interruption for some time. The cause of these vicissitudes is completely unknown; but they so closely resemble, in character, the changes of variable stars, that it seems impossible to exclude the sun from that category, spot-maxima corresponding with stellar light-maxima, and *vice versâ*.

Solar disturbances, however originating, are a sort of universal pulse-beat, with which the earth, and doubtless every other member of the solar cortège, throb in unison. The accompanying diagram (Fig. 2) shows how closely the magnetic needle sympathises with the variations in the state of the sun. The amplitude of its daily oscillations is represented by the dotted curve, while the smooth curve is constructed from the relative numbers of spots. The striking conformity in point of time-development, between two effects so disparate in their nature, extends to minute details. Violent commotions on the sun seldom fail to be reflected in magnetic storms and auroral manifestations on the earth; and exact correspondences have sometimes been observed; yet it does not seem possible to trace these simultaneous effects to the immediate magnetic action of the sun.

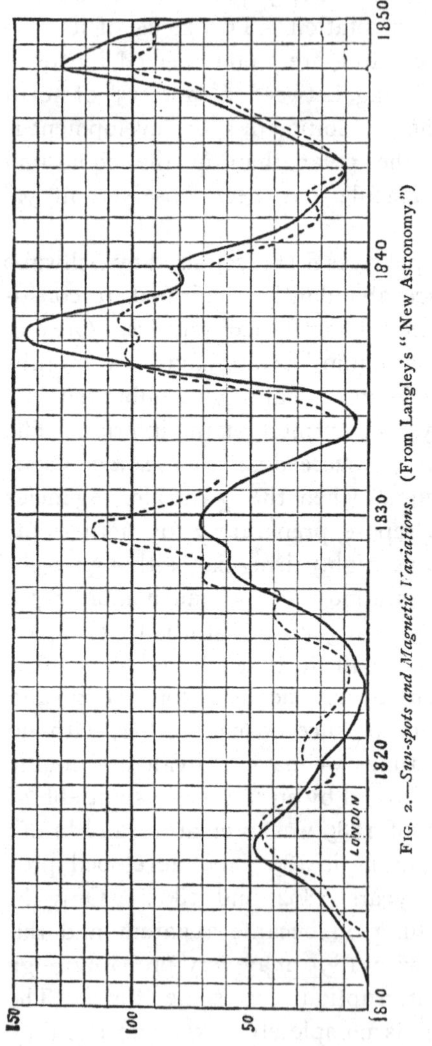

Fig. 2.—*Sun-spots and Magnetic Variations.* (From Langley's "New Astronomy.")

No meteorological cycle corresponding with the spot-cycle has yet been satisfactorily made out. The direct diminution of heat and light through the obscuration of a small part of the sun's photosphere amounts, at the utmost, to $\frac{1}{1000}$th of the whole. The spots are far from being totally dark or cool. Their blackest nuclei are really no less brilliant than lime-light; while about half as much heat is derived from them as from the surrounding disc when they are centrally situated, and 80 per cent. when they are near the limb.[1] Their dimming and cooling effects then are insignificant; they are probably more than compensated by the quickening of the sun's circulatory processes, and consequent increase of emission, through the disturbance of internal equilibrium of which outbreaks of spots are among the consequences.

The spot-zones do not always occupy the same positions. They shift with the progress of the eleven-year cycle. This curious circumstance, discovered by R. C. Carrington in 1856, illustrates, in his words, "the regular irregularity, and irregular regularity," distinguishing solar periodicity. At maxima, the mean latitude of the zones in question is about 16°; but they close down towards the equator as each wave of agitation dies out, its few latest products appearing in quite low latitudes. Then, when minimum is passed, a fresh start is made with the opening of a few small spots in 30° or 35° north or south latitude; and this newly-organised disturbance begins to descend as before, gaining strength as it proceeds. Thus, each impulse acts independently of the succeeding one.

The most cursory observation of sun spots suffices to show that the shining body marked by them rotates on an axis from west to east, in the same direction as the planetary revolutions. True, they emerge to sight on its eastern, and vanish at its western limb; but this is because we are located at its *backside*, and see their courses inverted. Attempts, however, to fix the sun's period of rotation were long baffled; for the spots, instead of being carried round as if attached to a rigid surface, gave signs of possessing "proper motions" of

[1] W. E. Wilson: *Monthly Notices*, vol. lv., p. 457.

uncertain and inconstant amount. The subject was first thoroughly investigated by Carrington; and he reached the unexpected conclusion that the sun has no uniform period, but gyrates in a composite fashion, quickest at the equator, and gradually slower towards the poles. From less than twenty-five days, he found the time of circuit to lengthen steadily to twenty-seven and a half in 50° of latitude. The axis round which this remarkably conditioned movement is performed makes an angle of 7° 15′ with the pole of the ecliptic; it inclines towards the earth's northern hemisphere from June to December, when the spots describe, in crossing the disc, paths curved downwards (to the eye of a northern observer); but the conditions being reversed between December and June, their paths are then curved upwards; while on June 3rd and December 5th, they pursue straight tracks, the earth being on those two days in the line of intersection between the sun's equatorial plane and that of the ecliptic.

Only a rough approximation, however, to the laws of solar rotation can be derived from spots. For they do not simply drift with the photospheric currents, but are subject to accelerations and retardations connected with their internal economy, as well as to mutual attractions and repulsions depending, it is supposed, upon their electrical condition. Fortunately, however, a method has been perfected by which these complications are abolished. Something has already been said as to spectroscopic determinations of motion in the line of sight. They are evidently applicable to the sun's axial movement. For, through its effect, his eastern limb is always advancing uniformly towards us, while the western limb is retreating at the same rate. Thus, the whole Fraunhofer spectrum is shifted slightly upward, or towards the blue, at the left-hand edge of the solar disc, and as much towards the red at the right-hand edge. The same lines of solar absorption, in fact, taken from opposite sides of the solar equator, and placed end to end, appear evidently notched, and can be distinguished at a glance from terrestrial absorption lines, which, having nothing to do with the sun's rotation, show no break at the junction of their sections. They in this way "virtually map" themselves, as Professor Langley proved experimentally in 1877.

In 1887-9, M. Dunér, of Upsala, succeeded in extending these delicate measurements to within fifteen degrees of the sun's poles, where the movement is so slow that it can only, by incredible refinements, be dealt with successfully. The upshot was to emphasise the law of slackening *angular* speed detected by Carrington and confirmed by Spoerer. From 25½ days at the Equator, the sun's period of rotation was found to become protracted to 38½ days at the seventy-fifth parallel of latitude. Its investigation from photographs of faculæ has been lately carried out by M. Stratonoff at Taschkent in Russia. The results of the three methods are collected in the following little table.[1]

THE SUN'S ROTATION.

Mean Solar Latitude.	Period from Faculæ. (Stratonoff.)	Period from Spots. (Spoerer.)	Period from Spectroscopic Measures. (Dunér.)
0°	24^{d}·66	25^{d}·09	25^{d}·46
15°	25 ·26	25 ·44	26 ·35
30°	25 ·48	26 ·53	27 ·57

These facts, although so various, are not necessarily discordant. They apply to different parts of the great solar machine, each one of which may rotate with a certain independence. The spots drift, more or less passively, *with* the photosphere. The faculæ are elevated above it, and appear to be everywhere accelerated relatively to its systematic currents. The strata originating the Fraunhofer lines, to which alone the spectroscope is applied, display, on the contrary, effects of retardation. "This peculiar law of the sun's rotation," Professor Holden remarks, "shows conclusively that it is not a

[1] *Observatory*, vol. xviii., p. 344.

rigid body, in which case, every one of its layers in every latitude must necessarily rotate in the same time. It is more like a vast whirlpool where the velocities of rotation depend on the situation of the rotating masses, not only as to latitude, but also as to depth beneath the exterior surface."

Solar chemistry progresses by successive interpretations; and the characters to be read are so multitudinous and so similar as to require very delicate discrimination. The work, carried on simultaneously in the sun and laboratory, becomes more arduous as it advances, and is still far from complete. Indeed, the difficulties attending detailed comparisons between the Fraunhofer lines and the innumerable components of terrestrial spectra, would be insuperable but for the aid of photography, here, as elsewhere, the versatile handmaiden of physical astronomy.

Here is a list of 36 solar elements published by Professor Rowland of Baltimore in 1891, and arranged according to the number of their representative lines in the solar spectrum.

Iron (2000 +)	Neodymium	Aluminium (4)
Nickel	Lanthanum	Cadmium
Titanium	Yttrium	Rhodium
Manganese	Niobium	Erbium
Chromium	Molybdenum	Zinc
Cobalt	Palladium	Copper (2)
Carbon (200 +)	Magnesium (20 +)	Silver (2)
Vanadium	Sodium (11 +)	Glucinium (2)
Zirconium	Silicon	Germanium
Cerium	Hydrogen	Tin
Calcium (75 +)	Strontium	Lead (1)
Scandium	Barium	Potassium (1)

Only two of these substances, carbon and silicon, are non-metallic, hydrogen ranking as a gaseous metal. Neither oxygen, nitrogen, nor argon, have yet spoken their "Adsum," but it is not impossible that they may do so in the future. Negative evidence, at any rate, is, in spectroscopic inquiries, absolutely inconclusive.

The spectra of sun-spots are, as might have been expected,

characterised by a great increase of absorption. There is a general darkening which extends far up in the ultra-violet, and is modified, in the green and blue, into remarkable dusky gratings made up of closely-set fine rays; and some of the ordinary Fraunhofer lines are besides thickened and blackened. The formation in spots of oxides is thought by Dr. Scheiner to be possibly indicated by these symptoms; "if so," he adds, "the presence of oxygen in the sun would thus be indirectly suggested."[1] Bright lines, too, flash out in the immediate neighbourhood of sun-spots, especially the "great twin brethren," "H" and "K," due to calcium, which stand in imposing breadth and strength at the violet end of the Fraunhofer spectrum, and are of corresponding importance as indexes to solar phenomena. With this pair, brilliant hydrogen rays are often associated, besides other "reversals," by which, upon the customary dark lines, flaming rays of identical wave lengths are superposed. But these signs of incandescence evidently belong to the facular stratum high up above the spot-umbra.

So long ago as 1769, the observations of Dr. Wilson of Glasgow were believed to have established, once for all, that spots are funnel-shaped depressions in the photosphere. But the perspective effects from which he argued are certainly not always, perhaps not very often, present. Mr. Howlett, after thirty-five years—1859 to 1895—devoted to testing the truth of the traditional conviction, has at last succeeded in shaking, if not in overthrowing, it. Most solar observers now admit that spots are of extremely various and extremely variable construction, so that the obscure umbra, at times a sort of pit or crater, in which vapours, cooled by expansion, well up from below, may, at another stage in the life-history even of the same spot, represent an actual accumulation of absorbent material above the brilliant solar cloud envelope. In any case, a spotted area appears to be an area of elevation. This might be due to a wide-spreading relief of pressure, or an accession of internal heat. The fact emerged clearly from a

[1] Frost-Scheiner: "Astronomical Spectroscopy," p. 177.

series of measurements of the sun's diameter executed by M. Sykora at Charkow, Russia, in 1895.[1]

The intensity of the agitations connected with sun-spots can be most fully appreciated from spectroscopic observations. Lines torn, displaced, and *branching*, testify to velocities in the line of sight of the matter surrounding or overlaying them up to three or four hundred miles a second! These tumultuous uprushes and downrushes are not of a systematic nature; they afford no insight, consequently, into the formative laws of spots. Of these we are indeed far more ignorant than Sir William Herschel supposed himself to be. Recent work on the sun has provided a grand store of facts ascertained with surprising skill and ingenuity. But they want *colligating*. No framework has yet been constructed that will hold them, each in its proper place. It has been truly said: "Considering the rapid progress which has been made in the observational or practical side of solar physics, it must be confessed that the theoretical side has been very imperfectly developed. Almost every student of solar physics has his own theory, and usually he himself is the only one who believes in it."

Since Sir John Herschel propounded his "cyclonic theory" of sun-spots in 1847, there has been a marked tendency to assimilate solar to terrestrial phenomena. But the circumstances of the two bodies are so utterly unlike that such attempts can only prove misleading. The earth is a solid globe warmed from without, hence, with hot tropical and frigid polar regions. This disparity is the prime motor in the circulation of its atmosphere and oceans; a circulation, essentially in latitude, directed towards the equalisation of temperature. The sun, on the contrary, is heated from within; there is no appreciable difference of temperature between its poles and equator; and its circulation is of the bodily kind belonging to fluid masses, and is carried on by vertical currents effecting exchanges of heat between the surface and the profundities beneath. Were these to stop, or even notably to slacken, the sun would promptly cease to shine, and lapse into the condition of a

[1] *Astronomische Nachrichten*, No. 3330.

"dark star." It is not then surprising that the drifting movements of the photosphere are *along*, not *across*, parallels of latitude. Solar meteorology, in a word, has almost nothing in common with terrestrial meteorology; and explanatory schemes, based upon an analogy which does not exist, must sooner or later be consigned to the limbo of vanities.

CHAPTER III.

THE SUN'S SURROUNDINGS.

"WHAT we ordinarily call the sun," wrote the late Mr. Ranyard, "is only the bright spherical nucleus of a nebulous body."[1] But it is only when the interposing moon cuts off the dazzling rays of the nucleus that we see directly anything of its nebular surroundings. Partial or annular eclipses are of little or no use for this purpose; the revelation belongs exclusively to the sombre, yet splendid moments of totality. No sooner has the last glint of sunshine vanished than the corona starts into view, encompassing the black lunar globe with a sort of "glory" of silvery streamers. Its radiated shape suggests vacillation of form and a flickering radiance; yet its immobility is absolute. The awe and wonder of the sight tend, for the moment, to supersede scientific curiosity, and they are enhanced by the perception, at the base of the corona, of the serrated scarlet "chromosphere" fringing the moon's circumference, while the towering "prominences" that are usually seen to spring from it produce the startling effect of a conflagration.

These marvellous appendages received no adequate notice until their disclosure during the total eclipse of July 8, 1842. Even the uninstructed crowds in the streets of Milan and

[1] *Knowledge*, vol. vi., p. 13.

Pavia shouted with amazement at what they saw; while by solar students the recurrence of similar opportunities has ever since been eagerly anticipated and diligently turned to account. The question that first pressed for solution related to the local habitation of prominences; for some unwisely persisted in attaching them to the moon. A decisive answer was given by photography at its first *effective* application to eclipses on July 18, 1860. From a comparison of negatives exposed at the beginning and end of totality, it became clearly apparent that the moon had, in the interval, moved *over* the prominences, uncovering, to a small extent, those on the west side and concealing those on the east.

Their solar connexion having thus been established by the camera, the spectroscope was called upon to determine their physical and chemical nature. An admirable opportunity for taking this further step was presented by the Indian eclipse of August 18, 1868. The result was decisive. The light of a huge spire of flame, 89,000 miles high, had no sooner passed through a prism than its gaseous origin declared itself. The spectrum consisted of several hydrogen-lines, and one unknown line in the yellow, slightly more refrangible than the sodium-pair D_1, and D_2, and hence called D_3. "Je verrai ces lignes-là en dehors des éclipses!" M. Janssen exclaimed, as they caught his eye; and on the following morning, at Guntoor in the Neilgherries, he actually started daylight spectroscopic work at the edge of the sun. He owed his success to a perfectly simple principle. The ordinary invisibility of prominences is due to the drowning of their light in reflected sunshine. But sunshine, because it is continuous—that is, made up of beams of all refrangibilities—can be weakened to almost any extent by dispersion, while the detached prominence-rays lose nothing by being separated. Hence, the result of passing the mixed light from near the solar limb through a train of prisms is that the tell-tale bright lines stand out distinctly from an *emaciated* prismatic background. The method was independently discovered by Mr. Norman Lockyer in England, and his and Janssen's communications on the subject were laid before the French Academy of Sciences on the same day of October, 1868. It has proved of inestimable value, and was further improved in 1869 by Dr. Huggins's device for viewing these objects in their proper shapes through an open slit, instead of building them up in narrow sections by successive observations through a narrow one. This was made possible by the intensity of their light. They can be observed in variously coloured images corresponding to the different rays they emit; but the least refrangible of the hydrogen series—the blood-red C (alias $H\alpha$)—is generally chosen as being the most brilliant and best defined.

The unrecognised substance giving the yellow prominence-line was named by Dr. Frankland "helium." It evidently existed near the sun in enormous quantities, and in close companionship with hydrogen. Yet no dark line corresponding to its absorption was to be found in the Fraunhofer spectrum, although it now and then emerged in spot-spectra. Conjectures were rife as to its nature and relations. It was generally believed to be specifically lighter than hydrogen, and some held it a product of its dissociation, and so of a different elemental standing. Everything about it, however, remained doubtful until, in March, 1895, Professor Ramsay produced a sample for inspection close at hand, extracted by heat from the rare mineral "clevite." The recognition-mark was its emission, when electrically excited, of the solar D_3, with which were associated several other chromospheric rays previously registered as of unknown origin, but now linked together as vibrations of the same molecules. A sudden and entirely unlooked-for advance was thus made in the chemistry of the sun's surroundings.

Helium is a colourless gas of about twice the density of hydrogen. Its peculiar qualities are shared only by argon, the new constituent of the earth's atmosphere. Both have unusual thermal relations; both are chemically inert. They refuse to combine with any other element, and thus stand apart from the round of multiform change involving the whole material world. Helium is nevertheless distributed freely throughout the universe. Hydrogen itself is scarcely more ubiquitous.

A considerable mass of information regarding the solar prominences was rapidly collected by means of the Janssen-Lockyer invention. They were at once divided into two classes. The "quiescent" kind occur in all solar latitudes; they change their shapes very gradually; they have no immediate relationship with spots. In form they resemble *pillared clouds* resting in banks like heavy cumuli, or floating, like expanses of thin cirrus, high above the chromosphere with which they are ordinarily connected by slender supports or conduit-pipes. But these are at times invisible or non-existent. Father Secchi occasionally watched isolated cloudlets form and grow spontaneously as if by condensation from saturated air; and on October 13, 1880, Professor Young made a confirmatory observation. About 11 A.M. he noticed a detached fiery mass at an elevation of 67,500 miles above the limb. "It grew rapidly, without any sensible rising or falling, and in an hour developed into a large stratiform cloud, irregular on the upper surface, but nearly flat beneath. From this lower surface pendent filaments grew out, and by the middle of the afternoon the object had become one of the ordinary stemmed prominences."[1] The size of these formations is enormous. They vary in height from about 10,000 to 100,000 miles; and ranges of them 450,000 miles in extent have been photographed during total eclipses.

The second class of prominences, known as "eruptive," are obviously manifestations of intense energy. In some of their forms they suggest geyser-like spoutings of incandescent vapours. They represent swords and scimetars, palms with twisted trunks composed of mounting flames, igneous vegetation of sundry types. Their chemistry is much more complex than that of the quiescent sort. Not only hydrogen and helium, but iron, magnesium, sodium, and a number of other

[1] "The Sun," p. 206, first edition.

metals enter into their composition. Belonging to the same order of disturbance with spots, they are closely conjoined with them, both in time and space. They conform to the sun-spot cycle, as well as to the "law of zones," showing that photospheric and chromospheric disturbances spring from a

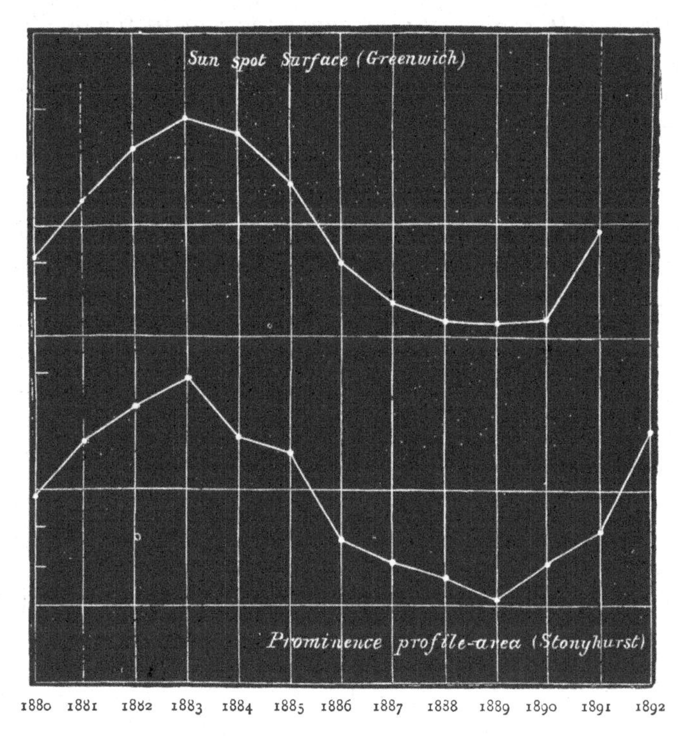

FIG. 3.—*Curves showing the development of Sun-spots and Prominences during the period* 1880 *to* 1891. (Sidgreaves.)

common cause. Fig. 3 (from the *Observatory* for March, 1893) embodies a comparison between the "spotted area" as determined at Greenwich 1880-1891, and the "profile area" of prominences (without distinction of kind) observed spectroscopically at Stonyhurst during the years 1880-1892. The agreement between the two curves is very striking; but the minimum of solar activity in 1889 is decidedly better represented by the prominence-tracing. Father Sidgreaves, director of the Stonyhurst Observatory, adds the important remark that wide-spreading elevations of the chromosphere attend spot-maxima, while depressions of equal extent occur at minima.

The chromosphere is a solar envelope, but not a solar atmosphere. It completely surrounds the sun to the depth of about 4,000 miles with a close tissue of scarlet flames, their filamentous or tufted summits swaying and intercrossing as if under the gusty sweep of fiery winds. Any of these summits which attain an unwonted height become "prominences," but it is a mere matter of convention when the change of nomenclature should take place. The chemical composition of the chromosphere does not differ essentially from that of prominences. Its permanent constituents were found by Professor Young to be hydrogen, helium, "coronium," and calcium, the last represented *only* by "H" and "K." But disturbances never failed to be indicated by the blaze of metallic lines, of which 273 in all have been determined by the same authority. Their appearance signified, without doubt, the injection from below of the corresponding vapours, chiefly those of iron, titanium, sodium, magnesium, strontium, barium, and manganese. At moments the reinforcement of the spectrum with bright rays was so extensive that it seemed as if the entire "reversing layer" had been uplifted bodily into the chromosphere.

The reversing layer lies quite close to the photosphere. It is scarcely more than 300 miles deep, and is hence invisible except during about a second at the beginning and end of total eclipses. Young was the first to be favoured with a sight of it, on December 22, 1870. No sooner was the direct solar spectrum intercepted by the moon, than "all at once, as suddenly as a bursting rocket shoots out its stars, the whole field of view was filled with bright lines, more numerous than one could count. The phenomenon was so sudden, so unexpected, and so wonderfully beautiful, as to force an involuntary exclamation."[1] It was afterwards frequently observed, and

[1] "Memoirs of the Royal Astronomical Society," vol. xli., p. 435.

at last satisfactorily photographed by Mr. Shackleton, a member of Sir George Baden-Powell's expedition to Novaya Zemlya, for the purpose of observing the total solar eclipse of August 9, 1896. The permanent record then secured was of peculiar importance as affording the means of confronting in detail the components of the vario-tinted flash at the eclipsed sun's limb with the dusky legion of the Fraunhofer lines. The correspondence is striking, and leaves no doubt that Young's stratum is the actual locality where the characteristic solar spectrum is produced. It may be described as an universal solar ocean of glowing metallic vapours, the rays emanating from which, although vivid when seen *off* the sun, are thrown out in dark relief by projection upon the white-hot photosphere. The existence of just such a heterogeneous absorbing layer had been predicted, on theoretical grounds, some years before it came into view.

The movements taking place in eruptive prominences are often of portentous speed. They are betrayed, so far as they coincide with the visual ray, by spectroscopic line-displacements; so far as they are directed *across* the visual ray, by immediate observation of the spectroscopic images. Thus, the up-and-downrushes of flaming hydrogen above spots on the disc reach velocities of 320 miles a second; and solar tornadoes (detected by Mr. Lockyer more than a quarter of a century ago) are often observed to whirl at rates which would be incredible were they less well authenticated. Vertical explosions at the limb, on the other hand, of still more unruly violence are rendered manifest by displacements, not of the emitted lines, but of the radiating substances themselves.

On September 19th and 20th, 1893, Father Fényi, director of the Kalocsa Observatory in Hungary, witnessed the development and dissolution of a pair of objects perhaps the most extraordinary in the astonishing record of solar phenomena.[1] They broke out within nineteen hours of each other, showed a close similarity of shape and structure, underwent analogous changes, and, strangest of all, were situated at

[1] *Astronomy and Astro-Physics*, vol. xiii., p. 122.

almost diametrically opposite points of the solar limb. The first was already, when first viewed at 2 P.M., 168,000 miles high ; within half an hour, it had sprung up to 224,000 miles (8′ 18″), and again subsided into a commonplace flame of the modest dimension of 13,650 miles (30″). The rate of ascent, directly measured (always necessarily through the medium of the spectroscope), was 132 miles a second. This vast, though transient construction, seemed to be formed of a multitude of distinct fiery tongues, each leaping and flaring independently. As a whole, it was also tongue-shaped, and "stood erect nearly in the direction of the sun's radius," travelling, meanwhile, towards the earth at an average rate of 186 miles a second.

The companion-prominence began to show at nine next morning, and, rising with a velocity of 300 miles per second, attained in twelve minutes to a height of 220,000 miles. This tremendous apparition was of the same "ragged" texture as its predecessor, and shone, even in its loftiest fragments, with the same intense glow. As might have been expected from its opposite position, its radial movement was *from* the earth. A prominence measured by the same observer, July 15, 1895, was diminishing its distance from the earth with the extraordinary velocity of 533 miles a second; and on September 30 of the same year, a colossal object resembling the bent and riven trunk of a great tree, was in the course of half an hour flung upwards to a minimum altitude of 313,000 miles, and had again faded out of sight. "The appearance," Father Fényi wrote, "of all the numerous great eruptions which I have observed has been such as would be produced by a kind of explosion over a spotted region, which, seizing upon a prominence already developed, hurls it upward from the surface, tears it to pieces, and brings it to a speedy end." The matter thus acted upon is of enormous volume, but negligeable mass.

Photographs of prominence-spectra, obtained by Dr. Schuster during the eclipse of May 17, 1882, brought out the remarkable predominance in their light of the

"H" and "K" emissions of calcium. It was re-discovered by means of spectrographs of those objects, taken in 1891 without an eclipse, by Professor Hale at Chicago, and by M. Deslandres in Paris. Both investigators promptly seized upon the advantage it offered for their chemical delineation in full daylight. The lines in question are dark and abnormally wide in the sun itself, bright and sharp in prominences. Thus, at these particular parts of the spectrum, the obliterating effects of scattered sunlight are non-existent. Just here, too, photographic sensitiveness is at its maximum. Hence, by working with either of these lines (K is preferable) nothing could be easier than to get impressions of the brilliant forms of prominences relieved against the background of solar absorption. (See Figures 4 and 5.) The thin, bright line is *sheltered* from daylight glare by the dusky, broad one. By the use of a "double slit," the method was completed. This, again, was simultaneously invented by Hale and Deslandres, although they had, without suspecting it, been anticipated by Janssen in 1869. The second slit is adjusted so as to exclude all but a single ray of the spectrum formed by dispersing the light admitted through the first. An unlimited power of selection is in this way afforded as to the quality of light to be employed; but for general purposes, K is not likely to be superseded.

In the Chicago spectroheliograph, two moveable slits, together with a powerful diffraction spectroscope, are attached to a twelve-inch refractor. With this instrument, monochromatic impressions of the sun with its spots, faculæ, and flame-garland are obtained without difficulty. To begin with, the solar disc is covered with a metal diaphragm, then the first slit is caused to traverse the artificially eclipsed image, the second following at such a rate that the K line alone always falls upon the sensitive plate. The result is a complete photographic record of the chromosphere and prominences. The diaphragm having been then removed, the return journey of the slits is very quickly made, so as to guard against the formidable actinic strength of even that small element of direct sunlight contained in the K line. The object of the second transit is to *insert* an autographic print of the sun itself into the space previously left blank to receive it. The entire operation occupies less than one minute. Portrayed thus in calcium light, the solar disc has a strange effect. It is entirely overspread with a reticulation of irregular bright markings, greatly emphasized over the spot-zones, and corresponding in general with the positions of faculæ. According to Professor Hale, these masses and wreathings of calcium vapour *are* faculæ. M. Deslandres regards them rather as gaseous formations connected with faculæ. Their extension and intensity are at times so great that M. Deslandres has actually succeeded, through the prevalence of their light, in photographing the sun as a "bright line star." The double slit method also affords the means of studying the distribution of each element of the reversing layer in the leisure of ordinary daylight, as M. Deslandres has shown by some preliminary experiments.[1]

To this extent astronomers have made themselves independent of eclipses. These momentous occurrences are, fortunately, not needed for researches concerned with distinct coloured rays separable by dispersion from diffuse sunshine. But with the corona it is different. For here we have a white glory to deal with. Coronal light is derived from three sources: from the original incandescence of solid or liquid particles, from sunshine reflected by them, and from gaseous emissions. The most conspicuous of these is a green ray of unknown chemical meaning. It proceeds from every part of the corona, even from the dark rifts separating its brilliant streamers, and the inconceivably tenuous substance to which it owes its origin has, accordingly, received the name of "coronium." The coronal spectrum includes many other bright lines, especially in the ultra-violet, photographed during eclipses; but the hydrogen, helium, and calcium lines which accompany them probably represent scattered chromospheric light.

The green coronal ray is much too faint to be isolated with

[1] *Comptes Rendus*, December 26, 1893.

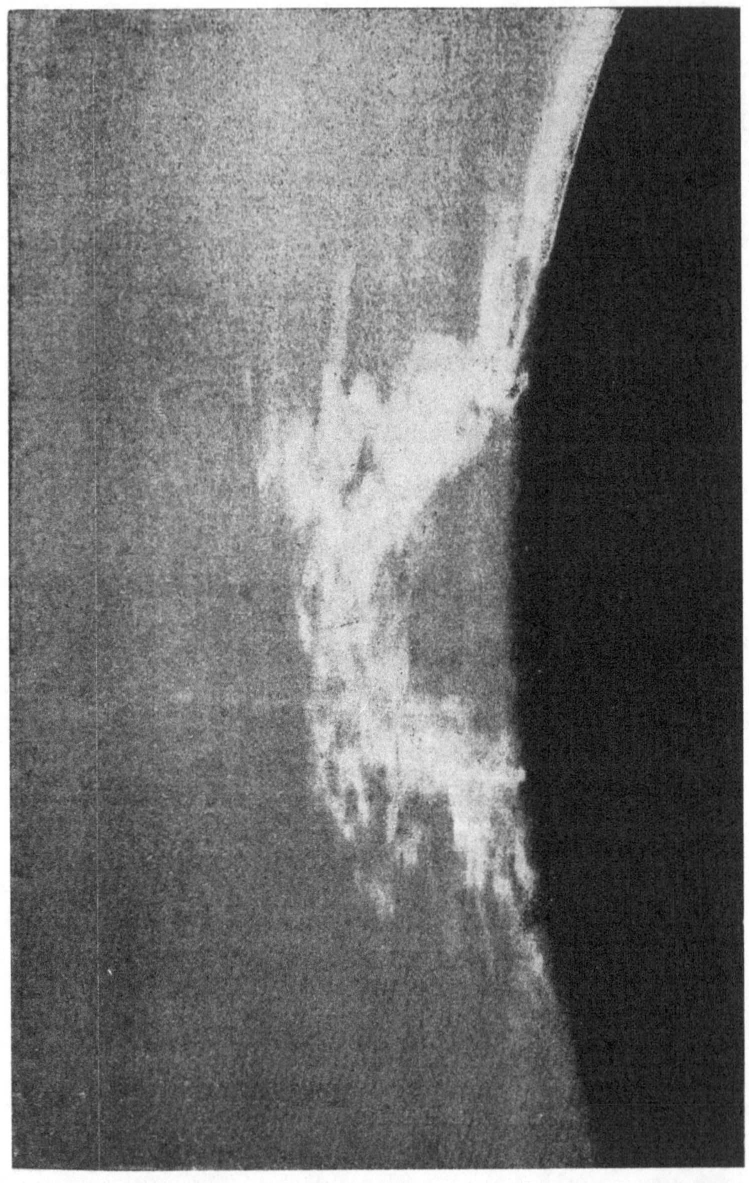

FIG. 4.—*Eruptive Prominence photographed by Professor Hale at the Kenwood Observatory, March 24, 1895, at 22h. 40m. Chicago mean time. (The photosphere is covered with a metallic disc.)*

FIG. 5.—*The same*, 18*m*. *later*.
(From the *Astrophysical Journal*, May, 1896.)

the spectroscope; but the continuous coronal spectrum has maxima of intensity compared with ordinary daylight, which suggested to Dr. Huggins, in 1882, a differential method of photographing the entire structure apart from eclipses. It has, however, as yet come to nothing, and Hale and

FIG. 6.—*The Eclipsed Sun, photographed at Sohag in Egypt, May 17, 1882. A Comet is almost involved in the Corona.* (From "Philosophical Transactions," vol. clxxv.)

Deslandres have been equally unsuccessful with their "double slit" apparatus. Hence, it is only by favour of the moon that this wonderful appendage can be investigated, and the available moments have not been allowed to pass in vain.

One result fully ascertained is that it changes in form

concurrently with the progress of the sun-spot period. The maximum coronal type is entirely different from the minimum type, and reappears in unmistakable connection with vehement solar disturbance. This cyclical relation was first pointed out by Mr. Ranyard. On July 29, 1878, a totality of 165 seconds was observed, under splendid conditions of weather, in the Western States of North America. No prominences worthy of note were visible, but the corona wore a most surprising aspect. A pair of enormous equatorial streamers stretched east and west of the sun to a distance of at least ten millions of miles. Indeed, they came to no definite end. They were best seen with the naked eye, and made no show on sensitive plates, but the application of low telescopic powers disclosed, near the base of the effusions, a mass of delicate and complex detail. The solar poles were as distinctively, although not so strikingly, garnished as the solar equator. Each was the centre from which diverged a dense brush of straight, electrical-looking rays. The sun was at the time in a state of profound tranquillity; and it was recalled that, at the previous minimum, in 1867, Grosch had delineated, at Santiago, just the same equatorial extensions, and just the same polar brushes. The connexion was emphasised during the maximum of 1882-4, by the substitution, when the moon covered the sun on May 17, 1882, and May 6, 1883, of a dazzling stellate formation for the winged corona of 1878. In Fig. 6 is reproduced a photograph by Dr. Schuster of the Sohag, or Egyptian corona, with the added embellishment of a comet hurrying up to perihelion, conspicuous to the eye at the time, but never seen again.

In 1889 the minimum type of corona reasserted itself. A drawing made by Miss M. L. Todd during the eclipse of January 1, gave the characteristic equatorial "fish-tails," reaching out on the west to four solar diameters.[1] And although the camera, owing to special difficulties, has not yet been able to pursue them so far, Professor Barnard's exquisite

[1] *Knowledge*, vol. iv., p. 105.

picture (Fig. 7), taken at Bartlett's Springs, California, with an exposure of $4\frac{1}{2}$ seconds, portrays the type to perfection, with its suggested indefinite expansions, "the soft feathery details of the inner corona, and the delicate fan-structures at the poles." Two minute notches mark the points where a couple of prominences have, by the intensity

FIG. 7.—*The Corona of January* 1, 1889, *photographed by Professor E. E. Barnard.*

of their actinic power, *eaten into* the black circumference of the lunar image.

Nine negatives were secured by the artist, but at a considerable personal sacrifice. "So impressive," he wrote, "was the magnificent spectacle upon the crowd that had gathered just outside our enclosure, that not a murmur was heard.

The frightened, half-whining bark of a dog, and the click-click of the driving-clock, alone were audible. When the sun suddenly burst forth, an almost instantaneous and highly-surprised cackling of the chickens, that had hastily sought their roosts at the beginning of totality, would have been amusing could one have shaken off the dazed feeling at the unexpectedly rapid termination of the semi-darkness. My own feelings were those of excessive disappointment and depression. So intent was I in watching the cameras and making the exposures, that I did not look up to the sun during totality, and therefore saw nothing of the corona."

On April 16, 1893, at the height of the last sun-spot maximum, a shadow-track crossed South America and Central Africa. Once more the coronal type had changed. Not a trace remained of the equatorial "wings"; not a trace of the polar "fans." Instead, the "compass-card" aureole of 1882 and 1883, shaped regardless of heliographic latitude, re-emerged from beneath the veil of daylight. That the sun's filmy "crown" follows, after its own inexplicable fashion, the general round of solar vicissitudes, no longer admitted of a doubt. The fact is thus stated by M. Deslandres, who observed the eclipse at Fundium, in the Senegal district.

"The form of the corona," he says, "undergoes periodical variations, which follow the simultaneous periodical variations already ascertained for spots, faculæ, prominences, auroræ, and terrestrial magnetism. This important relation, indicated by preceding eclipses, is strongly confirmed by the eclipse of 1893."[1]

Professor Schaeberle's photographs, taken on the same occasion at Mina Bronces in Chili, marked a decided advance in coronal portraiture. The sun's disc measured four inches on his plates, exposed with a photoheliograph forty feet in length; and the details of inner coronal construction came out accordingly with unprecedented perfection. The corona of August 9, 1896, reproduced the most striking features of

[1] "Rapport de la Mission envoyée an Sénégal," p. 31.

the corona observed August 29, 1886; and both corresponded to an intermediate epoch of the spot-cycle. The polar brushes were present without the equatorial extensions, while in both a protruding ray made an angle of some thirty or forty degrees with the solar axis. This distinctive trait imprinted itself with surprising emphasis on some of Sir George Baden-Powell's Novaya Zemlya photographs.

Researches, prosecuted under cover of eighteen eclipses, have greatly strengthened the visible analogy between coronal streamers, auroral coruscations, and comets' tails. The persuasion that electrical discharges in high vacua are concerned in all these phenomena is not easily resisted. Repulsive forces such as are at work in Crookes' tubes perhaps come into play, on the vast solar scale, to produce the strange and beautiful luminous forms revealed during eclipses. Their tenuity is certainly extreme. They probably contain very much less matter, volume for volume, than the incredibly exhausted tubes of modern physicists. The unresisted passage of comets through the corona demands this supposition, which is in complete accord with the fineness of the Fraunhofer lines. The corona shows no increase of density downwards, and the chromosphere very little. Hence neither can be a true solar atmosphere, weighing freely upon the sun's surface. For, under the immense power of solar gravity, the accumulated pressure of the superincumbent layers, even if there were only one hundred miles' thickness of them, could not be intelligibly conveyed in figures; how much less when the piling-up of the aerial strata is reckoned by thousands of miles!

To recapitulate. Starting from the photosphere, we meet first an envelope producing the *general* absorption, by which sunlight is enfeebled and reddened as if by the interposition of a slightly rufous shade. Next comes the reversing layer composed of mixed incandescent vapours, giving rise, by their *selective* absorption, to the Fraunhofer lines. No alterations in correspondence with the spot-cycle have yet been determined in either of these couches, which, close as they lie to the photosphere, remain, nevertheless, apparently indifferent to its agitations. They are overspread by the chromosphere and prominences; while above and beyond shines the mysterious corona; both chromosphere and corona strictly conforming, by manifest changes, to the sun's periodicity. One other solar appendage remains to be noticed.

After sunset in spring, and before sunrise in autumn, a mass of soft luminosity, often brighter than the Milky Way, may be seen tapering upward from the horizon along an axis approximating to the line of the ecliptic. Its more conspicuous visibility at those times just reverses the case of the harvest moon. As a rule, the apex of the cone barely reaches the Pleiades; but it does not really end here. Thrice during the present century, by Brorsen, Backhouse, and Barnard, the zodiacal "counterglow" has been independently discovered and studied. This is a hazy, luminous patch, ten to fifteen degrees across, and exactly 180° from the sun. It represents the *opposition aspect* of the Zodiacal Light, hence proved to be a formation in planetary space, extending considerably beyond the earth's orbit. Two plausible hypotheses as to its nature have been proposed. Professor Searle[1] holds it to represent the reflection of sunlight from "an infinite number of small asteroids." Professor Bigelow[2] considers it as an amassment in the plane of the sun's equator—"a place of zero potential"—of the particles electrically expelled from the poles. The Light is then, if this view be correct, an extension of the corona—a sort of "pocket or receptacle, wherein the coronal matter is accumulated and retained as a solar accompaniment." A continuous spectrum is derived from it; no element of original emission can be detected; so that the spectroscope "holds the balance even" between the two theories. If, however, the latter were true, the Zodiacal Light should spread out from the sun's equator; if the former, then its medial plane

[1] "Harvard Annals," vol. xix., part ii.; 1893.

[2] "The Solar Corona discussed by Spherical Harmonics;" Washington, 1889.

should deviate very slightly from that of the ecliptic, to which the fundamental, or "invariable" plane of the solar system is inclined only one and a half degrees. M. Marchand's observations from the Pic du Midi[1] appear to be decisive on the point. During three years, he mapped down the limits assigned by his observations night after night, to an emanation which, in that pure air, was seen to compass the entire sphere. The eventual comparison of his collected data showed its axis to be a great circle sensibly coincident with the sun's equator. All reasonable doubt as to the nature of the Zodiacal Light has thus been removed. It is a reservoir for the sun's waste matter—the sink, into which are daily flung the particles rejected through the agency of the aigrettes and streamers composing the wonderful eclipse-vision of the corona.

CHAPTER IV.

THE INTERIOR PLANETS.

THE Interior Planets are those which revolve within the earth's orbit. They are two in number—Mercury and Venus. Mercury, with a diameter of three thousand miles, is the smallest of the eight principal planets. It pursues a track, too, more eccentric and more highly inclined to the ecliptic than any other planetary orbit. The zodiac had of old to be made 16° wide in order to afford room for its excursions. These irregularities are, however, quite innocuous as regards the stability of the system, for the reason that they belong to a body of insignificant mass. The successive approaches to it of Encke's

[1] *Bulletin Astronomique*, April, 1896.

comet have afforded a means of ascertaining its gravitative power; and, according to the latest report from this filmy messenger, it is even less than had been supposed. Mercury, it appears, weighs little more than one-ten-millionth of the sun, or one-thirtieth of the earth. And since its volume is about one-nineteenth the terrestrial, the matter of which it is composed must be less dense in the proportion of 30 to 19. So that the planet would turn the balance against one equal globe of granite, or three and a half of water. We can hence easily calculate that gravity, at Mercury's surface, possesses less than one-fourth its power at the earth's surface. A man of sixteen stone transported thither, would find himself relieved of fully three-quarters of his habitual burthen.

The plane of Mercury's orbit makes an angle of 7° with the ecliptic, and he traverses it with a speed varying from 23 to 35 miles a second. The corresponding distances from the sun are 43½ and 28½ million miles, while the mean distance, or semi-major axis of the ellipse, measures just 36 millions. Independently then of what we call seasons, Mercury is subject, in the course of its year of 88 days, to considerable vicissitudes of temperature. At perihelion it receives nine times, at aphelion only four times, more heat than is imparted by the sun to an equal area of the earth.

The crucial point as regards the physical condition of a planet is the presence or absence of an atmosphere. And there is decisive evidence that Mercury is in this respect poorly provided. Certain luminous phenomena, often observed during its transits across the sun, appear to be of purely optical production, since they are less conspicuous with good than with indifferent telescopes; while, on the other hand, genuine refractive effects are absent. A corresponding indication is afforded by the low "albedo," that is, the slight reflective power of this planet. Of the light flooding its surface only 17 per cent.[1] is returned; 83 per cent. is absorbed. Now the albedo of clouds is about 72; a cloud-wrapt globe is little less

[1] According to G. Müller, *Potsdam Publicationen*, No. 30, p. 369, Zöllner fixed the albedo of Mercury at 0·13.

brilliant than if it were covered with fresh-fallen snow. Hence a high albedo accompanies a dense, vapour-laden atmosphere; a low albedo indicates a transparent one. And since Mercury, which sends back only about as much light as if it were made of grey granite, has the lowest albedo of any of the principal planets, it may be safely concluded to possess the thinnest aerial covering. Yet it is not, apparently, a totally airless globe. Spots upon its surface have been seen to become effaced as if by atmospheric veilings; and the spectroscope hints (although doubtfully) at aqueous absorption.

Mercury is "new" when nearest to the earth, and "full" when most remote from it. At both these periods, moreover, its position with regard to the sun renders it ordinarily invisible; so that it is usually seen as either gibbous or crescent shaped. The study of its phases has brought out a noteworthy circumstance. It is easy to understand that geometrical light-changes will not proceed by the same gradations upon a smooth and upon a rugged globe, where they are complicated by irregular shadows and illuminations. The laws of variation are quite different in each case, and their respective prevalence can be distinguished by steady observation. There seems no reason to doubt that the latter are obeyed by Mercury. After several years' watching of its phases, Professor G. Müller [1] of Potsdam concludes them to be such as characterise a broken and uneven surface.

Little or nothing was known about the rotation of Mercury when Schiaparelli of Milan undertook its determination in 1882. His observations were made in full daylight, in order to reduce atmospheric disturbances to a minimum; and he executed, in the course of a few months, a series of 150 Mercurian delineations upon which is founded the planisphere exhibited in Fig. 8. The surface of the planet, coloured light rose with a coppery tinge, was seen to be diversified by brownish-red markings which became effaced towards the limb as if through atmospheric absorption. Although evidently of a permanent nature, their outlines

[1] *Astr. Nach.*, No. 3171.

escaped precise definition. The most remarkable circumstance about them was that they showed no effects of rotation. During several consecutive hours of watching, they remained sensibly fixed in their places. The conclusion was finally arrived at that Mercury rotates on a nearly upright

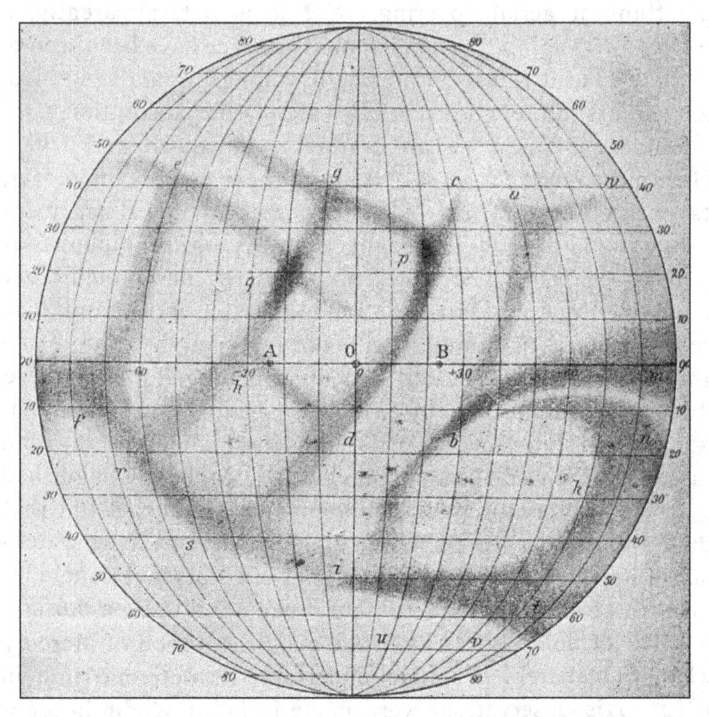

FIG. 8.—*Map of Mercury, by Schiaparelli.* (From *Astronomische Nachrichten*, No. 2944.)

axis in the same time that it revolves round the sun. Its day, no less than its year, is equal to 88 of our days. Consequently it turns at all times substantially the same face towards the sun ; and the "terminator," that is, the dividing-line between darkness and light, only "librates," without travelling right round the globe. The librations of Mercury are, however, extensive in proportion to the eccentricity of its orbit ; hence, five-eighths of its surface come in for some share of illumination during the Mercurian year. Over the remaining three-eighths darkness reigns supreme.

> "There is no light in earth or heaven,
> But the cold light of stars."

Satisfactory confirmation of this curious result was obtained by Mr. Percival Lowell at the Flagstaff Observatory in Arizona during the autumn of 1896.[1] In Schiaparelli's map, the axis of rotation lies in the plane of the paper, and the centre of the projected sphere thus represents the point on Mercury's surface where the sun is vertical at perihelion and aphelion; A and B, 23° 41′ to the east and west of it respectively, marking the places where the sun is vertical at the libration-limits. That formidable luminary oscillates from the zenith of A to the zenith of B and back in 88 days, occupying, in consequence of the planet's unequal motion, 51 in describing the arc from east to west (left to right), but only 37 in retracing it from west to east.[2]

The effects of these arrangements upon climate must be exceedingly peculiar. They cannot readily be traced in detail; but, thin as the Mercurian atmosphere is, it must be to some extent operative in modifying the contrast in temperature between the two hemispheres. Except in a few favoured localities, the existence of liquid water must be impossible in either. Mercurian oceans, could they ever have been formed, should long ago have been boiled off from the hot side, and condensed in "thick-ribbed ice" on the cold side.

Mercury is then, according to our ideas, totally unfitted to be the abode of organic life. Nor can it at any time have been more favourably circumstanced than at present. We need not hesitate to assert that its rotation was reduced to its actual minimum rate by the power of tidal friction. The brake was, moreover, applied by the sun. The attainment of rapid gyration was prevented by the resistance of solar tides raised on a plastic mass. Disruption was accordingly

[1] *Astr. Nach.*, No. 3406. [2] *Ibid.*, No. 2944.

rendered impossible. The planet was, by anticipation, deprived of satellites, and remained undivided and solitary.

Venus, the earth's nearest planetary neighbour, might be called its twin. Its diameter being 7,700 miles, it is of nearly the same size; it is not greatly inferior in mean density; gravity at its surface is of more than four-fifths its terrestrial strength, and it is supplied with an extensive atmosphere. Its movements are placid and well-regulated. In a period of 225 days it revolves at the rate of 22 miles per second in an almost circular track, deviating but slightly from the plane of the ecliptic. Its distance from the sun is 67,200,000 miles; hence it receives just twice as much heat and light as the earth. Moreover, it reflects at least 65 per cent. of the light incident upon it. Viewed in the same telescopic field with Mercury during a close conjunction in 1878, it shone, James Nasmyth reported, like burnished silver, while Mercury appeared as dull as zinc or lead. Yet Mercury is illuminated, on an average, three and a half times more intensely than its neighbour.

Atmospheric effects are conspicuous on Venus. At the beginning and end of transits, the part of the little black disc off the sun, has constantly been seen silver-edged through refraction; and when the planet, at inferior conjunction, passes above or below the sun, its whole circumference is not unfrequently bordered with a halo of solar rays, bent inwards as if by the action of a lens. Just in the same way, the *geometrical* rising of the heavenly bodies is *visually* anticipated, and their setting delayed on the earth, by the curvature of the beams refracted in passing through its atmosphere—or rather, through half of it; while we, as spectators of Venus from the outside, perceive the entire effect. Made on equal terms, the comparison is greatly to the disadvantage of the earth. Refraction, as directly measured on Venus, considerably exceeds its terrestrial amount; and the measurable refraction is only that produced in the higher part of the air surmounting the shell of clouds which constitutes the planet's visible surface. Thus, at the cloud-level a barometer would, by the lowest estimate, stand at 35 inches; while at the same altitude of, say, two miles, the column of mercury would, on the earth, drop to 21 inches. It is, indeed, very likely that the aerial envelope of Venus weighs twice as much as our own.

The occasional visibility of the dark side of Venus is still unexplained. The appearance is indistinguishable except in scale from that of the "old moon in the new moon's arms"; but illumination by earthshine, which is fully competent to produce the lunar effect, practically vanishes at the distance of Venus. The "ashen light," as it is called, ordinarily shows only when the planet figures as a narrow crescent; but M. Brenner of the Manora Observatory, who has a knack of being unprecedented, saw it in June, 1895,[1] during the gibbous phase. The appearances of this pale gleam follow no traceable law. They occur unsought; and are recalcitrant to vigilant expectation. Their closest analogy is with our auroræ. The "phosphorescence" of the dark side of Venus may quite reasonably be set down as of an electrical nature. But it does not seem, like terrestrial auroræ, to follow the lines of a magnetic system.

Distinct spectroscopic indications of aqueous absorption in the atmosphere of Venus were perceived, during the transits of 1874 and 1882, by Tacchini, Ricco, and Young. They accord well with the "snow-caps," which are one of the many puzzling Cytherean features. Since these can be resolved into groups of brilliant points, they represent, in the opinion of the late M. Trouvelot, mountainous formations penetrating the reflective stratum, and shining, lustrous with snow, in the clear upper air. They might almost equally well be cloud-like condensations of a permanent kind, called into existence by topographical peculiarities, and hence, after a fashion, *rooted in the soil.* On the other hand, Mr. Lowell questions their reality in any form; and his drawings represent extraordinarily sure seeing.

The only point regarding the planet's rotation upon

[1] *Astr. Nach.*, No. 3332.

which astronomers are agreed is that its axis is nearly perpendicular to the place of its orbit. As to its period, the divergence is enormous. It reaches all the way from 24 hours to 225 days. Bad as is the telescopic holding-ground on Mercury, that afforded by Venus is worse still. The disc falls off rapidly in brightness from the limb towards the terminator, and is sometimes diversified by filmy and indefinite markings, obviously of atmospheric origin (in Fig. 9 the shadings are much too pronounced). Attempts to use them as fiducial points are foredoomed to failure. The period, accordingly, of $23^h\ 21^m$ arrived at by forcing into artificial agreement the observations of Cassini at Bologna, of Bianchini and De Vico in Rome, obtained small credit. The subject lay, as it were, dormant until Schiaparelli made, in 1890, the provisional announcement that Venus rotates on the same plan as Mercury. A clamour of contradiction was immediately raised, and a large amount of evidence on both sides of the question has since been collected. It is curious to notice that, setting aside the opposite conclusions of Terby and Brenner, the Alps mark a dividing-line between the pros and the cons. Schiaparelli's period of 224·7 days (ratified by himself in 1895) is supported by Perrotin's observations both at Nice and Mont Mounier; by Tacchini's at Rome, Cerulli's at Teramo, and Mascari's at the complementary establishments of Catania and Mount Etna; while Niesten, Trouvelot, Villiger, Stanley Williams, and Flammarion, all under some disadvantage as regards climate, aver that the debated gyration is performed in "about" 24 hours. Now, in the first place, a period of 24 hours is in itself open to suspicion, since all delicate observations are liable to be affected by diurnal atmospheric variations; in the second, it is mainly, if not entirely, based upon supposed

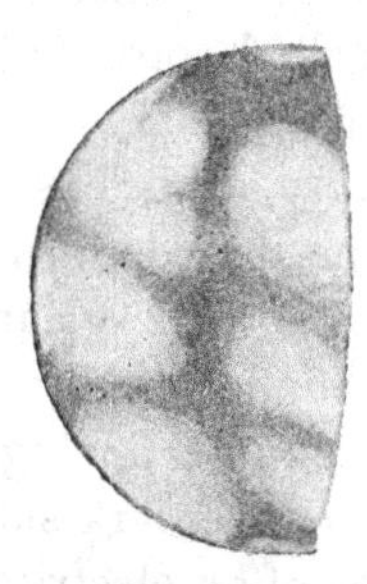

FIG. 9.—*Venus, from a drawing by Mascari.* (*Nature*, February 20, 1896.)

changes in almost evanescent shadings, while the long period of 224·7 days has been derived fundamentally from the immobility relative to the terminator, of definite and permanent topographical features. The perfect roundness of the disc of Venus affords independent proof of extremely slow rotation.

Spectroscopic evidence may before long become available. The quantity to be measured by the exquisite method of line-displacements is, indeed, at the most extremely small. The equatorial velocity of Venus would, with the 24-hour period, but slightly exceed a quarter of a mile a second; but this effect being doubled by reflexion from the planet, and doubled again by juxtaposition of light from its east and west limbs, could probably be made distinctly perceptible. In the negative case, the value of the support lent to the long-period hypothesis can only be appraised by the degree of refinement attained in the research.

The "long-period hypothesis" has, however, almost ceased to need such support. Schiaparelli's facts are inconsistent with any other; and they are scarcely controvertible. They have besides, as in the case of Mercury, been verified by Mr. Lowell's recent observations. Assuming, then, its truth, we may consider what it implies. Since the rotation and revolution of Venus synchronise, she always looks inwards toward the sun, perpetual day reigning on one hemisphere, perpetual night on the other. And these regulations are much more strictly conformed to than on Mercury. For the orbital motion of Venus is so nearly uniform that libratory effects count for very little. The equatorial breadth of the libration-zones, where light alternates with darkness, is only thirty-three miles. On the other hand, the atmospheric diffusion of sunshine is a powerful illuminating agency. The meteorology of the planet presents great difficulties. Its conditions are so remote from our experience that we can barely sketch out their results. The most obvious of these is the vehement aerial circulation which must proceed without ceasing between the hemisphere upon which the sun never rises and the hemisphere upon which the sun never sets. We should expect it to be accompanied by agitated conflicts of winds, and surgings of the atmosphere from its lowest to its highest strata, betrayed by rendings of the brilliant condensation-canopy, by the rapid transport of torn scuds, and wheeling vortices of clouds. But nothing of all this is telescopically visible. The aspect of the morning star suggests serenity rather than interior tumult.

One of the most remarkable instances of persistent optical illusion refers to a supposed satellite of Venus. It was first seen by Fontana at Naples in 1645; it was last seen by Horrebow at Copenhagen in 1768; and the intermediate observations were numerous, usually careful, and apparently authentic. Yet the body, of which they affirmed the existence, was purely fictitious; and it is a suggestive circumstance that it never ventured into the field of view of an achromatic lens.

Comparing the two planets nearest to the sun, the first spontaneous impression is of astonishment at their unlikeness. One travels in an almost circular, the other in a highly eccentric orbit. One possesses a dense and extensive atmosphere; the other is barely gauze-clad, and is hence exposed to almost unmitigated extremes of temperature, while the conformation of its solid surface is left open to telescopic scrutiny, impeded only by the inconvenient glare of the sun. That surface is of a reddish hue, and absorbs more than four-fifths of the light with which it is flooded; the disc of Venus being, on the contrary, of a dazzling whiteness, and little less reflective than a summer cloud. Yet these two globes, so dissimilar individually, have apparently had the same destiny prepared for them. Deprived of all but a remnant of their rotation by the frictional resistance of sun-raised tides, they were debarred from the production of satellites, and subjected to what we, in our ignorance, might be apt to call fantastic climatal conditions. With due reserve it may be added that they have thus apparently been rendered unfit to be the abodes of highly developed organisms. Why this has been so ordained we are unable to conjecture; we must wait to know.

CHAPTER V.

THE EARTH AND MOON.

THE earth occupies a critical position in the solar system. Its greater distance from the sun preserved it from the fate of Mercury and Venus. The influence of solar tidal friction fell short of predominance over the terrestrial future. All that it could do was to defer to the latest possible moment (so to speak) the separation of the moon, the comparatively large size of which was doubtless due to this postponement. For a viscous body, such as the earth must then have been, can bear much more rotational strain than a less coherent mass; but when the strain comes to be relieved, the needful sacrifice of material is proportionally greater. The process of fission, instead of being a mere incident, becomes a catastrophe. The most violent explosions are precisely those which are longest delayed.

Had the earth then been situated a few millions of miles nearer to the sun there would have been, so far as we can see, no moon; and the terrestrial day and year would have been of equal length. This equalisation was rendered impossible by lunar influence.[1] We are indebted to our satellite for the alternations of day and night which make life possible. How this came about is quite clear upon some brief consideration. Lunar tides are now about three times more effective than solar tides, and at their origin the disproportion was enormous. Their power might be called exclusive. Now, how was that power exercised? Primarily, in compelling an agreement between the duration of the month and day—that duration, to begin with, being of only a few hours. The day might, and in the long run did, fall short, but it could not possibly get ahead of the month. Hence the earth's rotation

[1] This was in principle suggested by Proctor in "The Old and New Astronomy."

was for ages protected against the destructive agency of solar tidal friction. By the time that the moon left it, as it were, to take care of itself, the plastic stage, during which alone rapid change could take place, had passed, and the earth was solid and secure.

Thus, the axial rotation of our planet in twenty-four sidereal hours is the outcome of a delicate balance of relations established in the "deep backward and abysm of time." Its shape matches, or has accommodated itself to the period, which has perhaps not varied much since the epoch when interior fires were first banked in by the formation of a rigid crust. The compression of rotating globes is so connected with the quickness of their spinning that one can be calculated from the other; and the earth's theoretical compression, or ellipticity, is found to be practically identical with its measured ellipticity of about $\frac{1}{293}$. Its mean diameter is 7,927 miles; the equatorial is 26 miles longer than the polar diameter; so that the globe is belted with a protuberance, 13 miles high, corresponding to the excess of centrifugal force at the Equator.

The heat by which it was originally maintained in a liquid condition is still in process of dissipation. A small part escapes year by year, but enough remains to keep the earth *alive* for ages to come. Were the supply exhausted, the oxygen of our air, and the water forming our oceans, would be rapidly absorbed, chemically and mechanically, and with them, vitality should disappear. Volcanic action, in some of its many forms, is accordingly a condition of existence. One unmistakable symptom of central fires still glowing is the increase of subterranean temperature. It averages one degree Fahrenheit for fifty-five feet of descent. Below two miles then, water can only remain liquid through the compulsion of the overlying strata, the slighest relaxation of which occasions it to flash explosively into steam; the devastating power of "super-heated" water being one of the chief causes of volcanic outbreaks. The growth of temperature downward cannot be supposed to proceed indefinitely; otherwise, a fabulous thermal state would be reached long before we got near the core of the globe; but the region of maximum heat depends upon an unknown quantity—that is, the lapse of time since the antique lava-globe began to crust over. Assuming it to be fifty million years, Lord Kelvin showed that the limiting temperature of about 5,400° F. is located not more than fifty miles from the surface. But 5,400° approaches the temperature of the electric arc, at which there is an all but universal vaporisation of material substances, and rocks liquefy while comparatively cool. Diabase, for instance, a typical basalt, is completely fluid at 2,200° F. On the other hand, the pressure at 50 miles beneath the earth's surface is of inconceivable power; and it is employed in resisting the expansive tendency of heat. The condition of matter subjected to these opposing and potent influences we are unable to divine, and have no means of ascertaining. We do, however, know from the results of various astronomical lines of enquiry that the earth is effectively as rigid as steel. Its mean density is about five and a half times that of water, the entire globe being more than twice as heavy as if made of the ordinary surface rocks. This, however, is not surprising, since oxygen enters largely into the composition of the exterior strata, while the subjacent materials are likely to be in large measure metallic.

The epoch of the earth's superficial solidification has again, quite lately, been under discussion. "The subject," Lord Kelvin wrote, "is intensely interesting. I would rather know the date of the *Consistentior Status* than of the Norman Conquest; but it can bring no comfort in respect to the demand for time in palæontological geology. Helmholtz, Newcomb, and another (Kelvin) are inexorable in refusing sunlight for more than a score, or a very few scores of millions of years."[1]

Improved data having been substituted, the problem was solved anew, with the result of very notably diminishing the "age of the earth." It is for the present fixed at twenty-four

[1] *Nature*, vol. li., p. 227.

million years, and upon such strong evidence as to "throw the burden of proof upon those who hold to the vaguely vast age derived from sedimentary geology." [1]

The earth is the largest of the terrestrial planets; and it is specifically the heaviest of all the planets. Its compactness is more likely to be a consequence of a particular relation between internal temperature and pressure, than of a difference in chemical constitution.

The mass of its atmosphere can be directly determined. We have only to look at a barometer in order to gain the information that our "cloud of all-sustaining air" weighs as much as a universal ocean of mercury thirty inches in depth. The corresponding depth of air, were it of the same density throughout, would be nearly five miles. But it is *not* of the same density throughout. With each three and a half miles of ascent, atmospheric pressure is halved; and the interval is lessened by making due allowance for decrease of temperature upwards. To the succession of these tenuous strata, no definite end can be assigned. The duration of twilight shows that, above forty-five miles, they cease to reflect light; yet meteors can be set ablaze at heights up to 120 miles, through the resistance offered to their motion by air reduced to $\frac{1}{250,000,000,000}$th its density at sea-level!

The cloud-bearing capability of the atmosphere has only of late been fully recognised. Ordinary cirrus float about five miles high. On December 4, 1894, an aeronaut, Dr. A. Berson, passed right through a bank of them at an altitude of five and a half miles, and was able to verify by actual contact their composition out of snow-flakelets.[2] But since 1885, still more delicate kind of floating formation has come within our acquaintanceship. "Luminous night-clouds" were first noticed by Ceraski; they have been systematically studied by O. Jesse of Berlin.[3] They appear long after sunset, be-

[1] Kelvin, *Nature*, p. 440; Clarence King, *American Journal of Science*, January, 1893.

[2] *Ciel et Terre*, 16th March, 1895.

[3] *Himmel und Erde*, Feb., 1889; *Astr. Nach.*, No. 3347; A. Battandier, *L'Astronomie*, 1894.

tween May and July, and derive their silvery radiance from the sun-rays which their elevated situation enables them to intercept, while all below is wrapt in darkness. Their height has been determined, from the comparison of photographs taken simultaneously at different places, to average fifty-one miles, and to range from fifty to fifty-four miles. They are an entirely new order of phenomenon.

This globe upon which we dwell is a great magnet. Its directive action upon the compass sufficiently proves the fact. But it is a magnet probably only by virtue of the electric currents which course round it. And since these currents originate from diverse interacting causes, the laws of terrestrial magnetism are necessarily complex. They are conditioned, yet not prescribed by the earth's rotation. The magnetic and geographical systems of co-ordinates approximate, but by no means coincide. The former is, indeed, both complex and variable.[1] The inclination, or "dip," of the needle does not vary in the same way as the declination, or horizontal position. There are two points on the earth's surface, called "poles of verticity," where a magnetic needle, freely swung, points vertically downward. One is situated in the arctic peninsula Boothia, the other on the antarctic continent within a few hundred miles of Mount Erebus. An intermediate line where the needle poises itself horizontally, corresponds roughly with the geographical equator. Each hemisphere contains besides two centres of maximum force, by the joint action of which magnetic deviations from true north and south are determined. Their mutual relations are highly intricate. The North American focus is stationary, the Siberian focus oscillates. Their relative and absolute intensity is probably also subject to fluctuations. Hence the inconstancy of magnetic directive influences. The variation of the compass varies.

It varies hour by hour, as well as year by year. The needle performs a diurnal oscillation, reaching an eastward maximum about eight A.M., and a corresponding westward

[1] Balfour Stewart: "Ency. Brit.," vol. xvi. pp. 164, 165.

maximum towards four P.M. Moreover, the range of this vibration increases concordantly with the growth of spotted area upon the sun, and falls off again as spots diminish (see Fig. 2). The cosmical relations of terrestrial magnetism are emphasised by the obvious connexion between a disturbed state of the sun and the occurrence of "magnetic storms." During these crises, the smooth progression and regression of the needle are superseded by violent and irregular movements. The photographic tracing in which they are recorded presents only a series of lawless zigzags; earth-currents are set up; telegraph-wires transmit messages without batteries; and the skies are at night draped with auroral streamers.

Auroræ are possibly a survival of our planet's original self-luminosity. If so, their dependence upon the terrestrial magnetic system is highly significant. They obey the magnetic period, they accompany magnetic disturbances, they illuminate magnetic lines of force. That they are immediately caused by electrical discharges in the high vacua of our upper air is no longer doubtful. In these latitudes, the auroral arch and crown are formed at a height of ninety to one hundred miles, in (about) $\frac{1}{1,000,000,000}$th of an atmosphere; but in the polar regions they approach much nearer to the earth. There, indeed, they more usually assume the form of a curtain, undulating in luminous folds, and traversed by vertical electric currents. That they are so traversed is demonstrated by the behaviour of the magnetic needle, the deviations of which change their sign as the auroral drapery crosses the zenith.[1] Auroræ seem to be confined to two zones of the earth, which, like the sun-spot zones, approach the equator as the solar cycle advances. Their frequency in temperate regions corresponds, accordingly, to a scarcity in high latitudes. The auroral spectrum consists of a number of bright rays, one of which is invariably present, and seems to be essential and fundamental. Its origin is unexplained.

The velocity of the earth in its orbit exceeds more than

[1] A. Paulsen: *Ciel et Terre*, 1 Juillet, 1895, p. 202.

sixty times that of a cannon-ball just leaving the muzzle of an eighty-ton gun. In other terms, the third planet from the sun travels at an average rate of $18\frac{1}{2}$ miles per second. Its albedo has been estimated—probably under-estimated—at 0·30. This would leave 70 per cent. of the solar emanations striking the upper surface of its atmosphere available for interior consumption. Most of this supply is absorbed or scattered in the atmosphere. The proportion sent back to space after reflection from the actual terrestrial surface must be extremely small. Very little topographical detail could be made out by telescopic scrutiny from the moon or Venus. At the most, the trend of some great mountain ranges, such as the Andes and Himalayas, and a dozen snow-clad peaks, could be visible. No sign of the teeming organic life brought forth by mother earth could be detected from without.

The more we know of the moon, the less inviting, from our point of view as animated beings, it appears. It is a harsh and inhospitable world, from which vital possibilities, if they were ever present, have plainly long ago departed. The diameter of our satellite is 2,162 miles. Its disc, so far as the most exact measurements tell, is perfectly round. This in itself indicates a slow rotation; and even casual observations suffice to show that they relate to only one lunar hemisphere. Rotation and revolution here again synchronise. In 27 days 8 hours (nearly), the moon executes one circuit of the earth, and one gyration on its axis. The coincidence was brought about in remote ages by the power of terrestrial tidal friction. The averted hemisphere does not, however, remain wholly invisible. Two-elevenths of it are, by the effect of librations, both in longitude and latitude, brought piecemeal into view. But the additional "lunes," thus thrown open to glimpses round the corner, are greatly foreshortened.

The area of the moon is somewhat less than one-thirteenth that of the earth. Yet room could be found there for the entire British Empire, with six million square miles to spare. Its volume is $\frac{1}{49}$th, its mass $\frac{1}{82}$th, the volume and mass of the earth. Hence the lunar materials are less dense than the terrestial in the proportion of about three to five. But this may be because they are under comparatively slight pressure.

At the moon's surface, gravity possesses only one-sixth its power here, so that a stone thrown upward with equal force would reach a six-fold height. Further, a projectile shot straight from our satellite with a velocity of one and a half miles a second would never return, while a speed of seven miles a second is just controllable by the earth, to say nothing of the immense efficacy of her dense atmosphere in hindering escape from her precincts. No terrestrial bomb, it may therefore be safely asserted, has ever been hurled into space, although volcanic ejecta may very well, in past ages, have made their way hither from the moon.

But lunar volcanoes are no longer active. Only their remains stand as records of a fiery past. In guiding a telescope across the scarred face of our satellite we seem to traverse a volcanic charnel-house. The evidence of ancient seismic action on the moon is overwhelming. Its surface is pitted all over with cones and craters. Nearly 33,000 are marked on Schmidt's map, and the list is very far from being exhaustive. The resulting chiaroscuro is obvious to the naked eye. Dante tried to explain it in the "Divina Commedia"; Galileo detected its cause and manner of composition. The chief facts about it are these.

The general albedo of the lunar surface is 0·17; but portions of the disc are as obscure as basalt or obsidian, while isolated spots glitter like snow-peaks. The former are usually admitted to be the oldest of conspicuous lunar formations, the latter to be comparatively recent. The dusky spaces too, are dead levels, if not depressions; they were formerly taken for seas, and retain the name of "Maria." One "ocean," extending over two million square miles, is included amongst them. This is the "Oceanus Procellarum" (see Fig. 10), which is five times larger than its nearest rival, the "Mare Nubium." The late Mr. Gwyn Elger regarded the lunar "seas" as lava outflows, by which certain earlier forma-

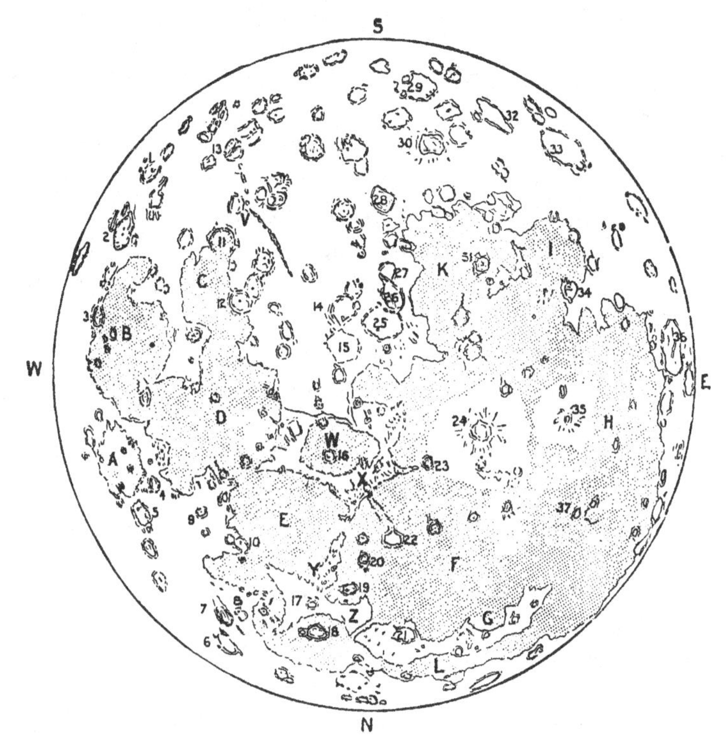

FIG 10.—*Map of the Moon.* (From Fowler's "Telescopic Astronomy.")

1. Furnerius
2. Petavius
3. Langrenus
4. Macrobius
5. Cleomedes
6. Endymion
7. Altas
8. Hercules
9. Römer
10. Posidonius
11. Fracastorius
12. Theophilus
13. Piccolomini
14. Albategnius
15. Hipparchus
16. Manilius
17. Eudoxus
18. Aristotle
19. Cassini
20. Aristillus
21. Plato
22. Archimedes
23. Eratosthenes
24. Copernicus
25. Ptolemy
26. Alphonsus
27. Arzachel
28. Walter
29. Clavius
30. Tycho
31. Bullialdus
32. Schiller
33. Schickard
34. Gassendi
35. Kepler
36. Grimaldi
37. Aristarchus

A. Mare Crisum
B. ,, Fercunditatis
C. ,, Nectaris
D. ,, Tranquilitatis
E. ,, Serenitatis
F. Mare Imbrium
G. Sinus Iridum
H. Oceanus Procellarum
I. Mare Humorum
K. ,, Nubium
V. Altai Mountains
W. Mare Vaporum
X. Apennine Mountains
Y. Caucasus ,,
Z. Alps

tions were all but obliterated. M. Suess explains them as areas where the primitive thin "slag-crust" re-melted. To the same category belong the vast "bulwark plains," the ramparts enclosing which are of so wide a sweep as to be, not merely "hull-down," but completely invisible to an imaginary spectator placed at their centres. Yet Pelions by the dozen are tumbled upon Ossas for their construction, with here and there an Olympus flung on the top. Typical examples are Ptolemaeus, 115 miles across; and Plato (near the Northern Pole), "sixty miles in diameter, with its bright border and dark steel-grey floor."[1]

The bottoms of lunar craters and "circuses" are nearly always depressed—sometimes thousands of feet—below the general level. Thus, the central peak of the great crater Copernicus towers to 11,300 feet above the depressed plain from which it rises, but surmounts by only 2,600 feet the average level of the moon.

Successive stages of activity have left ineffaceable marks upon this now stereotyped page. Groups of immense craters mutually encroach, and seem to have been scooped out of each others' flanks, like Kilauea from Mauna Loa; craters occur within craters, as Vesuvius inside the broken rampart of Somma; and the most recent are invariably the deepest and steepest. Cup-shaped depressions or "crater-pits" are innumerable; they result, according to Suess's theory,[2] each from a single explosion, the bursting of a "big bubble" of gas in a cooling lava-field. Mountain ranges are profusely strewn with them. These lunar Alps and Apennines appear to be as unmistakably igneous in their origin as Tycho or Aristarchus. They are colossal slag-walls. There are apparently no sedimentary deposits upon the moon. Aqueous action had no concern with its geological history. Yet on the earth water is essential to the production of volcanic phenomena. If they are to be developed without it, M. Angelot concludes, it must be by explosive escapes from

[1] Elger: "The Moon," p. 73.

[2] "Publications, Astronomical Society of the Pacific," vol. vii., p. 144.

solidifying materials, of gases absorbed by them when in a state of fusion.

The mountains of the moon are much higher, proportionally, than the summits of the Hindu-Kush, or of the Himalayas. Mount Everest, reduced to the lunar scale, would be a modest elevation of 8,200 feet; while pinnacles in the lunar Apennines spring up to 22,000 feet, and crater-peaks of eighteen or twenty thousand abound. The disparity is scarcely surprising when it is remembered that there the convulsive throes of cooling were restrained by gravity reduced to one-sixth the power it exerts here.

Among the puzzles of selenography are the objects termed respectively "rills" and "rays." The former are very numerous. Considerably more than a thousand of them have been mapped or photographed. They resemble the cañons of Colorado. Some few run to 150 miles; most are a couple of miles wide, and above a quarter of a mile deep. Their volcanic origin cannot be doubted. The "rays" diverge in extensive systems from such huge ring-craters as Tycho and Copernicus. They cast no shadows, and come out best at full moon, circumstances suggestive of their being immemorial lava-streams bleached by the chemical action of fumes from the interior. The whiteness of Aristarchus has been similarly explained; but accumulations of pumice and snow-like volcanic ashes perhaps enhance the effect. The flashing back by this wonderful peak, of earthshine at determinate angles of illumination, has often counterfeited the vivid glow of actual eruptions. Their possibility, however, belongs to the past. Nor have any of the rumoured alterations in lunar topography, which from time to time excited interest and raised controversy, made good their footing as solid facts. Agencies of change are certainly there, in tidal strains and alternations of temperature, but they work very slowly. There is no erosion by air or water; no grinding by ice; no transport of materials. Repose reigns apparently undisturbed. Lunar landscapes exhibit abrupt transitions from the blinding glare of crude sunlight to the blackness of absolute shadow.

Their aspect excludes any but the thinnest possible atmospheric remnant. To all intents and purposes, the moon is an airless globe. Occultations of stars afford a very refined test of this condition; and their instantaneousness alone suffices to demonstrate its reality. Spectroscopic evidence is to the same effect. Dr. Huggins watched, January 4, 1865, a *prismatic* occultation of the small star, ϵ Piscium. Had there been the slightest inequality of dispersion or absorption at the moon's limb, it could not have failed to be perceived. There was none. The spectrum remained unaffected, and vanished abruptly, all the colours together. And moonlight, analysed by the most powerful apparatus, varies not an iota from sunlight. It is reflected without the smallest selective change.

The absence of water is equally well attested. There are no river-beds to be seen, no rounded surfaces, no alluvial plains. A mosquito could not find a moist corner to lay its eggs in. There is nothing to show that this was otherwise in any past age, although it is not improbable that the lunar rocks contain large volumes of oxygen once free. As regards the earth, we can entertain no doubt that a goodly proportion of its original atmosphere and oceans is now permanently lodged in its bedded crust. But the geological histories of the earth and moon probably diverged from the first.

Indeed water, as such, could probably not exist upon the moon's surface. It would promptly take the form of ice. Professor Langley has shown that the temperature prevailing there, under vertical sunshine, is about that of frost; while it sinks, during the moon's long night of fourteen days, almost to absolute zero. This frigid state is due to the absence of atmospheric protection, leaving heat free to depart into space as fast as it is received. Thus, of the small quantity of heat contained in moonlight, nearly the entire comes to us by mere superficial reflection; a minute residuum only is absorbed previously to being emitted. The distinction is brought into view by comparing the solar and lunar heat-spectra, when moonlight is found to contain longer invisible heat-waves than can be detected in sunlight. Moreover, Professor Frank Very, through his experimental demonstration that the equatorial are slightly hotter than the polar regions, has established the fact of a slight retention of heat by the moon's substance. How slight the retention is, has been proved by Dr. Boeddicker's observations with the Rosse three-foot speculum, showing that, during total eclipses, moon-heat vanishes almost completely. Less than 1 per cent. survives. The thermal phases are not, however, identical with the luminous phases.

The eclipsed moon, on June 10, 1816, is said to have been utterly lost to sight; but, as a rule, with very few exceptions, our satellite traverses visibly the densest part of the earth's shadow. Even during "black eclipses," such as that of October 4, 1884, a dusky spot remains as an index to its locality; while in "red eclipses," the great craters and bulwark-plains can be easily distinguished with an opera-glass. Occasionally, the moon seems turned to blood, and the people cry out in the streets with fear. Such a phenomenon was witnessed by the writer at Florence, February 27, 1877. Its explanation is not difficult. The refractive power of the earth's atmosphere suffices to bring illumination to the lunar disc at the very middle of the shadow-cone. It is shut off from direct solar rays, not from those that are bent into convergence by the lens of our air. That they must be reddened by the process, sunset-effects on the earth tell plainly enough. But when the air is vapour, or dust-laden, and consequently opaque, little light is transmitted, and a scarcely mitigated eclipse ensues. That of 1884 is believed to have been darkened by the out-pourings from Krakatoa. A photograph by Professor Barnard, of the totally eclipsed moon, September 3, 1895, is reproduced in Fig. 11. It was one of a *search-series* for a lunar satellite. None was found; but the question of its possible existence was set at rest.

De la Rue's and Rutherfurd's plan of photographing the moon as a whole is no longer followed. Bit by bit photography, on a large scale, has superseded it. Splendid pictures of individual formations and separate regions have in this way been obtained, both at the Paris and the Lick Observatories; and their microscopic study has given some interesting results; yet it is undeniable that the "chemical retina" cannot here claim its usual superiority. "The best photograph of the moon ever taken," Professor W. H. Pickering

FIG 11.—*Photograph of the Totally Eclipsed Moon. By Professor Barnard. Exposure, 3 Minutes.*

avers,[1] "will not show what can be seen with a six-inch telescope, under favourable atmospheric conditions. For general outlines, for completeness of the coarser detail, and for purposes of future testimony, the photograph evidently stands without a rival; but as regards that which is really most interesting upon the moon—the finer detail and more delicate features—the photograph does not even hint at their existence." One of the most successful specimens of lunar photography forms the frontispiece to this volume. It was taken by MM. Loewy and Puiseux, with the large Coudé equatorial, February 14th, 1894, at 7^h 27^m Paris time, and cannot easily be surpassed in pictorial effect.

Atmospheric agitations are one cause of imperfection in

[1] "Harvard Annals," vol. xxxii., part i., p. 109.

lunar photographs. The eye can seize the instant of exquisite definition; the camera must take what comes. Then the disparities of actinic intensity in the various lunar formations are so wide that, in order to get an ideal picture, a different length of exposure should be given to each. What is enough for a plain—to take an example—is too much for the crater rising from it, or for the rampart enclosing it. Minute irregularities in the following motion of the telescope during the few seconds of exposure occasion further difficulties. A momentary shifting, by half a millimetre, of the image upon the sensitive plate, would suffice to blur the negative seriously, if not fatally. For this, as for several other lines of work, the instrument of the future may be of a type with which the equatorial has little in common. Professor Pickering considers it probable that "a horizontal telescope of three or four hundred feet focus, and twelve to fifteen inches aperture, would give the most satisfactory results. In such a case, it might be found best that the mirror should remain fixed during the exposure, while the plate was given an uniform motion by clockwork."

The suggestion is one among many signs that a revolution in the mounting of telescopes is at hand.

CHAPTER VI.

THE PLANET MARS.

THE furthest terrestrial planet from the sun is Mars, the "star of strength." No other heavenly body, except the moon, is so well placed for observation from our position in space. As a superior planet, it does not merely, like Mercury and Venus, oscillate about the sun, but is best seen when in opposition. It is then "full"; it crosses the meridian at midnight, and is at its least distance from the earth. These occasions recur every 780 days; but they are not all equally favourable. The opposition-distance of the planet varies, owing to the eccentricity of its orbit, from thirty-five to sixty-one million miles; so that the area of the disc is three times larger when a perihelion than when an aphelion-passage coincides with a midnight culmination. Under the best circumstances it is of the apparent dimensions of a half-sovereign 2,000 yards from the spectator.

The diameter of Mars is 4,200 miles; its surface is equal to two-sevenths, its volume to one-seventh those of the earth. But, in consequence of its inferior mean density, nine such spheres would go to make up the mass of our world. The superficial force of gravity on Mars, compared with its terrestrial value, is as thirty-eight to a hundred. A man could leap there a wall eight feet four inches in height with no more effort than it would cost him here to spring over a two-foot fence.

The planet's rotation is performed in 24 hours 37 minutes on an axis deviating from the vertical by 24° 50′. Hence its seasons resemble our own, except in being nearly twice as long, for the Martian year is of 687 days. They are modified, too, by the considerable elongation of the ellipse traversed by Mars, causing a difference of 26½ millions of miles in its greatest and least distances from the sun. These are respectively 155 and 128½ millions of miles, the mean distance being 141½ millions. A polar compression of $\frac{1}{220}$ is just what should be expected from its rotatory speed. When at quadrature, it is plainly gibbous; but our interior position with regard to it makes it impossible that it should ever take the crescent form. Its albedo, according to Zöllner, is 0·26—a figure intimating that sunlight is reflected from no cloud-canopy, but by the soil itself. This atmospheric transparency leaves the door open for researches into the condition of a very curious little world.

The disc of Mars is diversified with three shades of colour—reddish, or dull orange, dark greyish-green, and pure white. The last shows mainly in two diametrically opposite patches.

Each pole is surrounded by a brilliant cap, suggesting the deposition of ice or snow over the chilly spaces corresponding to our arctic and antarctic regions. Nor is this all. Each of the polar hoods shrinks to a mere remnant as the local summer advances, but regains its original size when wintry influences are again in the ascendant. Here, and nowhere else in the planetary system, we meet evidence of seasonal change; and seasonal change is associated with vital possibilities. Again, a globe upon which snow visibly melts, must contain water; hence the green markings cannot but image to our minds seas and inlets sub-dividing continents, the blond complexion of which may be caused by some native peculiarity of the soil. It is in no way connected with vegetation, since it neither fades nor flushes with the advent of spring; and an atmospheric origin is excluded by the circumstance that it becomes effaced by a whitish haze near the limb, just where the densest atmospheric strata are traversed by the line of sight.

The spots on Mars are by no means so sharply defined as lunar craters and *maria;* yet they are fundamentally permanent. Some can be recognised from drawings made over two hundred years ago; and these antique records have served modern astronomers to determine with minute accuracy the rotation-period of the planet. There is accordingly no doubt that "areography" has assured facts to deal with, although the facts are not quite as "hard" as they might be. Continents are somewhat vaguely outlined. Great tracts of them are of an uncertain and variable hue, as if subject to inundations. This peculiarity, thoroughly certified during the favourable opposition of 1892, makes a strong distinction between Mars and the Earth. Terrestrial oceans keep within the limits assigned to them. On the neighbouring planet—as M. Faye observed in 1892—"Water seems to march about at its ease," flooding, from time to time, regions as wide as France. The imperfect separation of the two elements recalls the conditions prevailing during the terrestrial carboniferous era.

The main part of the land of Mars is situated in the northern hemisphere. It covers two-thirds of the entire

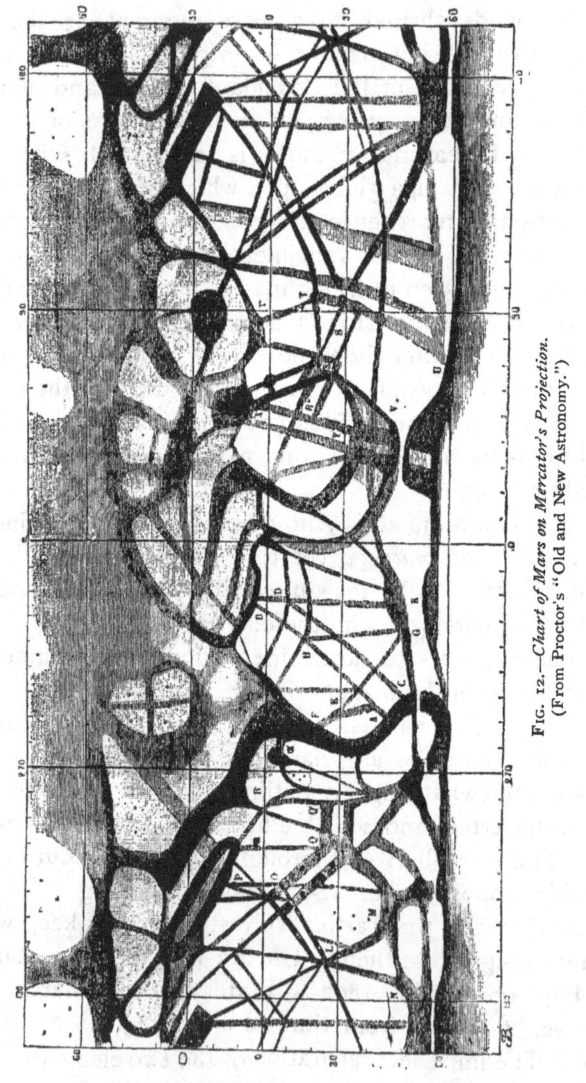

FIG. 12.—*Chart of Mars on Mercator's Projection.*
(From Proctor's "Old and New Astronomy.")

globular surface. Rather than land, indeed, it should be called a network of land and water. Fig. 12, from a chart by Schiaparelli, illustrates the remarkable fashion of their intermixture. The great continental block—so its orange tint declares it to be—is cut up in all possible directions by an intricate system of what appear to be waterways, running in perfectly straight lines—that is, along great circles of the globe—for distances varying from 350 to upwards of 4,000 miles. They are frequently seen in duplicate, strictly parallel companions developing thirty to three hundred miles apart from the original formations. This mysterious phenomenon is evanescent, or rather periodical. Canal-duplication is a recurrent change, depending upon the Martian seasons, and becoming obvious, according to Schiaparelli, chiefly near the equinoxes.

The canals invariably connect two bodies of water; hence they need no locks or hydraulic machinery; their course is on a dead level. The broadest of them are comparable with the Adriatic; those at the limit of visibility, stretching like the finest spider-threads across the disc, have a width of eighteen miles. "The canals," Schiaparelli says, "may intersect among themselves at all possible angles, but by preference they converge towards the small spots to which we have given the name of lakes. For example, seven are seen to converge in Lacus Phoenicis, eight in Trivium Charontis, six in Lunae Lacus, and six in Ismenius Lacus." [1]

These "lakes" evidently form an integral part of the canal system. They resemble huge railway-junctions; and the largest of them—the "Eye of Mars" (Schiaparelli's Lacus Solis)—seems, in Mr. Lowell's phrase, like the hub of a five-spoked wheel. It is depicted in Fig. 13 from a drawing made by Professor Barnard with the great Lick refractor, September 3, 1894. Mr. W. H. Pickering in 1892, and Mr. Percival Lowell in 1894, were amazed at their extraordinary abundance.

"Scattered over the orange-ochre groundwork of the continental regions of the planet," the latter wrote, "are any number of dark, round spots. How many there may be it is

[1] *Astronomy and Astro-Physics*, Nov., 1894, p. 718.

not possible to state, as the better the seeing, the more of them there seem to be. In spite, however, of their great number, there is no instance of one occurring unconnected with a canal. What is more, there is apparently none which does not lie at the junction of several canals. Reversely, all the junctions appear to be provided with spots."

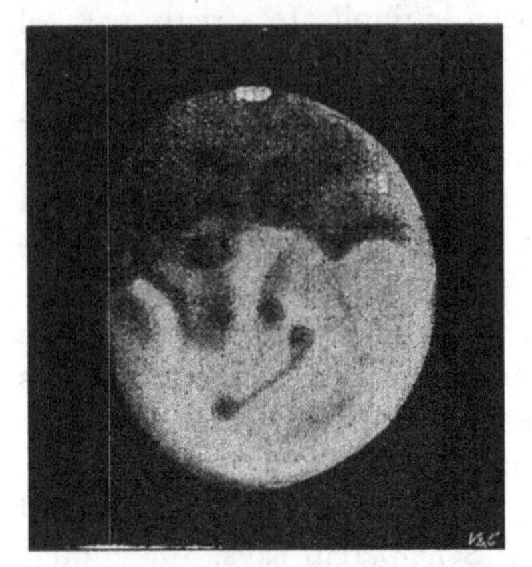

FIG. 13.—*The "Eye of Mars," drawn by Prof. Barnard, with the great Lick Refractor. The southern snow-cap is visible much shrunken by melting.*

Most of these foci are about 120 miles in diameter, and appear most precisely circular when most clearly seen. "Plotted upon a globe," Mr. Lowell continues, "they and their connecting canals make a most curious network over all the orange-ochre equatorial parts of the planet, a mass of lines and knots, the one marking being as omnipresent as the other. Indeed, the spots are as peculiar and distinctive a feature of Mars as the canals themselves."

Like the canals, too, they emerge periodically, and in the same but a retarded succession. They "are therefore, in the first place, seasonal phenomena, and, in the second place, phenomena that depend for their existence upon the prior existence of the canals." [1]

Mr. Lowell terms them "oases" (see Fig. 14), and does not shrink from the full implication of the term.

The most important result of the numerous observations of Mars, made during the oppositions of 1892 and 1894, was the recognition of a regular course of change dependent upon the succession of its seasons. Schiaparelli had long anticipated this result; he is commonly in advance of his time. These changes, moreover, when closely watched, are really

[1] "Popular Astronomy," 1895, p. 347.

self-explanatory. The alternate melting of the northern and southern snow-caps initiates, and to some extent determines them. As summer advances in either hemisphere, the wasting of the corresponding white calotte can be followed in every minute particular. "The snowy regions are then seen to be successively notched at their edges; black holes and huge fissures are formed in their interiors; great isolated fragments many miles in extent stand out from the principal mass, dissolve, and disappear a little later. In short, the same divisions and movements of these icy fields present themselves to us at a glance that occur during the summer of our own arctic regions."[1]

Indeed, glaciation on Mars is much less durable than on the earth. In 1894, the southern snow-cap vanished to the last speck 59 days after the solstice; and the remnant usually left looks scarcely enough to make a comfortable cap for Ben Nevis. An immense quantity of water is thus set free. The polar seas overflow; gigantic inundations reinforced, doubtless, from other sources, spread to the tropics; Syrtis regions of marsh or bog deepen in hue, and become distinctly aqueous; canals dawn on the sight, and grow into undeniable realities. We seem driven to believe that they discharge the function of flood-emissaries.

Mr. Lowell does not hesitate to pronounce them of artificial formation, and, on that large assumption, the purpose of their connexion with his "oases" becomes transparently clear. They bring to these Tadmors in the wilderness the water supply by which they are made to "blossom as the rose." The junction-spots, we are told, do not enlarge when the vernal freshet reaches them; they only darken through the sudden development of vegetation. These circular "districts, artificially fertilised by the canal system," are strewn broadcast over vast desert areas, the orange-ochreous sections of Mars, covering the greater part of its surface, but deep buried in the millennial dust of disintegrated red sandstone strata.

[1] *Astr. Nach.*, No. 3271 (Schiaparelli).

"Here, then," Mr. Lowell remarks,[1] "we have an end and reason for the existence of canals, and the most natural conceivable—namely, that the canals are constructed for the express purpose of fertilising the oases. When we consider the amazing system of the canal lines, we are carried to this conclusion as forth-right as is the water itself; what we see

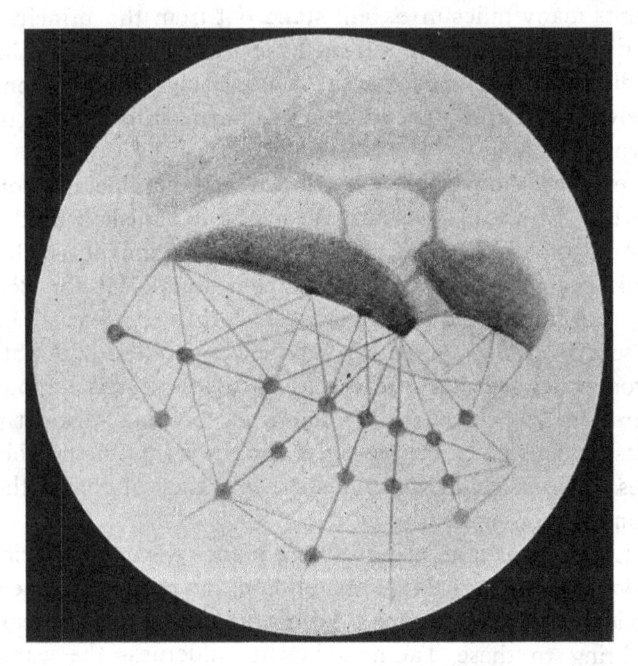

FIG. 14.—*The Oases of Mars. Drawn by Percival Lowell.* (From "Popular Astronomy," April, 1895.)

being not the canal itself, indeed, but the vegetation along its banks."

The idea that we see the water only by its effects along the shores of these prodigious troughs, originated with Professor W. H. Pickering. It is strikingly illustrated by the aspect of rivers from a balloon. Thus the Rhine, as M. Flammarion

[1] "Popular Astronomy," vol. i., p. 348.

attests,[1] seen from a perpendicular altitude of 8,000 feet, shows like a green thread drawn in the midst of a ribbon of meadow. The Martian canals, it is suggested, correspond to the "ribbon of meadow."

The hypothesis is seductive, but should not be hastily adopted. It gives no account of the doubling of the canals, yet the process takes place on a grand scale, at determinate epochs, and under fairly well ascertained conditions. It undoubtedly belongs to the series of vernal changes going forward upon the planet, and is accomplished with amazing rapidity. A single canal may be transformed into a double canal within twenty-four hours, and that simultaneously along its whole course. The two stripes, so curiously substituted for one, "run straight and equal with the exact geometrical precision of the two rails of a railroad."[2] The tendency is shared by the lakes or "oases." "One of these," we learn from the same authority, "is often seen transformed into two short, broad dark lines parallel to one another, and traversed by a yellow line."

This singular principle of subdivision offers at present no hold for profitable speculation. Schiaparelli trusts to the "courtesy of nature" for some ray of light by which, in the future, to penetrate the mystery; but wisely deprecates recourse being had to the intervention of intelligent beings. Such arbitrary modes of dealing with perplexing problems constitute, as he says, a grave obstacle to the acquisition of just notions concerning them. They raise prepossessions by which the progress of genuine research is impeded.

The proportion of water to land is much smaller on Mars than on the earth. Only two-sevenths of the disc are covered by the dusky areas, and of late the aqueous nature of some, if not all of these, has been seriously called in question. Professor Pickering was convinced by his observations, in 1892 and 1894, "that the permanent water area upon Mars, if it exist at all, is extremely limited in its dimensions."[3]

[1] *Scientific American*, Feb. 29, 1896.

[2] Schiaparelli: *Astronomy and Astro-Physics*, Nov., 1894, p. 720.

[3] *Astronomy and Astro-Physics*, August, 1894, p. 554.

He estimated it at about half the size of the Mediterranean. Professor Schaeberle is similarly incredulous. If the dark markings are seas, he asks, how explain the irregular gradations of shade in them?[1] How, above all, explain their apparent intersection by well-marked canals? Professor Barnard, observing with the Lick thirty-six inch in 1894, discerned on the Martian surface an astonishing wealth of detail, "so intricate, small, and abundant, that it baffled all attempts to properly delineate it."[2] It was embarrassing to find these minute features belonging more characteristically to the "seas" than to the "continents." Under the best conditions, the dark regions lost all trace of uniformity. Their appearance resembled that of a mountainous country, broken by cañon, rift, and ridge, seen from a great elevation. These effects were especially marked in the "ocean" area of the hour-glass sea.

Evidently the relations of solid and liquid in that remote orb are abnormal; they cannot be completely explained by terrestrial analogies. Yet a series of well-attested phenomena are intelligible only on the supposition that Mars is, in some real sense, a terraqueous globe. Where snows melt there must be water; and the origin of the Rhone from a great glacier is scarcely more evident to our senses than the dissolution of Martian ice-caps into pools and streams.

The testimony of the spectroscope is to the same effect. Dr. Huggins found, in 1867, the spectrum of Mars impressed with distinct traces of aqueous absorption, and the fact, although called in question by Professor Campbell of Lick, in 1894, has been re-affirmed both at Tulse Hill and at Potsdam. That clouds form and mists rise in the thin Martian air, admits of no doubt. During the latter half of October, 1894, an area much larger than Europe remained densely obscured. Whether or no actual rain was at that time falling over the Maraldi Sea and the adjacent continent, it would be useless to conjecture. We only know that with

[1] "Publ. Astro. Soc. of the Pacific," vol. iv., p. 196.

[2] *Monthly Notices*, vol. lvi., p. 166.

the low barometric pressure at the surface of Mars, the boiling point of water must be proportionately depressed (Flammarion puts it at 115° Fahrenheit), which implies that it evaporates rapidly, and can be transported easily.

If the Martian atmosphere be of the same proportionate mass as that of our earth, it can possess no more than one-seventh its superficial density. That is to say, it is more than twice as tenuous as the air at the summits of the Himalayas.[1] The corresponding height of a terrestrial barometer would be four and a half inches. Owing, however, to the reduced strength of gravity on Mars, this slender envelope is exceedingly extensive. In the pure sky scarcely veiled by it, the sun, diminished to less than half his size at our horizons, probably exhibits his coronal streamers and prominences as a regular part of his noontide glory; atmospheric circulation proceeds so tranquilly as not to trouble the repose of a land "In which it seemeth always afternoon"; no cyclones traverse its surface, only mild trade-winds flow towards the equator to supply for the volumes of air gently lifted by the power of the sun, to carry re-inforcements of water-vapour north and south. Aerial movements are, in fact, by a very strong presumption, of the terrestrial type, but executed with greatly abated vigour.

Brilliant projections above the terminator of Mars were first distinctly perceived at the Lick Observatory in 1890. They have been re-observed at Nice, Arequipa, and Flagstaff (Mr. Lowell's Observatory), coming into view, as a rule, when circumstances concur to favour their visibility. They strictly resemble lunar peaks and craters, catching the first rays of the sun, while the ground about them is still immersed in darkness;[2] and Professor Campbell[3] connects them with "mountain chains lying *across* the terminator of the planet," and in some cases possibly snow-covered. He calculates their height

[1] Campbell: "Publ. A. S. P.," vol. vi., p. 273.

[2] *Ibid.*, vol. ii., p. 248.

[3] *Ibid.*, vol. vi., p. 110.

at about ten thousand feet. Their presence was unlooked-for, since a flat expanse is a condition *sine quâ non* for the minute intersection of land by water, which seems to prevail on Mars.

Although the sun is less than half as powerful on Mars as it is here, the Martian climate, to outward appearance, compares favourably with our own. Polar glaciation is less extensive and more evanescent, and little snow falls outside the arctic and antarctic regions. Yet the theoretical mean temperature is minus 4° C., or 61° of Fahrenheit below freezing. This means a tremendous ice-grip. The coldest spot on the earth's surface is considerably warmer than this cruel average. Fortunately, it exists only on paper. Some compensatory store of warmth must then be possessed by Mars, and it can scarcely be provided by its attenuated air. Possibly, internal heat may still be effective, and we see exemplified in Mars the geological period when vines and magnolias flourished in Greenland, and date-palms ripened their fruit on the coast of Hampshire.

The climate of Mars, according to Schiaparelli,[1] "must resemble that of a clear day upon a high mountain. By day a very strong solar radiation hardly at all mitigated by mist or vapour; by night a copious radiation from the soil towards celestial space, and hence a very marked refrigeration; consequently, a climate of extremes, and great changes of temperature from day to night, and from one season to another. And as on the earth, at altitudes of from 17,000 to 20,000 feet, the vapour of the atmosphere is condensed only into the solid form, producing those whitish masses of suspended crystals which we call cirrus-clouds, so in the atmosphere of Mars it would be rarely possible to find collections of cloud capable of producing rain of any consequence. The variation of temperature from one season to another would be notably increased by their long duration, and thus we can understand the great freezing and melting of the snow, renewed in turn at the poles at each complete revolution of the planet round the sun."

[1] *Astronomy and Astro-Physics*, October, 1894, p. 640.

But the anomalies in the Martian domestic economy cannot thus easily be removed, and the only safe conclusion is Flammarion's, that "the general order of things is very different on Mars and on the earth."

The German astronomer, Mädler, searched in 1830 for a Martian satellite, and although his telescope was of less than four inches aperture, he satisfied himself that none with a diameter of as much as twenty-three miles could be in existence. As it happened, he was right. The pair of moons detected by Professor Asaph Hall with the Washington twenty-six refractor, August 11 and 17, 1877, are unquestionably below that limit of size. Neither of them can well be more than ten miles across. Their names, "Deimos" and "Phobos," are taken from the *Iliad*, where Fear and Panic are introduced as attendants upon the God of War. Deimos revolves in 30 hours and 18 minutes at a distance of 14,600 miles from the centre of Mars. And, since the planet rotates in 24 hours 37 minutes, the diurnal motion of the sphere from east to west is so nearly neutralised by the orbital circulation of the satellite from west to east that nearly 132 hours elapse between its rising and its setting. During the interval, it changes four times from new to full, and *vice versâ*. Professor Young estimates that Mars receives from it when full only $\frac{1}{1200}$th of full moonlight.

Phobos is more effective in illumination, both because it is larger, and because it is less distant. At the Martian equator, its brightness is equal to $\frac{1}{60}$th that of our moon, but beyond 69° of latitude it is permanently shut out from view by the curvature of the globe. This exclusion is an effect of its uncommon closeness to its surface, the interspace being only 3,700 miles, while its distance from the centre is 5,800. Moreover, the period of Phobos being only 7 hours 39 minutes, or less than $\frac{1}{3}$ the time of rotation of its primary, it rises in the west, sets in the east, and courses across the heavens in 11 hours, during which interval it accomplishes one entire cycle of its phases, and gets through half another. This is an unique phenomenon, and points to an unique origin for the little moon. No other known satellite revolves more quickly than its primary rotates, and the discovery of the fact has dealt a fatal blow to Laplace's method of planetary evolution. Were Phobos capable of raising any appreciable tide on Mars, its frictional effects would hence be of an opposite character to those of other tidal waves; and instead of being pushed outward, it would be drawn inward, and finally precipitated upon the planet. But it derives safety, on the one hand, from its small mass; on the other, from the insensibility of Mars to tidal action. The satellite is incapable of exerting the required influence; the planet is not in a state to respond to it, were it exerted. For the configuration of land and water upon its surface is such as effectually to prevent the flow of tides, were the compulsive power a thousand-fold that possessed by its pair of diminutive satellites.

CHAPTER VII.

THE ASTEROIDS.

BETWEEN the orbits of Mars and Jupiter is interposed a huge gap. On one side of it lie the terrestrial planets; on the other, the "major planets"—orbs belonging to a different order, both as to magnitude and as to constitution. The hiatus marks a change of front in planetary development, and its existence gravely compromises the symmetry of the solar system. Its inconsistency with Bode's law of planetary distances long troubled investigators. A member of the series had somehow dropped out; it was sought for under the form of a planet, and found, apparently, as its disintegrated constituents. The discovery of Uranus nearly at the distance indicated for it by the law roused astronomers to the necessity for a systematic chase; but before their organisation had got into full working order, the missing occupant of the vacant zone presented itself spontaneously. This was Ceres, the first asteroid, discovered by Piazzi at Palermo, January 1, 1801, the opening day of the present century.

A series of surprises followed. While watching its path, Dr. Olbers, March 28, 1802, came across an associated body. He named it Pallas, and it was at once proved by the calculations of Gauss to revolve practically at the same distance from the sun as Ceres. *Both* occupied nearly the position required by Bode's law. This double fulfilment was more than was bargained for; it was unprecedented and perplexing; but the anomaly was temporarily removed by Olbers's daring hypothesis of an exploded planet. The prediction based upon it that the acquaintance made with two specimen-products of the catastrophe would be followed by an introduction to many more, was strikingly verified by Harding's discovery of Juno, September 1, 1804, and by Olbers's of Vesta, March 29, 1807. By a further coincidence, both were at the time situated in the positions suggested as the most promising for a successful search—that is, near the line of intersection which should necessarily be common to orbits described by fragments of a single original mass.

The four asteroids received for many years no accession to their numbers. They were found to deviate, in several respects, from the example set them by the planets, properly so-called. They revolve, indeed, from west to east, thus following the current of systemic movement; but their paths are considerably eccentric and highly tilted. Each one of the quartette transgresses the zodiacal limits; and Pallas travels at an angle of no less than thirty-five degrees to the plane of the ecliptic.

Vesta, the brightest asteroid, can occasionally be seen with the naked eye; but the natural inference that it is the largest has lately been disproved. No trustworthy measurements of the real *discs* of the asteroids had been made until Professor Barnard in 1894 successfully performed the feat with a power of 1000 on the Lick refractor. The upshot has been to substitute Ceres for Vesta as the leading member of the group. Its diameter proved to be 485 miles, Pallas coming next with 304, while those of Vesta and Juno are respectively 243 and 118 miles. Now, Professor Edward Pickering, by comparing the brightness of the same bodies, and assuming for all indiscriminately an albedo equal to that of Mars, had arrived at a diameter for Vesta of 319, for Pallas of 169 miles. The disparity between his results and Barnard's can be reconciled only on the supposition of marked differences in reflective power. Their reality was established by G. Müller's photometric observations at Potsdam.[1] Thus Ceres is large and dull, Vesta comparatively small, but exceedingly bright—almost incredibly bright, indeed, since its albedo is estimated at 0·72, which represents a lustre midway between those of white paper and fresh-fallen snow. Ceres, on the other hand, is as obscure as Mercury, while Pallas throws back proportionately somewhat less, and Juno considerably more light than Mars.[2] The phases of these last two bodies progress besides in such a manner as to show that they are superficially uneven, and at quadratures flecked with profound shadows.

The facts thus arrived at are disconcerting to the views previously entertained. Few expected to meet with so much individuality in the asteroids. They were looked upon rather as loaves from the same batch. But now we find among them bodies as physically unlike as Venus and the moon. Ceres must be composed of rugged and sombre rock, unclothed probably by any vestige of air. Vesta displays a brilliant shell of clouds. And from Vesta alone among the asteroids, Vogel derived in 1873 some uncertain indications of atmospheric action upon the sun-rays reflected by it. There is, nevertheless, great difficulty in supposing a body of no more than one-thousandth the mass of Mars endowed with

[1] *Potsdam Publicationen*, No. 30, 1893.
[2] Barnard: *Monthly Notices*, vol. lvi., p. 55.

a dense atmosphere. Yet it must be dense and extensive in order to maintain the heavy cloud-layer implied, so far as our present knowledge goes, by an unusually high albedo. The difficulty is this. All gases tend, by their nature, to become indefinitely diffused through space. They can be restrained within a sphere of finite radius only through the exertion of some force capable of holding their elasticity in check. This force is gravity; none other suitable for the purpose is known. It acts as a counterpull to the translational velocities of the gaseous particles which, according to the dynamical theory of gases, constitute their elasticity. But if the confining power be insufficient, the roving particles will dart away, each on its own account, and will cease to form an atmosphere. This condition was adverted to some years ago by Dr. Johnstone Stoney, and he calculated the mass needed to secure to a heavenly body the lasting possession of an aerial envelope. It differs naturally for different gases; the lightest particles being affected by the swiftest movements, and hence being the readiest to escape. The earth, on this view, is impotent to retain hydrogen; since the critical velocity at its surface is seven miles a second, and hydrogen-molecules can, now and again, attain 7·4 miles, so that they would dribble away, one after another, until the whole original supply was exhausted. Mars (a projectile fired from which, with a speed exceeding three miles a second, would depart irrevocably), can but just hold oxygen, nitrogen, and water-vapour, all with more massive and sluggish molecules than those of hydrogen; while the moon has long ago been forsaken by whatever gaseous substances primitively belonged to it. The mass of Vesta, however, is only $\frac{1}{312}$ the lunar mass (supposing their mean densities the same); hence, if the relation just described holds good under all circumstances, its surface *ought* to be as bare and dry as any lunar volcano. The albedoes of the asteroids raise, then, questions of fundamental importance in planetary physics.

Endeavours to add to the asteroidal group, after having

been relinquished for over a score of years, were resumed, in 1830, by a retired Prussian post-master named Hencke. His watch was rewarded with the discoveries of Astraea, December 8, 1845, and of Hebe eighteen months later. Since then, every year has regularly brought its quota of detections. About forty astronomers devoted themselves systematically to the search, and some of them reckoned their trophies by the score. No less than eighty-five were credited, in 1893, to Palisa of Vienna; Peters of Clinton (N.Y.), whose career closed in 1890, owned forty-eight; Watson, another American professor, made testamentary provision for his twenty-two clients, lest, for lack of computational care, they should relapse into their former outcast condition. The task is, indeed, a heavy one of keeping guard over some hundreds of minute objects threading their way through a maze of orbits, amid throngs of stars, from which they are indistinguishable except by continuous observation, and the question, *Cui bono?* has been asked, and has only with hesitation been answered. But the business has, up to the present, been kept going; the registry and inquiry asteroidal office remains open at Berlin, and the almost overwhelming mass of calculations, necessary for identification, is punctually dealt with.

The work and responsibilities of this department have, of late, been alarmingly augmented. Until five years ago the telescope was the sole implement of research in connection with it, but on December 22, 1891, Professor Max Wolf of Heidelberg, discovered No. 323, afterwards named Brucia, on a sensitive plate exposed with a six-inch portrait lens, of thirty inches focus, and a field of seventy square degrees. Before the year 1892 had closed, his photographic discoveries of the same kind numbered eighteen, and they had, in January, 1897, run up to fifty-six, of which five were recorded on the same night. He picked up, besides, several "lost" or strayed asteroids. M. Charlois of Nice immediately adopted Wolf's method, and emulated his success. About ninety of these objects have already fallen to his share by telescopic and photographic means. In either case they are discriminated from stars solely by their motion; but on sensitive plates its effects are directly visible, fixed objects being represented by round dots, travelling objects by lines, the length of which is proportionate to the amount of displacement during the hour, or hours, of exposure.

About 440 asteroids are now established members of the solar system. It has long been thought that numerical identification is as much as they can properly claim; but the old and inconvenient system of mythological nomenclature is still pursued. Indeed, the supply of goddesses is running out, and has to be reinforced by apotheosis or invention. Already, to some extent, as Professor Holden remarks, the asteroidal catalogue "reads like the Christian names at a girls' school." Needless to say that the brightness of the objects annually registered is in steady course of decline. Very few of those now drawn to shore in the photographic net are likely to exceed twenty miles in diameter. Yet although mere planetary shreds, they are probably large compared with the grains of planetary dust, numberless as the sands of the seashore, which indiscernably revolve round the sun under analogous conditions.

Their aggregate mass is very small. Leverrier assigned for its superior limit one-fourth that of the earth, but the limit, we may rest assured, is very far from being attained. M. Niesten of Brussels estimated that the first 216 asteroids, including all the larger ones, amounted to $\frac{1}{4000}$th the earth's volume, and we may add, since they are beyond doubt specifically lighter, to about $\frac{1}{8000}$th the earth's mass. Mr. Roszl finds for the mass of 311 asteroids one-fortieth that of the moon.[1] Still later, M. Gustave Ravené has attempted to account for the superfluous movement of the perihelion of Mars by the gravitational influence of these bodies.[2] He computes the required mass to be two-thirds that of the moon.

[1] John Hopkins' *University Circular*, Jan., 1895.

[2] *Astr. Nach.*, No. 3359.

In other words, he assumes the group to be fairly represented by 500 globes as large as Juno (124 miles in diameter), and of terrestrial density. But he obviously puts some constraint on nature in order to secure the desired agreement.

The distribution of these dwarfed globes is not without significant features. It is such, at any rate, as absolutely to negative Olbers's hypothesis of their origin through the explosion of an already formed planet. They represent, on the contrary, the materials of a planet that never was, and never will be formed. They follow paths curiously intertwined. D'Arrest noticed forty-five years ago, as a proof of the intimate relation subsisting among the members of what was then a small group, "that, if their orbits are figured under the form of material rings, these rings will be found so entangled that it would be possible, by means of one among them taken at hazard, to lift up all the rest." They are not, however, scattered at random over the wide zone appropriated to them which, at its extreme limits, measures three times the radius of the earth's orbit. It includes blank spaces which seem as if cleared by some expulsive agency. That agency, as Professor Kirkwood divined in 1866, is the disturbing power of Jupiter. For the blank spaces occur where there would be commensurability of periods, and whence, accordingly, revolving particles should be ejected by accumulated perturbations. The clearing power was not exerted once for all; it is still active. But its effectiveness in modifying distribution is now perceived to be less complete than it seemed when our acquaintance with the bodies in question was more limited. It has produced in general only partial vacancies. M. Parmentier [1] analysed in 1895 the arrangement in space of 390 orbits, with the result of finding that some of the originally noted gaps had ceased to exist. The mean distances, for instance, corresponding to periods two-sevenths and three-sevenths the Jovian period, are fairly well frequented; while, on the other hand, there is an unmistakable thinning out where five revolutions are performed

[1] *Monthly Notices*, vol. lvi., p. 250.

while Jupiter accomplishes two. He found again that no asteroid circulates either in half, or in one-third the same dangerous period. Yet, even since he wrote, No. 401 has been detected occupying the former of these prohibited spaces. But this apparent breach of rule may turn out to result from a miscalculation, as in the case of Menippe, which has in consequence never been re-captured since she first presented herself in 1878, and was erroneously assigned a period two-fifths that of Jupiter. There is no doubt that the asteroids are collected most densely about the mean distance 2·8 of the earth's, just where conformity to Bode's law would place them. Nor is it less certain that Kirkwood's "rule of commensurability" has fundamentally influenced their distribution.

He further discerned among them groups of two or three moving in closely related orbits. Additional examples of this sort of connexion, which is far too close to be casual, have been pointed out by M. Tisserand and Mr. Monck, and eighty asteroids are at present known to have companions, their actual ties with which indicate, as Kirkwood held, original identity. Each group consists of fragments of a primitive nebular mass torn asunder by the unequal attraction of Jupiter shortly after its detachment from the great parent sphere eventually condensed to form the sun. As an example, we may take Juno and its twin Clotho. Both revolve at a mean distance from the sun 2·67 times that of the earth, in orbits of sensibly the same eccentricity, and of nearly the same inclination to the ecliptic, their major axes diverging, however, to the extent of ten degrees, obviously through unequal perturbations. As surely as corresponding scars on opposite cliffs vouch for their antique disruption, do these concurrent paths attest the primitive unity of the pair of planetules traversing them. And bodies similarly connected occur not in pairs only, but in triplets as well.

From whatever point of view the "planetary cluster" composed by the asteroids is regarded, the influence of Jupiter is perceived as dominant in the background.

The manner of planetary production underwent a marked change subsequently to the separation of his mighty mass. No interval of repose followed; but a constant shredding off of chips and shavings. This may safely be attributed (in accordance with Professor Kirkwood's surmise) to the tide-raising power of Jupiter at close quarters, by which strain in the central rotating mass was almost prevented, through the facility with which it was relieved. Hence the parent nebula long remained incapable of parting with any appreciable portion of its substance, and never resumed planet-making on the ancient scale. The asteroids then came into existence under Jupiter's auspices; they were, while still in an inchoate state, subdivided, or even pulverised by his disruptive influence, and scattered over the zone allotted to them under the compulsion of his perturbing power.

CHAPTER VIII.

THE PLANET JUPITER.

JUPITER is by far the most important member of the solar family. The aggregate mass of all the other planets is only two-fifths of his, which 316 earths would be needed to counter-balance. His size is on a still more colossal scale than his weight, since in volume he exceeds our globe 1,380 times. His polar and equatorial diameters measure respectively 84,570 and 90,190 miles,[1] giving a mean diameter of 88,250 miles, and a polar compression of $\frac{1}{16}$th. The corresponding equatorial protuberance rises to 2,000 miles, so that the elliptical figure of the planet strikes an observer at the first glance. This at once indicates rapid axial movement; and Jupiter's rotation is accordingly performed in nine hours

[1] Barnard: *Astr. Journal*, No. 325, 1894.

and fifty-five minutes, with an uncertainty of a couple of minutes. The cause of this uncertainty will presently appear.

The numbers just given imply that this great planet is of somewhat slight consistence, and its mean density is in fact, a little less than that of the sun. The sun is heavier than an equal bulk of water in the proportion 1·4 to 1, Jupiter in the proportion of 1·33 to 1. The earth is thus more than four times specifically heavier than the latter globe. Three Jupiters would keep in equipoise four equal globes of water, while the earth would turn the scale against five and a half aqueous models of itself. This low density, an unfailing characteristic of all the giant planets, is charged with meaning. It at once gives us to understand that, in crossing the zone of asteroids, we enter upon a different planetary region from that left behind. The bodies revolving there are on an immensely larger scale of magnitude than those on the hither side; they are of solar, rather than terrestrial, density; they rotate much more rapidly, and are in consequence of a more elliptical shape; they display, and most likely possess, no solid surface; they are attended by retinues of satellites.

Jupiter circulates round the sun in 11·86 years, in an orbit deviating by less than one and a half degrees from the plane of the ecliptic, but of thrice the eccentricity of the ellipse traced out by the earth. With a mean distance from the sun of 483 millions of miles, it accordingly approaches within 462 at perihelion, and withdraws to 504 millions of miles at aphelion. And since the heat and light received from the sun are inversely as the squares of these numbers, it follows that Jupiter is better warmed and illuminated when at the near than when at the far extremity of its orbit, in the proportion of 109 to 100. Seasons it has none worth mentioning; nor could they be of much effect even if they were better marked. At its mean distance of 5·2 "astronomical units"—that is, radii of the earth's orbit—the sun's potency is reduced to $\frac{1}{27}$th what it is here; we might accordingly have expected to meet in this planet the conditions of a frozen world. But this anticipation has been singularly falsified.

Under propitious circumstances Jupiter comes within 369 million miles of the earth. These occur when he is in opposition nearly at the epoch of his perihelion passage. His maximum opposition distance, on the other hand, is 411 million miles. He is then at aphelion. Thus, at the most favourable opposition, he is 42 million miles nearer to us than at the least favourable. The effect on his brightness is evident to the eye. When his midnight culmination takes place in October, he in fact sends us one and a half times more light than when the event comes round to April. We need only recall the unusual splendour of his appearance in September and October, 1892, when his lustre was double that of Sirius. His opposition period, as we may call it, is 399 days.

The intrinsic brilliancy of his surface is surprising, especially when we consider that it is somewhat deeply tinged with colour. According to Müller's determination (relative to Mars), it actually returns 78 per cent. of the incident light. But this would imply self-luminosity, the presence of which is negatived by trustworthy evidence. Hence Zollner's absolute albedo of 0·62 seems preferable. In either case, Jupiter does not fall far short of being as reflective as white paper.

The minimum diameter of the visible disc considerably exceeds the maximum of that of Mars. The latter never measures more than 25″; Jupiter at conjunction, when (in round numbers), 600 million miles distant from us, presents a surface 32″ in diameter, widened at a favourable opposition to 50″. Even with a low power it thus makes a beautiful and interesting telescopic object. Its distinctive aspect is that of a belted planet, the belts varying greatly in number and arrangement. As many as thirty have, on occasions, been counted, delicately ruling the disc from pole to pole. They are always parallel to the equator, but are otherwise highly changeable, and cannot be too closely studied as an index to the planet's physical constitution. Two in particular are remarkable. They are called the north and south equatorial belts, and enclose a lustrous equatorial zone. The poles are shaded by dusky hoods.

This general scheme of markings, however, when viewed with one of the great telescopes of the world, is so overlaid with minor particulars as sometimes to be scarcely recognisable. One cannot see the wood for the trees. Lovely colour-effects, too, come out under the best circumstances of definition and aerial transparency. The tropical belts may be summarily described as red; but they are of complex structure, and their subordinate features and formations are marked out, under the sway of alternating and tumultuous activities, by strips and patches of vermilion, pink, purple, drab and brown. The intermediate space is divided into two bands by a line, or narrow riband, pretty nearly coinciding with the equator, and rosy, or vivid scarlet in hue. The polar caps are sometimes of a delicate wine-colour, sometimes pale grey.

Professor Keeler made an elaborate study of the planet with the Lick 36-inch in 1889, and executed a series of valuable drawings, one of which we are privileged to reproduce (Fig. 15). With a power of 320, the disc, he tells us, "was a most beautiful object, covered with a wealth of detail which could not possibly be accurately represented in a drawing." Most of the surface was then "mottled with flocculent and irregular cloud-masses. The edges of the equatorial zone were brilliantly white, and were formed of rounded, cloud-like masses, which, at certain places, extended into the red belt as long streamers. These formed the most remarkable and curious feature of the equatorial regions. They are the cause of the double or triple aspect which the red belts present in small telescopes."

Near their starting-points the streamers were white and sharply defined, but became gradually diffused over the ruddy surface of the belts. When at all elongated, they invariably flowed backward *against* the rotational drift, and were inferred to be cloud-like masses expelled from the equatorial region, and progressively left behind by its advance. This

[1] "Publ. A. S. P.," vol. ii., p. 286.

hypothesis was confirmed by the motion of some bright points, or knots, on the streamers. "The portions of the equatorial zone surrounding the roots of well-marked streamers were somewhat brighter," Professor Keeler continues, "than at other places, and it is a curious circumstance that they were

FIG. 15.—*Jupiter, October* 3, 1890. *Drawn by Professor Keeler with the great Lick Refractor. The Red Spot is visible.*

almost invariably suffused with a pale olive-green colour, which seemed to be associated with great disturbance, and was rarely seen elsewhere."

Now, if the material of the streamers had been simply a superficial overflow, it should have carried with it into higher latitudes an excess of linear rotational speed, and should hence have pushed its way onwards as it proceeded north and south. But, instead, it fell behind ; its velocity was less, not greater than that of the belts with which it eventually became incorporated. What are we to gather from this fact? Evidently that the currents issuing north and south were of eruptive origin. Their motion, in miles per second, was slow, because they belonged to profound strata of the planet's interior. Their backward drift measured the depth from which they had been flung upward.

The spots, red, white, and black, constantly visible on the Jovian surface, excite the highest curiosity. They are of all kinds and qualities, and their histories and adventures are as diverse as they are in themselves. Some are quite evanescent; others last for years. At times they come in undistinguished crowds, like flocks of sheep, then a solitary spot will acquire notoriety on its own account. White spots appear in both ways; black spots more often in communities; and it is remarkable that the former frequent distinctively, though not exclusively, the southern, the latter the northern hemisphere. Red spots, too, develop pretty freely; but the attention due to them has been mainly absorbed by one striking specimen.

The Great Red Spot has been present with us for at least nineteen years; and it is a moot point whether its beginnings were not watched by Cassini more than two centuries ago. Its modern conspicuousness, however, dates from 1878. Then of a full brick-red hue, and strongly-marked contour, it measured 30,000 by nearly 7,000 miles, and might easily have enclosed three such bodies as the earth. It has since faded several times to the verge of extinction, and partially recovered; but there has never been a time when it ceased to dominate the planet's surface-configuration. More than once it has been replaced by a bare elliptical outline, as if through an effusion of white matter into a mould previously filled with red matter; and just such a sketch was observed by Gledhill in 1870. The red spot is attached, on the polar side, to the southern equatorial belt. It might almost be described as jammed down upon it; for a huge gulf, bounded at one end by a jutting promontory, appears as if scooped out of the chocolate-coloured material of the belt to make room for it.

Absolute contact, nevertheless, seems impossible. The spot is surrounded by a shining aureola, which seemingly defends it against encroachments, and acts as a *chevaux de frise* to preserve its integrity. The formation thus constituted behaves like an irremovable obstacle in a strong current. The belt-stuff encounters its resistance, and rears itself up into a promontory or "shoulder," testifying to the solid presence of the spot, even though it be temporarily submerged. The great red spot, the white aureola, and the brownish shoulder are indissolubly connected.

The spot is then no mere cloudy condensation. Yet it has no real fixity. Its period of rotation is inconstant. In 1879-80, it was of 9 hours, 55 minutes, 34 seconds; in 1885-86, it was longer by 7 seconds. The object had retrograded at a rate corresponding to one complete circuit of Jupiter in six years, or of the earth in seven months.[1] It is not then fast moored, but floats at the mercy of the currents and breezes predominant in the strange region it navigates. A quiescent condition is implied by the approximate constancy of its rotation-period during the last ten years. With the paling of its colour, its "proper motion" slackens or ceases. This must mean that, at its maxima of agitation, it is the scene of uprushes from great depths, which, bringing with them a slower linear velocity, occasion the observed laggings. It is not self-luminous, and shows no symptom of being depressed below the general level of the Jovian surface. A promising opportunity was offered in 1891 of determining its altitude relative to a small dark spot on the same parallel, by which, after months of pursuit, it was finally overtaken. An occultation appeared to be the only alternative from a transit; yet neither occurred. The dark spot chose a third. It coasted round the obstacle in its way, and got damaged beyond recognition in the process. Its material, as Mr. Stanley Williams observed, "was diverted and forced bodily southwards, and obliged to pass round the southern side of the red spot as if it were an island projecting above a stream."

[1] Maunder: *Knowledge*, vol. xix., p. 5.

Jupiter has no certain and single period of rotation. Nearly all the spots that from time to time come into view on its disc are in relative motion, and thus give only individual results. The great red spot has the slowest drift of all (with the rarest exceptions), while the black cohorts of the northern hemisphere outmarch all competitors. Mr. Stanley Williams,[1] as the upshot of long study, has delimitated nine atmospheric surfaces with definite periods. They are well marked, and evidently have some degree of permanence, yet the velocities severally belonging to them are distributed with extreme irregularity. Thus, two narrow, adjacent zones differ in movement by 400 miles an hour. This state of things must obviously be maintained by some constantly acting force, since friction, if unchecked, would very quickly abolish such enormous discrepancies. The rotational zones are unsymmetrically placed ; there is no correspondence between those north and south of the Jovian equator ; and, although the equatorial drift is quicker than that of either tropic, it is outdone in 20° to 24° north latitude. The stability of this anomalous mode of rotation was remarkably illustrated by Dr. Rambaud's measurements of the "Garnet Spot" of October, 1895. Its movement proved to be strictly conformable to that of the zone in which it was situated (10° to 20° north latitude), and to agree, moreover, within a fifth of a second with the value deduced by Schröter in 1787 for that of a spot in the same "zenographical" district.[2]

Jupiter's equatorial rotation, as indicated by observations of spots, is accomplished in 9 hours 50 minutes ; but Bélopolsky's and Deslandres' spectrographic determinations gave rates of approach and recession falling somewhat short of the corresponding velocity.[3] Possibly the spots forge ahead in the medium that sustains them ; or it may be, as M. Bélopolsky suggests, that the planetary sphere itself has been measured too large, owing to refraction in its atmosphere.

[1] *Monthly Notices*, vol. lvi., p. 143.

[2] "Scientific Proceedings, R. Dublin Society," vol. viii., p. 398.

[3] *Astro.-Phys. Journal*, May, 1896, p. 394 ; "Rapport de l'Observatoire de Paris," 1895, p. 22.

However this be, the rotation of the great planet, albeit ill-regulated (if the expression be permissible), is distinctly of the solar type. It is itself a "semi-sun," showing no trace of a solid surface, but a continual succession of cloud-like masses belched forth from within. Each series, in fact, of certain classes of markings, such as the equatorial "port-holes," plainly owes its origin to the rhythmical activity of a solitary, deep-buried focus.[1] Jupiter's low mean density, considered apart from every other circumstance, suffices to demonstrate the primitive nature of his state. Under the enormous pressure reigning in his interior, the same materials should be vastly more massive, specifically, than within our own small globe; their fourfold expansion gives us to understand the intensity of that heat by which pressure has been so much more than neutralised. Moreover, the agitations due to the cooling of a fluid globe make their mark on its turbulent surface. On a solidified body like the earth, circulation is kept up by heat received from without, and is purely atmospheric, and essentially horizontal. In a sun-like body, the circulation is bodily and vertical. That the processes going on in Jupiter are of this kind is beyond question. Exchanges of hot and colder substances are effected, not by surface-flows, but by up and down rushes. The parallelism of his belts to his equator makes this visible to the eye. An occasional oblique streak[2] betokens a current in latitude, but it is exceptional, and might be called out of character.

Jupiter's true atmosphere encompasses the disturbed shell of vapours observed telescopically. Its general absorptive action upon light is betrayed by the darkening of the planet's limb—another point of resemblance to the sun; while its special, or selective, absorption can only be detected with the spectroscope. The arresting effect of water-vapour was early noticed by Huggins and Vogel, and they measured a strong line in the red of unknown origin, but contained in banded

[1] Proctor: "Old and New Astronomy," p. 584.

[2] "The subject of slant-markings," Mr. Stanley Williams remarks (*loc. cit.*), "has only just begun to be investigated."

star-spectra. Atmospheric absorption is strongest above the ruddy equatorial belts, which are hence concluded to be placed at a lower level than the white surface.

Planetary photography was set on foot by Dr. Gould of Boston, in 1879, when he obtained some promise of success with Mars, Jupiter, and Saturn; and Dr. Lohse prosecuted the subject in 1883. The actinic power of Jupiter's light is very remarkable. It surpasses that of moonlight nine times, and that of Mars twenty-four times. Dr. Lohse further ascertained that the southern hemisphere is twice as chemically effective as the northern.[1] This superiority is doubtless connected with the greater physical agitation of the same region. A series of photographs of Jupiter, taken in 1891 with the great Lick refractor, were the first of any value for purposes of investigation. Each is one inch in diameter; the image of the planet having been enlarged eight times before being received upon the plate. Mr. Stanley Williams found them full of interesting detail. Figure 16 shows an enlargement of a striking photograph taken by Professor E. C. Pickering.

Jupiter's satellites were the first trophies of telescopic observation. They are, indeed, bright enough for naked eye perception, could they be removed from the disc which obscures them with its excessive splendour; and the first and third have actually been seen, in despite of the glare, by a few persons with phenomenally good eyesight. The mythological titles of the Galilean group—Io, Europa, Ganymede, and Calypso (proceeding from within outward) have been superseded by prosaic numbers. The change was unlucky, but is now probably irremediable.

The Jovian family presents an animated and attractive spectacle. The smallest of its original members (No. II.) is almost exactly the size of our moon; the largest (No. III.), with its diameter of 3,550 miles, considerably exceeds the modest proportions of Mercury. Satellite I. revolves in 42½ hours at the same average distance from Jupiter's surface that our moon does from that of the earth. No. II. has a period of

[1] "Jupiter and his System," by Ellen M. Clerke, p. 43.

3 days 13 hours, and its distance from Jupiter's centre is 415,000 miles. Both these orbits are sensibly circular; and Nos. III. and IV. travel in ellipses of very small eccentricity, the one at a mean distance of 664,000, the other at 1,167,000

FIG. 16.—*Photograph of Jupiter. Exposure*, 87 *seconds.*
(From *Knowledge*, November, 1889.

miles, in periods respectively of 7 days 4 hours, and 16 days 16½ hours. All four revolve strictly in the plane of Jupiter's equator.

They constitute a system bound together by peculiar dynamical relations, in consequence of which they can never be all either eclipsed, or seen aligned at one side of their primary, at the same time. They can all, however, be simultaneously hidden behind it, or in its shadow; although this moonless condition is looked out for as a telescopic rarity.

The varied phenomena of eclipses, occultations, and transits, offer the interest, not only of predictions fulfilled, but sometimes of discrepancies detected. The three inner satellites plunge through the huge neighbouring shadow-cone at every revolution; the fourth, owing to its greater distance, escapes eclipse when the shadow makes an appreciable angle with the plane of its orbit. When Jupiter is in opposition or conjunction, occultations, but no eclipses, of his moons take place; at other periods, the two kinds of obscuration merge into, or succeed each other. "Time cannot stale their infinite variety."

From observations of the eclipses of Jupiter's satellites, Olaus Römer gathered, in 1675, the first intimations of the finite velocity of light. He noticed that their visibility was alternately retarded and accelerated as the earth withdrew from, and approached the scene of their occurrence; and he designated half the extreme difference, or the time occupied by light in travelling from the earth to the sun, the "equation of light." Its value is 500 seconds; and until recently, no other measure was available of that fundamental constant of nature—the rate of luminous transmission.

The transits of the satellites across the Jovian disc present many curious appearances, due to complicated and changeable effects of light and shade both upon the planetary background, and upon the little circular objects self-compared with it. These, in the ordinary course, show bright while near the dusky limb, then vanish during the central passage, and re-emerge again bright at the opposite side. But, instead of duly vanishing, they now and then darken even to the point of becoming indistinguishable from their own shadows, by which they are preceded or followed. This difference of behaviour cannot be attributed wholly to varieties of lustre in the sections of the disc transited; otherwise, it could be predicted. But this has never been attempted; "black transits" come when least expected. The third and fourth satellites are those chiefly subject to these phases; the second has never been known to exhibit them; and they but slightly affect the first. A drawing by Professor Barnard of one of its bright transits with an attendant shadow that Peter Schlemyl might have envied, is reproduced in Figure 17. Its belted appearance,

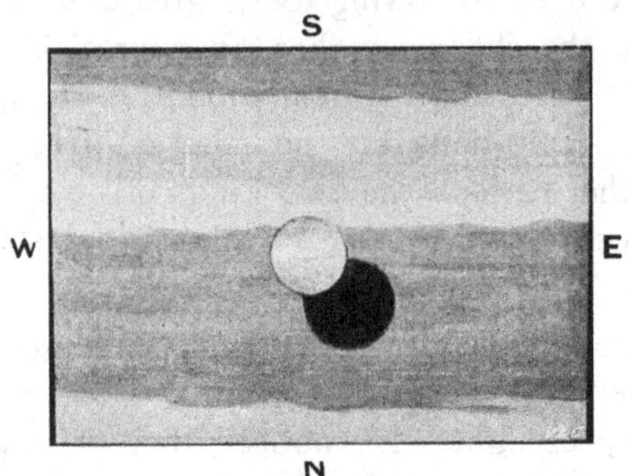

FIG. 17.—*Transit of Jupiter's first Satellite, with Shadow, drawn by Prof. Barnard, November* 19, 1893. (From *Monthly Notices*, January, 1894.)

detected by that eminent observer, will be noted. Indeed, all the satellites, except perhaps No. II. are striped or spotted; and this leads to seeming deformations in their shape, as well as fluctuations in their brightness, the markings being evidently of atmospheric origin, and hence changeable. Their distinct and accurate perception has been made possible by the excellence of the Lick thirty-six inch refractor.

Jupiter's moons seem to resemble him in constitution. The three first possess the same high reflective power. No. II., is as bright as the planet's brightest parts, so that its albedo cannot fall short of 0·70. And even No. IV. (formerly designated "Calypso" in reference to its frequent obscurations) exactly matches, during its darkest phases, the blue-grey polar hoods of its primary. On an average, too, the satellites seem to be of about the same mean density as Jupiter, No. I. being considerably the lightest for its bulk; and their spectra, according to Vogel's observations in 1873, are composed of solar rays modified in precisely the same way as those reflected by the planet. Nothing is known quite certainly about their rotation-periods. Sir William Herschel concluded them to be of the same length with their periods of revolution; but recent work throws some doubt upon the reality of this agreement.

The discovery, September 9, 1892, of Jupiter's "fifth satellite" was one of the keenest astronomical surprises on record. An accession to a system so symmetrically arranged, so complete, to our judgment, as it stood, appeared superfluous, and, considering the eager scrutiny devoted to it during 282 years, well-nigh incredible. But the extra member was in truth out of reach until it was found; original discovery being, as every one knows, a greatly more arduous feat than subsequent verification. Nor could it have been casually detected. Professor Barnard seized the opportunity, lent by the specially favourable opposition of 1892, to rummage the system for novelties. Keeping the telescopic field dark by means of a metallic bar placed so as to occult the gorgeous planetary round, he sought, night after night, for what might appear. At length, on September 9, he caught the glimmer he wanted, and made sure, September 10, that it truly intimated the presence of a new satellite.

This small body revolves in a period of 11 hours, 57 minutes, 23 seconds, at a mean distance of 112,160 miles from Jupiter's centre, or 67,000 from his bulged equatorial surface. Hence, it should by right be called "No. I." instead of "No. V." The major axis of the ellipse in which it circulates advances so rapidly, owing to the disturbance caused by Jupiter's spheroidal figure, as to complete a revolution in five months. The implied eccentricity of its orbit, as M. Tisserand has shown,[1] very slightly exceeds that of the orbit of Venus, yet it has been made obvious by Barnard's observations of the differences between its east and west elongations. Its orbital velocity of 16½ miles a second far surpasses that of any other satellite in the

[1] *Comptes Rendus*, t. cxix., p. 581.

solar system. Close vicinity to a mass so vast as Jupiter's demands counter-balancing swiftness. Its period of revolution being, however, longer by one hour than Jupiter's period of rotation, it so far conducts itself normally as to rise in the east and set in the west. On the other hand, since its progress over the sphere is measured by the difference between the two periods, it spends five Jovian days in journeying from one horizon to the other, running, in the meantime, four times through all its phases. Yet it never appears full. Jupiter's voluminous shadow cuts off sunlight from it during nearly one-fifth of each circuit.

It is an exceedingly elusive telescopic object. There is no chance of catching a glimpse of it except with a powerful and perfect telescope at its "elongations," or furthest excursions of about eight seconds of arc on either side of the planet. For the most part, it lurks within the blaze as closely as Teucer behind the shield of Ajax. It is far too small to be discerned in projection upon the disc, which, viewed from it in mid-transit, is *full* with a diameter of 42° 2′, and an area 6,440 times that of our moon. Yet, since its intrinsic lustre is less in the proportion of 2 to 15, the light shed by Jupiter upon the "fifth satellite" equals the joint radiance of no more than 860 full moons.

The new satellite is indistinguishable in aspect from a star of the thirteenth magnitude. And its neighbour No. I. being of 5·6 magnitude, we receive from it 910 times more light than from the stranger. If both be equally reflective, the diameter of the latter is $\frac{1}{30}$th the diameter of the former, or, approximately, 80 miles. But its albedo is unlikely to exceed that of Mars. By a rough estimate, therefore, this interesting object measures 120 miles across, and 9000 such miniature globes would go to the making of one full-sized Jovian attendant. Instead of being a late addition to the system, or, so to speak, an afterthought, it may be presumed, from the perceptible eccentricity of its path, to be the senior member of the family. But the subject of its origin is not yet ripe for discussion.

CHAPTER IX.

THE SATURNIAN SYSTEM.

NEARLY twice as far from the sun as Jupiter revolves a planet, the spacious orbit of which was, until 1781, supposed to mark the uttermost boundary of the solar system. The mean radius of that orbit is 886 millions of miles; but in consequence of its eccentricity, the sun is displaced from its middle point to the extent of 50 million miles, and Saturn is accordingly 100 million miles nearer to him at perihelion than at aphelion. The immense round assigned to the "saturnine" planet is traversed in 29½ years, at the tardy pace of six miles a second. His seasons are thus twenty-nine times more protracted than ours, and are nominally more accentuated, since his axis of rotation deviates from the vertical by 27°. But solar heat, however distributed, plays an insignificant part in his internal economy. In the first place, its amount is only $\frac{1}{91}$th its amount on the earth; in the second, Saturn, like Jupiter—even more than Jupiter—is thermally self-supporting. The bulk of his globe comparatively to its mass suffices in itself to make this certain. The mean diameter of Saturn is 71,000 miles, or nine times (very nearly) that of the earth; if of equal density, its mass should then be nine cubed, or 729 times the same unit. The actual proportion, however, is 95; hence the planet has a mean density of only $\frac{95}{29}$, or between $\frac{1}{7}$th and $\frac{1}{8}$th the terrestrial, and being thus composed of matter as light as cork, would float in water. Professor G. H. Darwin has moreover demonstrated, from the movements of its largest satellite, that its density gains markedly with descent into the interior, so that its surface-materials must be lighter than any known solid or liquid.

When at its nearest to the earth, Saturn is as large as a sixpence held up at a distance of 210 yards.[1] But instead of

[1] G. H. Darwin: *Harper's Magazine*, June, 1889.

being round like a sixpence, it is strongly compressed—more compressed even than Jupiter. The spectra of the two planets are almost identical. Both are impressed with traces of aqueous absorption, and include the "red star line." About the albedo of Saturn there is some uncertainty. Zöllmer made it 0·50, a very probable value; Müller of Potsdam determined it at 3·3 times that of Mars, the unit of his scale. For the value of the unit, the only authority is Zöllner, who found Mars to give back 0·26 of the light dispensed to him. Multiplying then 0·26 by 3·3 we get for the albedo of Saturn 0·86, an impossible number for a non-luminous body, the albedo of "untrodden snow" being, as already stated, 0·78.

Saturn resembles to the eye a large, dull star; its rays are entirely devoid of the sparkling quality which distinguishes those of Jupiter. But it shows telescopically an analogous surface-structure. Its most conspicuous markings are tropical dark belts of a greyish or greenish hue; the equatorial region is light yellow, diversified by vague white spots; while the poles carry extensive pale blue canopies. The apparent tranquillity of the disc may be attributed in part to the vast distance from which it is viewed; yet not wholly. For lack of fiducial points, no attempt was made to determine the planet's rotation until 1794, when the elder Herschel, by following an identified irregularity in a complex banded formation, arrived at a period of 10 hours 16 minutes. The first possibility of checking this result offered itself to Professor Hall of Washington, after fourteen years of vain expectation, in the emergence of a white spot just north of the equator, the movement of which gave for the length of the Saturnian day, 10 hours, 14 minutes, 24 seconds. In 1891-2, Mr. Stanley Williams made observations upon a good many such objects; and their discussion by Mr. Denning afforded a mean period two seconds longer than Hall's. Individual variations, however, to the extent of 14 seconds were brought out by it, proving that Saturnian, like Jovian, spots have "proper motions," and cannot be depended upon to give the true rotation of the planet.

Its compound nature may be suspected, but has not yet been proved.

From measures executed by Barnard in 1895, it appears that the equatorial diameter of Saturn is 76,470, its polar diameter 69,770 miles, giving a mean diameter of 74,240, and a compression of about $\frac{1}{12}$. Gravity, at its surface, is only one-fifth more powerful than on the earth.

Thus, Saturn not only belongs to the same celestial species as Jupiter, but is a closely-related individual of that species. There is no probability that either is to any extent solid. Both exhibit the same type of markings; both betray in-

FIG 18.—*Saturn and its Rings. Drawn by Prof. Barnard, July 2 1894.*

ternal tumults by eruptions of spots which, by their varying movements, supply a measure for the profundity of their origin; both possess identically constituted atmospheres, and are darkened marginally by atmospheric absorption.

Saturn is, however, distinguished by the possession of an unique set of appendages. Nothing like them is to be seen elsewhere in the heavens; and when well opened (as in Fig. 18) they form, with the globe they enclose, and the retinue of satellites in waiting outside, a strange and wonderful telescopic object. The rings, since they lie in the plane of Saturn's equator, are inclined 27° to the Saturnian orbit, and 28° to the ecliptic. The earth is, however, comparatively to Saturn, so near the sun, that their variations in aspect, as viewed from it, may in a rough way be considered the same as if seen from the sun. They correspond exactly with the Saturnian seasons. At the Saturnian equinoxes, the rings are illuminated edgewise, and disappear, totally or approximately; at the Saturnian solstices, sunlight strikes them nearly at the full angle of 27°, first from *below*, then from *above*. At these epochs, we perceive the appendage expanded into an ellipse about half as wide as it is long. Two concentric rings (generally called A and B) are then very plainly distinguishable, the inner being the brighter. The black fissure which separates them is called "Cassini's division," because that eminent observer was, in 1675, the first to perceive it. A chasm known as "Encke's division," in the outer ring (A), is a thinning-out rather than an empty space; and temporary gaps frequently appear in A, while B is entirely exempt from them. There are then two definite and permanent bright rings, and no more; but with them is associated the dusky formation discovered by W. C. Bond, November 15, 1850, and described by Lassell as "something like a crape veil covering a part of the sky within the inner ring." It is semi-transparent, the limb of Saturn showing distinctly through it.

The exterior diameter of the ring-system is 172,800, while its breadth is 42,300 miles.[1] The rings A and C are each 11,000 miles wide; while B measures 18,000, Cassini's division 2,270, and the clear interval between C and the planetary surface somewhat less than 6,000 miles. Each ring, C included, is brightest at its outer edge; but there is no gap between the shining and the dusky structures, B shading by insensible gradations up to C, yet maintaining distinctness from it. The earliest exact determinations of the former were made by Bradley in 1719, since when they

[1] Barnard, *Monthly Notices*, vol. lvi., p. 163.

have been affected by no appreciable change.[1] The theoretically inevitable subversion of the system is progressing with extreme slowness.

The thickness of the rings is quite inconsiderable. They are flat sheets, without (so to speak) a third dimension. For this reason, they disappear utterly in most telescopes, when their plane passes through the earth, as it does twice in each Saturnian year. Only under exceptional conditions, a narrow, knotted, often nebulous, streak survives as an index to their whereabouts. On October 26, 1891, Professor Barnard,[2] armed with the Lick refractor, found it impossible to see them projected upon the sky, notwithstanding that their shadow lay heavily on the planet. It was not until three days later, that "slender threads of light" came into view. The corresponding thickness of the formation was estimated at less than fifty miles. The phenomenon of the disappearance of the rings will not recur until July 29, 1907.

The constitution of this marvellous structure is no longer doubtful. It represents what might be called the fixed form of a revolving multitude of diminutive bodies. This was demonstrated by Clerk Maxwell in the Adams Prize Essay of 1857. His conclusion proved irreversible. The pulverulent composition of Saturn's rings is one of the acquired truths of science. An incalculable number of tiny satellites, revolving independently in distinct orbits, in the precise periods prescribed by their several distances from the planet, are aggregated into the unmatched appendages of Galileo's *tergeminus planeta*. The local differences in their brightness depend upon the distribution of the component satelloids. Where they are closely packed, as in the outer margins of rings A and B, sunlight is copiously reflected; where the interspaces are wide, the blackness of the sky is barely veiled by the scanty rays thrown back from the thinly scattered cosmic dust. The appearance of the crape

[1] Lewis: *Observatory*, vol. xviii., p. 379.
[2] *Monthly Notices*, vol. lii., p. 419.

ring as a *dark* stripe on the planet results—as M. Seeliger has pointed out—not from the transits of the objects themselves, but from the flitting of their shadows in continual procession across the disc.

The albedo of these particles is so high as to render it improbable that they are of an earthy or rocky nature, such as the meteorites which penetrate our atmosphere. The rings they form are, on the whole, more lustrous than Saturn's globe; but this superiority is held to be due to the absence of atmospheric absorption. Their spectrum is that of unmodified sunlight.

An eclipse of Japetus, the eighth Saturnian moon, by the globe and rings, November 1, 1889, was highly instructive as to the nature of the dusky appendage. The satellite was never lost sight of during its passage behind it; but became more and more deeply obscured as it travelled outward; then, at the moment of ingress into the shadow of ring B, suddenly disappeared. Certainty was thus acquired that the particles forming the crape ring are most sparsely strewn at its inner edge—which is, nevertheless, perfectly definite—and gradually reach a maximum of density at its outer edge. Yet, while there is not the smallest clear interval, a sharp line of demarcation separates it from the contiguous bright ring. Professor Barnard was the only observer of these curious appearances. The distribution of the ring-constituents, like that of the asteroids, was governed by the law of commensurable periods, Saturn's moons replacing Jupiter as the perturbing and regulating power. Kirkwood showed in 1867, that Cassini's division represents a region of peculiarly strong disturbance; since a body revolving there would have a period connected by a simple relation with the periods of no less than *four* satellites. Encke's division, too, as Dr. Meyer has indicated, and other lines of scanty occupation and occasional vacancy, coincide with districts of space where similar combinations occur.

The "satellite-theory" of Saturn's rings has received confirmation from apparently the least promising quarters.

Professor Seeliger of Munich showed, from photometric experiments in 1888, that their constant lustre under angles of illumination ranging from 0° to 30° was proof positive of their composition out of discrete small bodies.[1] And Professor Keeler of Alleghany, by a beautiful and refined application of the spectroscopic method, arrived at the same result in April, 1895.[2] "Under the two different hypotheses," he remarked, "that the ring is a rigid body, and that it is a swarm of satellites, the relative motion of its parts would be essentially different." The former would necessarily involve increasing velocity *outward*, the latter, increase of velocity *inward*, just for the same reason that Mercury moves more swiftly than the earth, and the earth than Saturn; while the sections of a solid body, which could have but one period of rotation, should move faster, *in miles per second*, the farther they were from the centre of attraction. The line of sight test is then theoretically available; but it was an arduous task to render it practically so. The difficulties were, however, one by one overcome; and a successful photograph of the spectra of Saturn and its rings gave the required information in unmistakable shape. From measurements of the inclinations of five dusky rays contained in it with reference to a standard horizontal line, rates of movement were derived of 12½ miles per second for the inner edge of ring B, and of 10 miles for the outer edge of ring A. The agreement with theory was, as nearly as possible, exact; the components of the rings were experimentally demonstrated to be moving, each independently of every other, under the dominion of Kepler's laws.

For the globe of Saturn, Professor Keeler obtained, by the same exquisite method, a rotational period of 10 hours, 14 minutes, 24 seconds, in precise accordance with that indicated by the white spot of 1876, which thus seems to have had no proper motion, but to have floated on the ochreous equatorial surface as tranquilly as a water-lily upon a stagnant

1 "Abhandlungen Akad. der Wissensch." München, Bd. xvi., p. 403.

2 *Astro-Physical Journal*, May, June, 1895.

pool. The result, so far as it goes, hints that Saturn may be really, as well as apparently, less ebullient than Jupiter.

Seers into the future of the heavenly bodies consider that the rings of Saturn, like the gills of a tadpole, are symptomatic of an early stage of development ; and will be disposed of before he arrives at maturity. They cannot be regarded otherwise than as abnormal excrescences. No other planet retains matter circulating round it in such close relative vicinity. It was proved by Roche of Montpellier that no secondary body of importance can exist within less than 2·44 mean radii of its primary; inside of that limit, it would be rent asunder by tidal strain. But the entire ring-system lies within the assigned boundary; hence, being *where* it is, it can only exist *as* it is—in flights of discrete particles. Will it, however, always remain where it is?

"Clerk Maxwell," wrote Mr. Cowper Ranyard,[1] "used to describe the matter of the rings as a shower of brickbats, amongst which there would inevitably be continual collisions. The theoretical results of such impacts would be a spreading of the ring both inwards and outwards. The outward spreading will in time carry the meteorites beyond Roche's limit, where, in all probability, they will, as Professor Darwin suggests, slowly aggregate, and a minute satellite will be formed. The inward spreading will in time carry the meteorites at the inner edge of the ring into the atmosphere of the planet, where they will become incandescent, and disappear as meteorites do in our atmosphere."

Yet it may be that collisions are infrequent in this conglomeration of "brickbats." There is the strongest presumption that they all circulate in the same direction, in orbits nearly circular, and scarcely deviating from the plane of the Saturnian equator. Those pursuing markedly eccentric tracks must long ago have been eliminated. Thus, encounters can only occur through gravitational disturbances by Saturn's moons, and they must be of a mild character, depending upon very small differences of velocity. The first sign of a

[1] "Old and New Astronomy," p. 640.

"spreading outwards" should be the formation of an exterior "crape ring," of which no faintest trace has yet been perceived.

Saturn's rings are entirely invisible from its polar regions, but occasion prolonged and complex eclipse-effects in its temperate and equatorial zones. They have been fully treated of from the geometrical point of view by Mr. Proctor in "Saturn and its System."

Of this planet's eight satellites, the largest, Titan (No VI.), was discovered first (by Huygens in 1655), and the smallest, Hyperion (No. VII.), last (by Lassell and Bond in 1848).. The five others were detected by J. D. Cassini and William Herschel. Titan, alone of the entire group, equals our moon in size. It measures, according to Professor Barnard, 2,720 miles across. Its period of revolution is nearly sixteen days, its distance from Saturn's centre, 771,000 miles. The orbit of Japetus (No. VIII.) is the largest, and its period the longest of any secondary body in the solar system. It circulates in 79⅓ days at a distance of 2,225,000 miles, equal to 59½ of Saturn's equatorial radii. Hence its path is of about the same *proportional* dimensions as that of our moon. Japetus is remarkable for its variability in light. It is capable of tripling or quadrupling its minimum lustre. Sir William Herschel noticed that these maxima coincided with a position on the western side of the planet, and inferred rotation of the lunar kind. "From the changes in this body," he argued in 1792,[1] "we may conclude that some part of its surface, and this by far the largest, reflects much less light than the rest; and that neither the darkest nor the brightest side is turned towards the planet, but partly one and partly the other, though probably less of the bright side."

This explanation, however, he admitted to be incomplete. There was, and is, outstanding variability, which seems to intimate the presence of an atmosphere and the formation of clouds. But no positive knowledge has yet been gained regarding the physical state of Saturn's moons. We may

[1] "Phil. Trans.," vol. lxxxii., p. 17.

nevertheless conjecture that, since tidal friction has destroyed the rotation (as regards Saturn) of the remotest member of the family, it has not spared those more exposed to its grinding-down action. All presumably rotate in the same time that they revolve.

The five inner satellites move in approximately circular orbits; the three outer in ellipses about twice as eccentric as the terrestrial path. All, Japetus only excepted, keep strictly to the plane of the rings. And since this makes an angle of 27° with the planet's orbit, eclipses are much less frequent here than in the Jovian system. They can only occur when Saturn is within a certain distance (different for each) from the node of the satellite-orbit. Even Mimas (No. I.), although it wheels round the ring at an interval of only 34,000 miles, often slips outside the obliquely-projected shadow-cone. Its distance from Saturn's centre is 118,000 miles, and it completes a circuit in 22½ hours. Perpetually wrapped in the glare of its magnificent primary, it is a very shy object, only to be caught sight of in its timid excursions by the very finest telescopes. Like all the Saturnian moons, except Titan, and, by a rare conjuncture, Japetus, it is far too much contracted to be visible in transit across the disc.

The movements of these bodies have been carefully studied, and their mutual perturbations to some extent unravelled. They have proved exceedingly interesting to students of celestial mechanics. Titan has, in this department, chiefly to be reckoned with. He exercises in the Saturnian system a similar overpowering influence to that wielded by Jupiter in the solar system. Mr. Stone finds its mass to be $\frac{1}{4600}$th that of Saturn, showing that its density is nearly equal to that of our moon. This seems to indicate an advanced stage of cooling. On the other hand, its albedo is evidently very high. The other satellites appear in the largest telescopes as mere stellar points.

CHAPTER X.

URANUS AND NEPTUNE.

THE four giant planets, closely allied as they are, and strongly distinguished in physical constitution from the terrestrial planets, divide again of themselves into two sub-groups. Jupiter and Saturn have much more in common than either has with Uranus or Neptune; while Uranus and Neptune present peculiar analogies. Conclusions concerning one may almost be said to apply to the other. Their enormous distance, it is true, tends to efface minor differences; yet it is insufficient to obliterate similarities of a peculiar kind.

Uranus is a globe 32,000 miles in mean diameter, and decidedly elliptical in shape. Mädler and Schiaparelli agreed in assigning to it a compression of $\frac{1}{11}$; Barnard, in 1894, uninformed of their results, noticed the disc to be more oval than Saturn's. The indicated rotational movement must be very swift; and a lucid spot watched by MM. Perrotin and Thollon at Nice in 1884, seemed to fix it at about ten hours. This was, however, only a vague estimate. Faint equatorial belts, too, have with difficulty been seen. Remembering, indeed, that the object they diversify is just large enough to be *annularly eclipsed* by a cricket ball two miles off, there is little cause for surprise at the indistinctness of its surface-markings. They probably consist, like those of Jupiter and Saturn, in dusky polar hoods, a brilliant equatorial zone, and obscure intermediate bands. The last were seen as "the merest shades on the planet's surface," and under a somewhat deformed aspect, by the Lick observers in 1890 and 1891.[1] By Professor Young in 1883, on the other hand, and by the MM. Henry at Paris in 1884, they were observed to be symmetrically placed, parallel one to the other, and of what might be called the normal type for great planets. That they

[1] "Publications Astr. Soc. of the Pacific," vol. iii., p. 284.

constitute, with the bright space they enclose, an equatorial scheme of marking, was proved by Barnard's comparison of the trend (or position angle), determined for them by Young, with the direction of the shortest axis of the little disc they traverse.[1] Their considerable foreshortening in 1894 was, doubtless, the reason why Barnard, with his acute vision, was compelled to rely upon earlier observations, brought up to date by computation. Unless, indeed, the markings are intrinsically variable.

This was suspected at Nice in 1889, when a thirty-inch refractor was available for their scrutiny.[2] Dusky rulings were obvious on a strongly compressed spheroid; and they ran parallel to the major axis of the spheroid—that is, to the planet's equator. But their appearance varied, and their width seemed irregular. At the same establishment, but with a fourteen-inch telescope, Uranus was observed, under particularly favourable circumstances, March 18, 1884.[3] An unexpected resemblance to Mars was apparent. The ordinarily sea-green disc was divided into a sombre north-western and a bluish-white south-eastern hemisphere. Dark spots were visible, and a conspicuous white one at the limb simulated a snow-cap. But ulterior observations resolved the spots into belts, and showed the shining patch to be, not polar, but equatorial. It was presumably of an eruptive nature.

The axis upon which Uranus rotates is very much bowed towards the plane of its orbit. Its seasons are hence abnormal; but their vicissitudes can scarcely be sensible at a distance from the sun more than twice that of Saturn. This, as Mr. Proctor noticed, is the only case in which the ratio of one to two is exceeded in the radii of two adjacent planetary orbits. The radius of the Uranian track, pursued at the leisurely pace of $4\frac{1}{5}$ miles a second, is 1,782 millions of miles, or more than 19 astronomical units. It consequently receives

[1] *Astr. Journal*, No. 370.

[2] Perrotin: "Vierteljahrsschrift Astr. Ges.," Jahrg. xxiv., p. 267.

[3] "Annales de l'Observatoire de Nice," t. ii., 1887.

from the sun 370 times less warmth and light than the earth does. Area for area, it is true, the sun shines with the same intensity there as here; the difference lies in its apparent size. Instead of the broad eye of day to which we are accustomed, the luminary of Uranus presents a surface only $2\frac{1}{4}$ times that of Jupiter, as seen from the earth at an *unfavourable* opposition; and although Uranus is 166 millions of miles nearer to the sun at perihelion than at aphelion, no conspicuous difference would mark the passage from one to the opposite point. This is accomplished in 42, the entire round in 84 years.

In point of size, as Professor Young remarks, Uranus compares with the earth very much as the earth compares with the moon. For its surface exceeds the terrestrial surface about sixteen times, and its volume amounts to sixty-six times the terrestrial volume. Its mass, however, is less than fifteen times that of the earth, whence its density is represented (in round numbers) by the fraction $\frac{15}{66}$. The large globe is then nearly five times less dense than the small one, its materials exceeding the weight of an equal bulk of water by only one-fifth. Gravity is actually less at its surface than at the sea-level on the earth. Every ton of coal, for instance, delivered in that remote globe would fall short by two hundred pounds. The albedo of Uranus differs little from that of Jupiter; if anything, it is somewhat higher, and is nearly represented by the brilliancy of white paper.

The spectrum of Uranus indicates an emphatic departure from the planetary conditions so far met with. This body is obviously surrounded by a powerfully absorptive atmosphere, of a constitution foreign to our experience. The greenish hue of the light which has traversed some of its strata gives a preliminary indication of the manner in which it has been affected. This its spectrum, first inspected by Secchi in 1869, expounds in detail. He noticed a number of heavy dark bands in the red, while the green and blue sections remaining open gave to the planet its characteristic colour. A couple of years later, Huggins and Vogel executed concordant measurements of six pronounced bands, besides some faint streaks; and on June 3, 1889, the former obtained, with two hours' exposure, a beautiful spectrographic impression extending far up into the ultra-violet. A corroborative, though less comprehensive, photograph was taken by Mr. Frost at Potsdam, April 23, 1892. Both included many Fraunhofer lines, the presence of which demonstrates that the light of Uranus, although more powerfully stamped with original absorption than that of the rest of the planets, consists essentially of reflected solar rays. Professor Keeler's admirable series of visual observations with the Lick refractor were undertaken in 1889 to test the truth of a suggestion that this peculiar spectrum consisted of bright bands upon a dark ground, and not of dark bands upon a bright ground. His decision in favour of the latter alternative was without appeal.

Of the six principal dark bands representing the arresting action upon light of the planetary atmosphere, four are quite distinctive; the fifth is the "red-star line" common to the spectra of Jupiter and Saturn; the sixth is the hydrogen "F" (Hβ)—not definite and narrow as it is seen in the solar spectrum, but hazy, and graduating in darkness towards the middle, an undoubted outcome of native absorption.[1] Now, this is a fact that implies a great deal. It gives direct evidence of a very high temperature. Free hydrogen ceases to be present in a body upon which water can form—given, of course, the presence of oxygen, which it would be in the highest degree arbitrary to exclude. At one epoch of its development, the earth must have been surrounded by immense volumes of hydrogen. But with the diminution of heat, union with oxygen became possible, and the gas vanished to reappear in the form of liquid oceans, with their related hydrographic and cloud-systems. Uranus is presumably—almost certainly—still too hot to permit the combination of hydrogen and oxygen; and the absence from its spectrum of the slightest trace of aqueous absorption, strengthens this inference. Doubtless, the time will come

[1] Keeler: *Astr. Nach.*, No. 2927.

when the two elements will no longer be held at arms' length; their affinities will come into play; the familiar, all-important terrestrial liquid will be formed, and the geological history of Uranus will begin.

Uranus is attended by four moons. They are named Ariel, Umbriel, Titania and Oberon. Titania—the third in order of distance from the primary—is the brightest of the group, and has a diameter of possibly one thousand miles. Oberon is slightly inferior. Both were detected by Herschel in 1787. Ariel and Umbriel, captured by Lassell at Malta in 1851, are insignificant bodies in themselves—their dimensions probably differing but slightly from those of Hyperion, the seventh and least Saturnian moon, estimated to measure five hundred miles across. They are among the most difficult of telescopic objects, since they circulate about as close to Uranus as Mimas and Enceladus do to Saturn, are physically smaller, and more than twice as remote from the earth. Both were believed variable by Lassell, and Newcomb obtained in 1875 plausible, though not convincing, evidence that Ariel, at any rate, is subject to light-changes in the period of its orbital circulation, showing that, here again, tidal friction has done its work of synchronising rotation and revolution.[1] None of the four orbits are appreciably eccentric; they all lie in the same plane, and are described in periods ranging from 2½ to 13½ days.

The position of that plane is, however, exceedingly remarkable. It is tilted at an angle of 98° to the ecliptic. This means that the satellites move *backward*, against the succession of the zodiacal signs. For direct becomes retrograde motion automatically, so to speak, by turning the plane in which it is performed beyond the limit of the vertical. The same fact is merely expressed in two different ways by saying that the bodies in question travel from west to east at an angle of 98°, or from east to west at an angle of 82° to the ecliptic. The planes of the ecliptic and of the Uranian orbit deviate, it should be mentioned, by only two-thirds of a

[1] Gregory: *Nature*, vol. xl., p. 236.

degree. The disturbance by which the Uranian system was set topsy-turvy did not in the least affect the motion of Uranus itself.

Another unusual circumstance about that system is that the satellite-plane departs widely from the equatorial plane. Our own moon, it is true, is similarly circumstanced; but, on the Uranian scale, it is nearly eight times farther from its primary than Ariel, and 2·6 times farther than Oberon; while the enormous equatorial protuberance of Uranus almost seems to impose conformity upon bodies revolving so close to it. Conformity, none the less, is absent. The direction taken by the equator of Uranus, as we have seen, is indicated in a two-fold manner: first, by the trend of the belts; secondly, by the lie of the major-axis. And these indications agree. Supposed discrepancies between them have been reconciled by improvements in the conditions of observation. But with the equatorial line the plane of satellite-revolution cannot be brought to coincide. The angle of divergence is uncertain, but may be put roughly at 20°. This would give 78° for the inclination of the Uranian equator, so that the rotation of the planet is likely to be direct. If so, the extraordinary anomaly is here met with of a satellite-system circulating in a direction opposite to that of its primary's rotation.

Uranus can at times be perceived with the naked eye. Indian traditions of an eighth "dark" planet have been thought to refer to it, and its slow course among the stars had been noted by savage tribes long before Herschel singled it out from them by its tiny disc. It is about three times brighter than Vesta; and Mr. Proctor stated that "in the summer of 1887 they were comparable under favourable conditions," when both, in the transparent skies of Florida, were "quite conspicuous without telescopic aid." Twenty chances of discovering Uranus were missed before it came to Herschel's turn. So many times it had been located or catalogued as a fixed star by astronomers far from indifferent to immortal fame.

Neptune is much nearer to the sun than it ought to be. Both Leverrier and Adams assumed that Bode's law would hold good for the planet still below the horizon of knowledge; they could do no otherwise; yet the rule played them false. Some have even asserted paradoxically that the planet found was not the planet sought. In point of fact, the distance of the theoretical Neptune is thirty-eight, that of the real Neptune thirty astronomical units. The mean radius of its orbit measures 2,792 million miles. Hence the sun is reduced to $\frac{1}{900}$th its terrestrial brilliancy, and could be replaced by 687 full moons. "As seen from Neptune," Professor Young remarks, "the sun would look very much like a large electric arc lamp at a distance of a few feet. It would give about forty-four millions the light of a first-magnitude star."[1] Accordingly, Neptune does not circulate by any means in outer darkness. His orbit, although very slightly eccentric, brings him at perihelion fifty millions of miles nearer to the sun than at aphelion. It makes an angle of less than 2° with the ecliptic, and is traversed, at the rate of 3⅓ miles a second, in a period of 165 years.

Neptune, being fainter than the eighth stellar magnitude, is quite inaccessible to unaided vision. But a good telescope at once displays the seeming star in the guise of a small planetary nebula with a diameter of 2″·433. This mean value, reduced to the mean distance of the planet from the sun, was afforded by Barnard's measures in 1895 with a power of 1,000 on the Lick refractor.[2] It corresponds to a linear diameter of 32,900 miles. Neptune accordingly, although only 17 times more massive than the earth, is 72 times more bulky, and composed of materials 4·2 times specifically lighter. Gravity at its surface has almost precisely its terrestrial power. The albedo of Neptune, combining Zöllner's with Müller's results, is 0·65; and its spectrum appears identical with that of Uranus. It may be inferred that this planet also is too hot to contain water.

[1] "General Astronomy," p. 372.
[2] *Astronomical Journal*, No. 342.

Its satellite is believed to be of about the size of the moon; but since it is 12,000 times more distant, it can be distinguished only with the most powerful telescopes as a star of the fourteenth magnitude. The radius of its orbit measures 225,000, that of our moon 238,000 miles; but Neptune's attendant completes a circuit in 5 days 21 hours; and it is through this rapidity of movement that the large mass of its primary has been learned. It resembles the moon besides in being solitary, so far as can be ascertained by the most diligent researches; and it is beyond doubt that if any companion-bodies exist they are comparatively small or obscure. That they do exist, appears probable on the face of it.

The one Neptunian satellite emphasises the problems set by the Uranian four. These problems are concerned with the origin and early mechanical relations of the solar system. Here, at its utmost verge, we encounter a decided reversal in the direction of systemic motion—a reversal prepared for, as it might seem, by the nearly vertical position of the Uranian plane of satellite-revolution. This diversity is in no sense "accidental," as some have unwisely asserted, invoking impacts of comets, and such like futile devices, to account for it; it belongs fundamentally to the design of planetary evolution. Laplace's scheme has no room for it; Faye's, constructed expressly to include it, requires that Uranus and Neptune, instead of being the first, should have been the latest formed of all the solar train. And their obviously rudimentary condition favours the suggestion. Neptune's satellite revolves from east to west in a quasi-circular path, inclined to the ecliptic at an angle of 35°; or, putting it otherwise, it revolves from west to east at an angle of 145°.

As the only member of the solar system exempt from perturbations by a third body (the sun being too remote to cause perceptible deflections), it seemed admirably fitted to discharge the functions of a standard celestial clock, greatly needed, but nowhere to be found in our system.[1] But in 1886 Mr. Marth drew attention to certain divagations of this

[1] Tisserand: *Astronomy and Astro-Physics*, vol. xiii., p. 291 (1894).

"ideal time-keeper" resulting from conspicuous changes in the position and plane of its orbit. They were explained almost simultaneously in 1888 by M. Tisserand,[1] late director of the Paris Observatory, and by Professor Newcomb.[2] The disturbance, which, in its mode of production, is analogous to the precession of the equinoxes, results from the polar compression of the Neptunian globe combined with a deviation of the satellite's motion from its equatorial plane. By the action of the protuberant girdle, a slow gyration of the secondary body's orbital plane is produced, its inclination to the primary's equator remaining unchanged. Viewed under a different aspect, the same phenomenon may be described as a retrograde movement, in a period of at least five hundred years, of the pole of the satellite's orbit round the pole of the planet's equator. The radius of the circle described cannot be less than 20°, implying a flattening of the Neptunian globe of $\frac{1}{85}$th, and may easily amount to 30°, with which an ellipticity of $\frac{1}{115}$ should be associated. But before the centre of this circle—that is, the pole of Neptune's axial movement—can be satisfactorily located, several centuries must elapse. At present we may affirm with reasonable certainty: first, that the rotation in question is retrograde, like the satellite's revolution; secondly, basing the inference upon the comparatively slight ellipticity of Neptune's figure, that it is much slower than the vertiginous spinning of Jupiter, Saturn, and Uranus.

Uranus and Neptune are, as has been said, companion globes. In bulk and density they differ very slightly; their albedoes are virtually the same, their spectra indistinguishable. They seem perfectly alike in chemical and physical constitution, and to be situated at precisely the same stage of development. Both govern retrograde systems. In Uranus the peculiarity appears as if in an incipient form; in Neptune, strongly accentuated.

Viewed from the position of Neptune, all the planets are morning and evening stars. They are tethered to the

[1] *Comptes Rendus*, t. cvii., p. 804.

[2] *Astronomical Journal*, No. 186.

chariot-wheels of the sun, instead of having the run of the sky. "The four terrestrial planets," Professor Young writes, "would be hopelessly invisible, unless with powerful telescopes, and by carefully screening off sunlight. Mars would never reach an elongation of three degrees from the sun; the maximum elongation of the earth would be two, and that of Venus about one and a half degrees. Jupiter, attaining an elongation of about ten degrees, would probably be easily seen somewhat as we see Mercury. Saturn and Uranus would be conspicuous, though the latter is the only planet of the whole system that can be better seen from Neptune than it can be from the earth." [1]

To a spectator retreating with the velocity of light, all the planetary cortège would in a few hours disappear, and the sun would shine alone. No sign would remain that his office is purely ministerial—that he exists only to enlighten, rule, and vivify the relatively minute globes shred from his mass in the beginning, maintaining by his attractive power the adjusted movements of the complicated piece of mechanism they constitute. The skies perhaps hold millions of his stamp; every solitary star telescopically visible may be the centre of a planetary scheme like our own; or, on the other hand, our own may, quite conceivably, have no counterpart in the wide universe.

CHAPTER XI.

FAMOUS COMETS.

In the fourth year of the 101st Olympiad (373 B.C.), the Greeks were startled by a celestial portent. They did not, at that time, draw fine distinctions, and posterity would have remained ignorant that the terrifying object was a great

[1] "General Astronomy," p. 372.

comet but for the description of it left by Aristotle, who saw it as a boy at Stagira. It was mid-winter when it flared up from due west at sunset, its narrow, definite tail running "like a road through the constellations" over a third of the heavens. Diodorus relates that it cast shadows like the moon, which implies a very unusual, yet not impossible, degree of brightness. The prompt engulfment by an earthquake and its attendant tidal wave of the Achaean towns, Helice and Bura, justified the apprehensions it aroused. It never came back to retrieve its reputation. During at least two thousand subsequent years, such objects lay under the ban of popular superstition; and the counts upon which they were accused of malefic influence were so many and so vague that acquittal was impossible. Their respect of persons was notorious; nor were they consistent in their dealings with the great, to whom alone they paid individual attention. A comet marked the apotheosis of the great Julius; a comet announced the death of Constantine; a comet illuminated the cradle of Napoleon.

The very word "comet" takes us back to the Stagyrite; for it is derived from the Greek word κόμη, hair, and signifies a *hirsute* star. Shakspeare's "crystal tresses" represent what we now, in homely fashion, call the "tail," while the "nucleus" and "coma" make up the "head." The nucleus, in great comets, shines like a star of the first magnitude, sometimes indeed surpassing the brilliancy of Jupiter. It is usually of measurable dimensions, often of granular texture. The planetary disc, round which the filmy appendages of the comet of December 1618 were displayed, was observed by Cysatus, a Jesuit astronomer at Ingolstadt, to become transformed into the semblance of a star-cluster; Hevelius noticed a double nucleus in the comet of 1652; and modern instances of the same kind abound. There is indeed no likelihood that substantial globes are ever included in the construction of comets.

The coma is of immense volume, and extreme tenuity.

The rays of faint stars traverse, undimmed and unrefracted, strata of it tens of thousands of miles in thickness. Yet strong lines of structure develop in it through the influence of forces emanating from the sun. As they approach our system out of the depths of space, comets are scarcely distinguishable from round nebulæ, and they relapse into a similar quiescent condition on leaving it. Their temperature must then be very near the absolute zero of cold, since they cannot be supposed either to contain stores of native heat, or to retain stores of borrowed heat. Thus the rapidly augmenting power of solar radiation, as they rush with accelerated velocity nearer and nearer to its source, produce upon them stupendous effects. The nucleus blazes out into a coruscating star; the coma, violently driven off from it, forms multiple envelopes like thin gauze veils, one outside the other, flung round the nucleus on the side next the sun, separated by intervening dark spaces, and diversified by brilliant jets and sectors. The tail is the outcome of a double repulsion. Matter expelled by the nucleus towards the sun is, at a certain point, thrown back to form an immense, oppositely directed appendage, usually convex on the forward side. Some tails resemble hollow cones, being bright at the edges, and dark within: others are traversed by a shining *backbone;* many, perhaps all, are composite. The magnificent object first seen by Klinkenberg at Haarlem, December 9, 1743, was supplied with six, varying in length from 30° to 44°, each, according to the extant representations, being separately *rooted* in the head. Grouped into a lustrous fan, they presented a very beautiful and surprising appearance, not again to be displayed until the world and humanity have undergone some unlooked-for changes. For the period of the comet was computed to be one hundred thousand years! Tails, less obviously and splendidly multiplex, are rather the rule than an exception. Or rather, closer observations, chiefly photographic, have made it manifest that the single efflux of nebulous stuff generally designated as a comet's tail can be analysed into bundles of fibres, into straight rays and curved plumes of light, or into knotted and branching emanations. Homogeneous outflows, such as are seen in drawings, do not really exist. Tails pointing *towards* the sun have also been occasionally noticed; but they are always feeble. Olbers recorded, however, that, during eight days of January, 1824, the comet then visible had a solar tail of 7°, while its anti-solar tail was only $3\frac{1}{2}$° long.

The great comet of 1680 will always be memorable for having had its orbit calculated by Newton on gravitational principles. It was not unworthy of the distinction. Approaching the sun almost in a straight line, it penetrated the corona at the rate of 370 miles a second, and passing within 140,000 miles of the photosphere, escaped by means of its extraordinary velocity from those perilous precincts. Resulting internal commotions became evident through the rapid development of a tail more than a hundred million miles in length. Newton calculated that particles from the head reached its extremity in two days. He assigned to the comet a highly elliptical orbit traversed in six centuries. But, since its speed might be called parabolic, millenniums may be nearer the mark than centuries. It cannot, therefore, be identified with any earlier apparition.

The comet of 1682 was Halley's, the predicted return of which, in 1759, was unprecedented and memorable. At its apparition in 1835, valuable observations of a physical kind were made upon it by Bessel at Königsberg, and by Sir John Herschel at the Cape. They were facilitated by the circumstance that this far-travelling body, the perihelion-distance of which is 55 million miles, and the aphelion-distance $2\frac{1}{2}$ times that of Neptune, approached the earth on this occasion within $4\frac{1}{2}$ million miles. It was remarkable for singular and sudden changes of aspect. To Bessel the nucleus seemed like a burning rocket. Divergent flames issued from it towards the sun, and he took especial note of a blazing "sector," which swung like a pendulum to and fro, in a period of $4\frac{3}{5}$ days. These emanations, accumulating at the surface where the solar balanced the cometary repulsive force, were then swept back, as if by a tempestuous wind, to form a tail, which, on October 15, measured at least 24°. The conviction was forced upon him that the body in which these wonderful processes were going on was affected by opposite polarities; and he fully concurred with Olbers in the opinion that tail-production was a purely electrical phenomenon.

During some time before and after its perihelion-passage on November 16, the comet wore the disguise of a star. All its hairy appendages had vanished. On the 23rd of January, 1836, it was sharply stellar; twenty-four hours later it had acquired, besides a twenty-fold increase of light, a disc like that of the planet Neptune, enclosed in a nebulous sheath of about four-fold breadth. Later in its career, Sir John Herschel[1] observed the nucleus under the form of "a miniature comet, having a nucleus, head, and tail of its own, perfectly distinct, and considerably exceeding in intensity of light the nebulous disc or envelope" containing it, which was, properly speaking, the "head" of the comet. At last, on May 5, through the progress of distension, the last thin shred of its substance melted into the sky. The next return of Halley's comet, somewhat accelerated by Jupiter's influence, is looked for in the year 1910.

The "vintage comet" lingered in northern skies during 510 days—from March 26, 1811, until August 17, 1812. It was attentively observed by Sir William Herschel, who gathered from it the then new truth that comets are self-luminous bodies. "The quality of giving out light," he acutely remarked, "is immensely increased by an approach to the sun." But he failed to persuade his contemporaries or successors. His inference had to wait for spectroscopic demonstration. The nucleus of the comet of 1811 he found to measure 428 miles. It showed a ruddy hue, and was eccentrically placed within a greenish-blue "planetary body" 127,000 miles in diameter. This was again enclosed in a shining atmosphere about four times as wide, round which

[1] "Observations at the Cape of Good Hope," p. 396.

was flung an envelope of a yellow tint, forming a thin hemispherical shell on the side next the sun, and continued indefinitely away from the sun as the hollow cone of the tail. Owing to this mode of construction, the space between the head and the hemispherical sheath, as well as the central part of the tail, appeared dark. The latter extended, in October, over 100 million miles of space, and was 15 million miles broad. Its soft radiance resembled that of the Milky Way, side by side with which it ran on November 9, 1811. The comet's path lay entirely outside the earth's orbit, and Argelander assigned to it a period of 3,065 years. The restriction was needless. Between a period of infinite length, and one of 3,000, or 1,000 years, no valid distinction can, where comets are in question, be drawn. The short sections of their tracks observable from the earth might belong equally well to parabolas or to the far-stretching ellipses which such protracted periods imply.

The apparition of 1811 suggested to Olbers the "electrical theory" of comets' tails. The uncommon impressiveness with which it displayed not uncommon phenomena, was perhaps a result of its considerable distance from the sun, owing to which the *interior* force obtained an advantage over the *exterior*, and the locus of equilibrium between solar and cometary repulsion was pushed back further than usual from the nucleus.[1] He calculated that the materials of the tail spent 11 minutes in the journey from its root to its tip, indicating ejection by a force greatly more powerful than the opposing force of gravity. Olbers anticipated the modern view that chemical differences determine the shapes of comets' tails, the various species of matter being diversely acted upon by electrical repulsion. The long, straight ray, for instance, issuing from the comet of 1807, must, he perceived, have been composed of particles much more energetically repelled than those aggregated in the inflected plume with which it was associated. The curvature of these appendages, in fact, depends upon the relation between the orbital

[1] "Monat. Correspondenz," Bd. xxv., pp. 3-22, 1812.

velocity of the comet and the velocity of ejection imparted to their constituent molecules. It has to be borne in mind, however, that while curved tails may appear straight in projection, straight tails can never appear curved.

Olbers's classification of comets is still of great significance. He divided them into:

(1.) Comets which develop no matter subject to solar repulsion. These are without tails, and may be regarded as simple nebulosities devoid of solid nuclei.

(2.) Comets showing no trace of nuclear, while subject to solar repulsion. They throw out no matter *towards* the sun; the heads are consequently left bare of envelopes, and are of simple structure. The comet of 1807 was of this kind.

(3.) Comets manifesting the effects of both species of action. They are characterised by the presence of a dark hoop round the head, and of a dark rift in the tail, by which it may be judged to be a hollow conoid.

On February 28, 1843, a "short, dagger-like object" blazed out at an interval of only fifty-two minutes of arc from the sun's limb. It was viewed with amazement in various parts of the world; and spectators in Italy, by shielding their eyes from the direct mid-day glare, were able to discern a tail already several degrees long. The proportions of the appendage rapidly grew. On March 3, it measured twenty-five degrees; on March 11, an adjunct to it shot out, within twenty-four hours, to nearly twice the apparent length of the main structure, conveying, as Sir John Herschel said, "an astounding impression of the intensity of the forces at work." It was first seen in this country after sunset on March 17, as "a perfectly straight, narrow band of white cloud, thirty degrees in length, and about one and a half in width." On the following night, Sir John identified this "luminous appearance" as the tail of a grand comet, stretching over an extent of space (as it afterwards proved) of no less than two hundred millions of miles.

The movements of this body were as surprising as its aspect. It rushed past perihelion with a speed of 366 miles a second, leaving an interval of 100,000 miles between its centre and the sun's surface, and swinging through two right angles in two hours and eleven minutes. The northern part of its course was finished in two hours and a half; hence, it was a "southern" comet. Very curiously, it seems to have remained obscure throughout its journey towards the sun, reserving its outburst for the day *after* perihelion. Periods were assigned to it ranging from seven to six hundred years.

Strangest of all, it turned out to be but one member of a whole family of similarly-conditioned bodies. The "great southern comet" of February, 1880, seemed like its ghost. It had no perceptible nucleus, but an inordinately extended train, which rapidly faded; and it scarcely deviated by a hair's breadth from the track of its predecessor. That is to say, so far as could be ascertained; for the object was so indefinite as to elude exact observation. Its period could not even be conjectured. The nature of the relationship between the comets was thus left uncertain.

But after the lapse of two years and a half, the question was reopened by the appearance of the leading constituent of the group. Like the comet of 1843, the "great September comet" of 1882, was first seen close beside the sun. At Ealing, shortly before noon, on September 17, Dr. Common was struck with the astonishing spectacle of a brilliant comet hurrying up to perihelion. A transit was evidently imminent, but clouds veiled the scene. Its completion was, however, fortunately witnessed six thousand miles away by Mr. Finlay and Dr. Elkin at the Cape Observatory. The comet was watched by them "right into the boiling of the limb," which it had no sooner touched, than it utterly disappeared. This cannot have been through the absence of contrast; for although its intrinsic brilliancy was excessive, it must either have shown bright against the sun's dusky margin, or dark when projected upon his dazzling centre. Since neither effect was produced, it can only be inferred that the object was translucent owing to insubstantiality. That it had not passed *behind* the sun was later fully ascertained. During three subsequent days the "blazing star near the sun" drew popular attention in the southern hemisphere, and many parts of Europe. Nothing quite so extraordinary had ever been seen before. The spectacle of 1843 was renewed, but outdone.

Meanwhile, an astonished public hung on the dicta of perplexed astronomers. The speculation which obtained most currency was that the three successive southern comets were accelerated returns of the same body, destined, after a few short, spiral circuits, to make fiery shipwreck in the glowing solar ocean. The effects upon terrestrial life were unwarrantably described as likely to prove disastrous ; but only an abortive panic ensued. Data, however, to serve as the basis of a determinate conclusion, were on this occasion collected in abundance. The comet of 1882 was not lost sight of until June 1, 1883, when its distance from the earth was more than five astronomical units—the greatest at which any previous comet except that of 1729 had been observed. Hence the general character of its orbit became thoroughly known. It proved to deviate somewhat from the tracks pursued by the comets of 1843 and 1880; it gave the sun a slightly wider berth ; above all, its period had unmistakably a duration of several centuries. There could then be no further question of its being a return of either, or both of those bodies, although its close connexion with them was assured. This can be most rationally explained by supposing them to have primitively constituted a single body. According to Professor Kreutz's able and exhaustive research, the period of the September comet is 772, that of the comet of 1843, between five and six hundred years; and the relative situation of their orbits indicates that the supposed catastrophe of their disruption took place at perihelion, where a large incoherent mass could scarcely fail to yield to the strain of the sun's unequal attraction at the excessively close quarters it was brought into by the conditions of its movement. The comet of 1880 is another splinter from the same trunk ; and yet one more fragment presented itself to M. Thome at Cordoba, January 18, 1887, when he observed literally a "nine days' wonder" in the guise of a shadowy ray, thirty-five degrees in extent, following the lead of the other "southern comets," and taking rank (so far) as the last and least of their company.

A tendency to still further disaggregation was evident in the comet of 1882. It did not pass with impunity through the fiery ordeal of its visit to the sun; internal agitations supervened; abnormal appendages of rarefied texture, but prodigious dimensions, issued from it sunward; the nucleus broke up into six spherules like strung pearls; and it was noticed in October to be surrounded by detached nebulous masses, just launched perhaps on independent cometary careers. The tail was two-fold. It consisted of a dim, straight ray which temporarily attained a length of a couple of hundred millions of miles, and a massive forked appendage, strongly luminous and unusually permanent. Fig. 19 shows

FIG. 19.—*Great Comet of September*, 1882. *Photographed at the Royal Observatory, Cape of Good Hope.* (From Clerke's "History of Astronomy," 3rd ed.)

one of a series of photographs of this comet taken with an ordinary portrait-lens under Dr. Gill's direction in October, 1882. The observations of its transit proved to be of great importance. Having been made just before perihelion, they availed to demonstrate that no loss of motion had been suffered in its plunge through the corona. This incontrovertible fact implies an inconceivable degree of rarity in the solar surroundings.

So long ago as 1831, Clausen pointed out that many comets are grouped together after the manner incomparably exemplified later by the southern comets. An analogous system, composed of only two known members, is formed by the comet of 1807, and Tebbutt's comet of 1881. The former, made by Bessel the subject of a masterly investigation, was not again due at perihelion until the remote epoch 3346 A.D., so that the announcement of a re-appearance so exceedingly premature was startling. But when the new comet was also found to have a period of several thousand years, it became clear that no return had been observed, but only a companion recognised. Tebbutt's comet was a beautiful object. Its head, adorned with interlacing arcs of light, was an overmatch for Capella, while so translucent that a star of the seventh magnitude seemed rather to gain than to lose brightness by shining centrally through it. As the upshot of these singular experiences, the difficulty of identifying comets has been increased tenfold. Their aspects were always perceived to be well-nigh interchangeable, but their movements were held to be distinctive; now their very orbits are found to be, to a considerable extent, common property.

A small, glimmering nebulosity descried at Florence by Donati, June 2, 1858, gave little promise of coming splendour. Yet few more picturesque celestial effects have been witnessed than it presented, October 5, when Arcturus blazed undimmed through the denser part of the tail, in brilliant conjunction with the equal splendour of the nucleus. The ineffable grace with which the comet spread its luminous plumage was set off by the juxtaposition, as if for the purpose of determining the amount of its curvature, of a long, perfectly straight ray. The aspect of this beautiful object on October 3, is represented in Fig. 20; some idea of its rapid development in size and brilliancy can be gathered from an inspection of the Frontispiece to this Section. The apparition lasted, to the naked eye, for 112 days, and will not again be visible for 2,000 years. So that Donati's comet may be reckoned an "irrevocable traveller."

Twice during the present century the earth has traversed, with impunity, the tail of a comet. First, on June 26, 1819,

FIG. 20.—*Donati's Comet, October* 3, 1858. (From Langley's "New Astronomy.") *The Star to the left of the Comet's head is Arcturus.*

when a comet passed invisibly between us and the sun, sending its tail our way. Again on June 30, 1861. The sun had scarcely set that evening when a yellowish disc became apparent at the horizon, from which issued an enormous double train, enclosing our planet within its folds. The closing-up and withdrawal of the "outspread fan" to which they were compared was accomplished in a few hours. The head of the comet had as many envelopes as a Chinese puzzle.

The first recognised "short-period" comet approached within one and a half million miles of the earth, July 1, 1770. Had it possessed $\frac{1}{5000}$th the mass of the globe which rushed by it with entire indifference, a preceptible lengthening of the year should have ensued; and its gravitational insignificance was confirmed by the fact that it passed, in 1779, right through the Jovian system without troubling the mutual relations of its members. Lexell (with whose name it has continued to be associated) fixed its period of revolution at five and a half years; yet it had never been seen before. Astronomers, in fact, caught it on its trial trip along a fresh orbit to which it had been transported in 1767 by the disturbing power of Jupiter, and whence it was removed by the same influence in 1779. An intermediate return in 1776 had doubtless occurred; but circumstances precluded its observation. Further encounters with the giant planet may, however, bring back the vagrant, and the possibility was thought to have been realised when the history of a comet discovered by Mr. Brooks of Geneva, N.Y., July 6, 1889, came to be inquired into. Its return about the predicted time in 1896 afforded an opportunity for revising the laborious inquiry, with the result of disproving the case for identity.

A comet, lost under very different circumstances, was picked up February 27, 1826, by an Austrian officer, Wilhelm von Biela. His calculations led him to the unlooked-for discovery that it travelled in an orbit with a period of 6½ years, and had already been observed in 1772 and in 1805. On its return in 1832, when it had become reduced to the status of a telescopic object, Sir John Herschel watched its conjunction with a knot of minute stars, the rays of which traversed it without the smallest obstruction. It had neither tail nor nucleus; its aspect was that of the commonest type of nebula. On December 29, 1845, however, a curious change was seen to have affected it. The comet had split into two, each of which immediately assumed the characteristic cometary shape, by providing itself with a tail and bright nucleus. Thus divided and regenerated, the pair advanced side by side, 157,000 miles apart, without the least trace of mutual action through gravity, but displaying vivid interchanges of brightness, reasonably attributed to the play of electrical forces.[1] They re-visited the sun in 1852, but have never since, and most probably will never again, be seen. Their end came through senile decay. It was that predicted by Newton for all such bodies. *Diffundi tandem et spargi per universos cœlos.*

The most rapidly-revolving comet of our acquaintance was investigated in 1819 by Johann Franz Encke, of the Seeberg Observatory, who assigned to it a period of 3⅓ years, and predicted its return in May, 1822. It was punctually recaptured at Sir Thomas Brisbane's Observatory in New South Wales. Encke traced back its appearances to 1786, and identified it with a comet detected by Caroline Herschel in 1795. At its last return in 1894-5, it was just at the limit of naked eye visibility. It fluctuates, however, considerably, at successive apparitions. M. Berberich[2] has sought to associate these perplexing changes with solar vicissitudes; but his arguments are not entirely convincing. Encke's comet, even if 45,000 billion times less dense than air at atmospheric pressure—the consistence attributed by Babinet to cometary matter—would still weigh twelve hundred tons.[3] Its excessive rarefaction is a matter of ocular proof. On October 21, 1881, Barnard observed a central passage of this comet, then more than usually bright and condensed, over a

[1] Fessenden: *Astro-Physical Journal*, vol. iii., p. 40.
[2] *Astr. Nach.*, No. 2837.
[3] Guillemin: "The World of Comets," p. 282.

ninth magnitude star, which "remained so remarkably distinct during the entire progress of occultation, that it formally impressed me with the idea of a transit of the star *across* the comet—a pearly point floating between me and the bright mass of vapour."[1]

This object signally exemplifies the cometary peculiarity of contracting near perihelion, and re-expanding after the critical point has been passed. Thus, it measured 312,000 miles across, October 28, 1828, when 135 million miles from the sun, but only 14,000 on December 24, when its distance had been reduced to 50 millions; and in passing perihelion, December 17, 1838, at an interval of 32 millions, its diameter had shrunk to 3,000 miles. It fulfils, as regards Mercury, the function of spying upon the planets, assigned to comets by Airy; for, only through the Mercurian disturbances of its motion has the Mercurian mass been at all definitely ascertained; and a residual acceleration, which, at each circuit, brings it back to perihelion a couple of hours before the appointed time, has long been regarded as an index to the condition of planetary space. Encke explained this shortening of period by the action of an hypothetical "resisting medium" augmenting in density towards the sun; but accumulated facts have swept it out of existence. The southern comets performed for our benefit, one after the other, an *experimentum crucis* in the matter. The chief of them, on September 17, 1882, swept through a region where Encke's medium should be *two hundred thousand* times denser than it is at the perihelion distance of Encke's comet; yet suffered no appreciable loss of motion. Nor has the comet itself of late complied with the requirements of the theory it suggested. At its return to the sun in 1868, the acceleration had fallen to one-half its customary, and until then, constant value. And the change has proved to be permanent. But the influence of the postulated medium is evidently incapable of diminution. Thus, the movements of Encke's comet still remain problematical.

[1] *Astr. Nach.*, No. 2437.

CHAPTER XII.

NATURE AND ORIGIN OF COMETS.

COMETS reflect sunlight, and also emit light of their own. But the combination was scarcely thought of as possible until the spectroscope gave its verdict. The first analysis of cometary rays was made by Donati at Florence, August 5, 1864. They were dispersed by his prisms into a yellow, a green, and a blue band, with wide intervals between. Their chemical interpretation was afforded by Dr. Huggins in 1868. The subject of his experiments was Winnecke's comet, an insignificant object with a period of five and a half years. He found it to be composed—at least in part—of acetylene, or some other hydro-carbon gas. The coloured bands agreed precisely in position with those in the spectrum of the blue light at the base of a candle-flame, or of a gas-jet. The spectra of the immense majority of comets is of this pattern, with more or less of continuous light added. A portion of this is borrowed, a portion inherent. A photograph of the spectrum of Tebbutt's comet (1881, III.), taken by Dr. Huggins, June 24, 1881, demonstrated by its distinct impression with several Fraunhofer lines the presence of solar radiance; the association of which with native emissions of the continuous sort has been made evident in various comets by sudden outbursts of white light.

Comets do not then consist entirely of carbon-compounds; but their remaining constituents make no distinctive show in their spectra unless when sun-raised agitation is particularly vehement. Thus, an approach within five million miles of the sun evoked in comet Wells (1882, I.), sodium-luminosity, detected by Dr. Copeland at Dunecht, June 17, 1882. The blaze was so vivid that a crocus-tinted image of the entire head with the beginning of the tail was visible, like a solar prominence, through the open slit of the spectroscope. The same observer witnessed an outbreak of both sodium and iron lines in the September comet (1882, II.). In both cases, the newly-kindled emissions effaced the old, and, after a time, were replaced by them. This mode of procedure is characteristic of electrical action, and combines with other symptoms to assure us that cometary illumination is produced by interior electrical disruptive discharges due to solar induction.

Olbers's felicitous conjecture has been developed into a plausible theory of comets' tails by M. Bredichin, late director of the Pulkowa Observatory. He divided them into three "types," distinguished by the values of the repulsive forces employed severally in their production. Those belonging to type I. imply the exertion of a counter-influence fourteen times stronger than gravity. They are long, straight rays, the constituent particles of which are carried, in a torrent too swift to be deflected, to the observed extraordinary distances. Their outward velocity of five miles a second to start with is, we must remember, constantly accelerated, and finally becomes enormous. Halley's comet and the great comets of 1811 and 1861 had tails of this type. Donati's great plume exemplified the second, in which the average strength of repulsion exceeds that of gravity one and a half times. Tails of the third type correspond to a ratio varying from three-tenths to one-tenth. Solar attraction is, in them, only partially neutralised. They are short, strongly-bent, brush-like appendages, seldom seen apart from those of a more striking kind.

These three types have a physical meaning of great interest. The attractive force of gravity varies as the mass, the repulsive force of electricity as the surface of the molecules they sway; hence the ratio of repulsion is inversely as the ratio of molecular weight, the lightest particles being the most violently driven away from the sun. Assuming them to be hydrogen-molecules, Bredichin found that the atomic weights of hydro-carbon gases and iron would correspond fairly well with the speed of projection signified respectively by the curvatures of the second and third types of tail. Materials of other kinds are not excluded; their presence is, indeed, demanded by the width of these appendages, which obviously consist of bundles of emanations differently influenced, and presumably of a different chemical nature. Bredichin's theory works admirably from a geometrical point of view. All the varieties of cometary trains can be constructed by strict calculation from the basis it supplies. Yet there are spectroscopic difficulties in the way of accepting it unreservedly. No evidence is at present forthcoming of any connexion between the chemistry of tails and their shapes; and hydrogen-rays are conspicuously absent from cometary spectra.

"Short period," or "planetary" comets may be defined as those revolving in periods of less than eight years. They have much more in common, however, than the quickness of their successive returns to the sun. All move from west to east; they show some preference for the plane of the ecliptic; and none of their orbits are excessively elongated. Thus, they tend towards conformity with the regular ordinances of the solar system, which its less accustomed visitants completely ignore. All, too, have a *used-up* appearance. This is easily understood. They have wasted their substance spinning out nebulous appendages—*sicut bombyces filo fundendo*, as Kepler said—at their frequent returns to perihelion. They are thus visibly effete bodies. Before long, they will drop out of individual existence, and survive obscurely, reduced to the "dust of death." Yet the supply is not likely to become exhausted. Discovery proceeds faster than disappearance.

"Lost comets" belong, without exception, to this class. Two typical instances have already been mentioned in the disaggregation of Biela's, and the removal of Lexell's comet. The fate of Biela may have been shared by Brorsen's, a comet with an established period of five and a half years, which has, nevertheless, remained submerged since 1879. It is believed by Dr. Lamp to have exploded through internal forces in 1881, and he recognises as one of its fragments a faint comet detected by Mr. Denning at Bristol, March 26,

1894. The adventures of displaced comets, such as Lexell's can be traced only by arduous and delicate inquiries. They depend upon a single cause. Unsettled comets are those which pass near Jupiter's orbit, and are subject to encounters with his mighty mass. And since they must necessarily return to the point of disturbance, the series of their vicissitudes can come to an end only by their being driven off finally from the solar system along a hyperbolic path.

The condition of these bodies might be described by saying that, in the regular course of things, they revolve round the sun disturbed by Jupiter; while, during brief but energetic crises, they revolve round Jupiter disturbed by the sun. Their abnormal condition results from the situation of their aphelia close to the Jovian track. This is the case, in a minor degree, with many comets of comparatively settled habits. They escape eviction and exile, and suffer only disquietment. Such are Winnecke's, D'Arrest's, Faye's comets, which, having been continuously observed during half a century, are, as Mr. Plummer expresses it, "well under control."[1]

Short-period comets, with the solitary exception of Encke's, appear to be inevitably connected with Jupiter. The peculiarity is rendered more significant by the circumstance that the other great planets are also provided with cometary clients. The Jovian group is the largest; it includes more than two dozen recognised individuals. Saturn claims nine, Uranus eight, and Neptune five. Halley's comet belongs to the Neptunian family. Another of its members was discovered by Pons in 1812, and re-discovered by Brooks in 1883, so that it has a period of 71 years. And the re-appearance in 1887 of a comet first seen by Olbers in 1815, bore reassuring testimony to the regularity with which Neptune's comets conduct themselves during their long periods of invisibility.

The nature of these planetary relationships was at once conjectured. It seemed an open secret that the comets had been taken prisoners by the attractive force of the great globes they flitted past on their way to the sun. But

[1] *Knowledge*, Feb., 1896, p 41.

astonomers can take nothing for granted ; and preliminary mathematical inquiries served rather to discredit the first and easy surmise. The case had to be thoroughly sifted ; and it was only through the profound researches of Tisserand, Callandreau, and Newton of Yale, that the "capture-theory" has taken its place as a highly probable truth. With an unstinted allowance of time and *comets*, it can perform all that is required of it. "Captures" are not effected all at once ; the lasso is thrown many times over the escaping body before it is definitively secured. Moreover, at each such effort, the chances are even of its being made in the wrong direction. We observe only the outcome of the hits ; the misses are beyond our reckoning. A multitude of happy accidents have led to the domestication in our system of Faye's, Tuttle's, Winnecke's, D'Arrest's comets. Mr. Plummer has adverted to the likelihood that we are indebted to some slight but well-directed pulls from Mercury for the permanent addition of Encke to the solar company ; and Neptune exerted itself ages ago with similar success as regards Halley's comet, yet under great difficulties, since retrograde comets, and those with highly inclined orbits are, as a rule, exempt from capture. This is one of the reasons why short-period comets show some degree of conformity to planetary modes of motion.

These investigations remove all doubt as to the foreign origin of comets. Those that are in the solar system are not of it. They assuredly remained unaffected by the gradual processes of its development. Yet they, as well as the multitude of parabolic comets, belong to it in a wider sense. That is to say, they accompany its march through space. Otherwise, as M. Fabry has demonstrated, most of their orbits should be strongly hyperbolic ; and no such cometary orbits are known. They should, besides, if casually encountered, present themselves chiefly along the line of the sun's way ; they arrive, on the contrary, indifferently from all quarters of the heavens. They are then subject to the same mysterious influences which govern his motion, and drift with the

FIG. 21.—*Photograph of Swift's Comet. Taken by Prof. Barnard, April* 6, 1892. *Exposure*, 1*h*. 5*m*.

FIG. 22.—*Photograph of Swift's Comet. Taken by Prof. Barnard 24h. later. Exposure, 50m.*

cosmic current which bears the solar family along, we know not how or whither.

Comet-photography became possible only through the introduction of highly-sensitive gelatine plates; and even with them, exposures of an hour and upwards are necessary in order to obtain the desired results. But these results are of such importance as to deserve the closest attention. For investigating either the forms or the spectra of comets, the camera is unrivalled. Its systematic employment for these purposes dates from 1892. It can also serve as an engine of discovery. On October 12, 1892, a comet so faint that, had it not been photographed, it would most likely never have been seen, appeared as a nebulous trail on a plate exposed by Professor Barnard to the Milky Way in Aquila. It proved to be one of Jupiter's dependents, pursuing, in a period of 6·3 years, a track so closely resembling the orbit of Wolf's comet in 1884, that Schulhof regarded them as the offspring of one parent body.

In the year 1892, seven comets were detected; and all, by one of those picturesque coincidences with which nature loves to entertain her devotees, were, towards its close, visible in the sky together. One of them was first noticed by Lewis Swift—a specialist in that line—and passed perihelion April 6.[1] The head competed in brightness with a third-magnitude star; the tail was 20° long, and came out, in a photograph taken by Mr. Russell at Sydney, on March 22, self-analysed into eight perfectly distinct rays. *No such structure could be seen with the telescope.* Figs. 21 and 22 reproduce two pictures of this object obtained by Professor Barnard, April 6 and 7 respectively. During the interval, a striking change had occurred. In the first photograph, the tail is sharply separated into two branches, and shows traces of further indefinite subdivisions. The uneven, knotty texture of the main stream is obvious. The matter composing it seems as if it had rushed in a torrent over a rocky bed, whirling and foaming round the

[1] For an account of its spectral changes, see Campbell in *Astr. and Astr.-Physics*, vol. xi., p. 698.

obstacles it encountered. Twenty-four hours later, this powerful emanation left scarcely a trace on the plate. Its dwindled remnant had split up into two faint streaks, while the almost negligeable offset of the previous night had sprung into unlooked-for prominence. A unique feature was added in the apparent development of a secondary comet two degrees behind the head. The anomalous enlargement brightened gradually inwards, and can readily be seen upon the plate to be the centre of an entirely new system of tails.[1]

Owing to moonlight and clouds, the autobiography of this planetary *bud* unfortunately remained a fragment; and since Swift's comet has an indefinitely long period, it will never again exhibit for our benefit any of its caprices of change.

FIG. 23.—*Photograph by Prof. Barnard of Holmes' Comet near the Andromeda Nebula.*

On November 8, 1892, Professor Barnard secured a very perfect representation (shown in Fig. 23) of a peculiar-looking comet grouped with the great Andromeda and its attendant nebula. Discovered only two days previously by Mr.

[1] Barnard, *Knowledge*, vol. viii., p. 229.

Edwin Holmes of London, it presented a great round disc with definite edges visible to the naked eye. This contained a tail in embryo, which subsequently opened out into a feeble brush, the head being then pear-shaped, and granulated like a remote star-cluster.[1] A strictly continuous spectrum was derived from it. "Its appearance," Professor Barnard wrote, "was absolutely different from that of any comet I had ever seen. It was a perfectly circular and clean-cut disc of dense light, almost planetary in outline. There was a faint, hazy nucleus."[2] A photograph taken by him, November 10, showed, distant about one degree to the south-east, "a large irregular mass of nebulosity covering an area of one square degree or more, and noticeably connected with the comet by a short, hazy tail."

This object underwent extraordinary vicissitudes of aspect. From a seeming planet it quickly degenerated by distension into the thinnest of nebulosities; then suddenly, on January 16, 1893, gathered itself together into an ill-defined star of the eighth magnitude. This evanescent outburst was simultaneously observed in several parts of the world. After some minor rallies and relapses, the comet finally, on April 6, 1893, melted into the sky-ground. Jupiter is responsible for its introduction into the solar system, and it will again be due at perihelion in May, 1899. Yet its reappearance is considered doubtful.

It was perhaps caught sight of during a temporary crisis of internal agitation, which may not recur. Certainly it could not, if as bright as when discerned by Mr. Holmes, have remained many nights unnoticed. Nevertheless, it had passed the sun five months previously. Its orbit is more nearly circular than that of any previously observed comet, and it revolves wholly within the asteroidal zone. That is to say, its perihelion lies outside the orbit of Mars, its aphelion inside that of Jupiter. Hence, it ought to be visible like a planet, at every opposition. Professor Barnard, however,

[1] Denning: *Astronomy and Astro-Physics*, vol. xii., p. 371.
[2] *Astroph. Journal*, Jan., 1896, p. 42.

sought vainly for it, when thus situated. The apparition was in many ways enigmatical.

A comet discovered by Brooks, October 16, 1893, was photographed by Barnard three nights later, when a tail was disclosed, 3½° long, and flowing off in two branches with a spine-like ray attached to each. A series of impressions were fortunately taken, and that of October 21 (reproduced in Fig. 24) proved to be of peculiar interest. Since the night before, the tail had apparently met with an accident. It imprinted itself upon the plate shattered, deformed, and affected by a double curvature. A collision with some external body was at first suggested as the cause of this untoward state of things; but, knowing all that we do about the violent interior paroxysms of comets, it seems more rational to attribute it to extreme irregularities in the quantity and direction of effluences from the nucleus. The following night's photograph gave evidence of a partial return to normal conditions. Yet the appendage still looked badly damaged; and an elliptical fragment, wrenched from it during the convulsion, showed no tendency towards reunion. At the time of this incident, Brooks' comet was situated well outside the orbit of the earth.

The facts already collected by the photographic study of comets are concordant, and easily interpreted. One obvious inference from them is "that the matter of a comet's tail is driven away from the nucleus in a very irregular and spasmodic manner."[1] At certain crises, outflows are only accomplished by convulsions, compared by Mr. Ranyard to the explosions of terrestrial volcanoes, or solar prominences. Moreover, capricious as cometary forms are to the eye, they are still more inconstant as recorded chemically. "The appearance one day," Professor Hussey says, "affords no indication as to what it may be the next. The most radical changes of form have been observed in almost every reasonably bright comet that has been photographed; and they

[1] Ranyard: *Knowledge*, vol. ix., p. 159.

FIG. 24.—*Brooks' Comet, photographed by Prof. Barnard, October* 21, 1893. *Exposure,* 35*m.*

sometimes take place so rapidly as to become conspicuous in an hour or two."[1]

Comets' tails appear very different in structure photographically and visually. On the sensitive plate, they are perceived to be composed of innumerable, distinct filaments, sometimes tied up, as it were, into sheaves. The filaments, or streamers may, however, according to the same authority, "leave the coma in a single compressed bundle, or they may spring from it in widely divergent and loosely connected groups; they may be smooth, and straight, and distinct, or they may be lumpy, crooked, interlacing, and spirally twisted; or again, they may be broken into fragments, and scattered as though they were smoke driven by the wind." And these effects often swiftly succeed each other in the same comet.

In photographs of Swift's and Rordame's comets in 1892 and 1893 (taken by Barnard and Hussey respectively), the effects of a spiral outward movement in the grouped streamers of the tail can be plainly recognised. They are indistinguishable from "the twisted forms produced by an electrical discharge in a magnetic field."[2] Another much more common peculiarity of such appendages brought into prominence by chemical portraiture, is the occurrence upon them of knots, or condensations. These are evidently accumulations of outflowing matter. Again, in most of the comets recently photographed, the tails start directly from the nuclei, which appear destitute of genuine envelopes. This is the precise criterion of Olbers' first cometary division, in which solar repulsion acts alone, nuclear repulsion being ineffective, or non-existent. It comes out remarkably in Barnard's photographs of Gale's comet in 1894.

We may now resume in a few words what we have learned about comets. To begin with, they are of such small mass that no gravitational effects from their closest vicinity have ever yet been detected. Their bulk, on the other hand, is enormous. The great comet of 1811 comprised a nebulous

[1] "Publications Astr. Pac. Society," vol. vii., p. 166.

[2] Hussey: *loc. cit.*, p. 171.

globe $2\frac{1}{2}$ times larger than the sun, with a tail many thousand times more voluminous. Hence the extraordinary tenuity of such bodies. They must indeed contain solid matter; otherwise they could not hold together even in the imperfect way that they do; but it is probably in a state of very loose aggregation. Their permeability to light may thus be accounted for. The visibly granular texture of their nuclei is confirmatory of the supposition. If, then, the nuclei of comets are essentially "meteor-swarms," all the constituent particles must revolve round the centre of gravity of the whole, in a common period, but with a velocity directly proportional to distance from the centre—that is, increasing outward. And the joint mass being so small, the utmost speed attained would perhaps rarely exceed a couple of hundred yards a second. Moreover, towards the centre, where the components of the swarm would crowd most closely together, motion would become so slow as to be scarcely perceptible. Hence collisions would be infrequent and of slight effect; while the probability of their occurrence should diminish with the comet's approach to the sun, which, by its unequal attraction, would draw the revolving particles asunder, and amplify their allowance of space. Internal collisions may then fairly be left out of the account in considering the phenomena of comets. The expansion of their nuclear parts, due to tidal forces, is, however, usually disguised by the contraction, near perihelion, of their nebulous surroundings. The latter effect can be explained by the immense predominance at that conjuncture of solar over cometary electrical repulsion.

That the light-emissions of comets are largely of electrical origin is no longer doubtful; so that the present rush-ahead in this branch of knowledge cannot but help to elucidate many of the still mysterious circumstances connected with these strange visitants from the uttermost verge of the sun's empire. The tie of allegiance hangs loosely there; but by the persevering efforts of the great planets it is sometimes drawn closer, with the result of domiciling under their control a train of dilapidated comets, verging towards dissolution.

Carbon, sodium, and iron, are the only substances directly known to exist in these bodies. Spectroscopic evidence also suggests the presence of nitrogen or hydrogen; and a number of chemical elements which make no show in their light doubtless enter into their composition. The state of comets when remote from the sun can only be surmised. Their gaseous constituents may be solidified by cold. They can, in any case, scarcely be other than obscure and inert bodies.

CHAPTER XIII.

METEORITES AND SHOOTING STARS.

AT Madrid, on the morning of February 10, 1896, the sunshine was at 9.30 overpowered by a vivid flash of bluish light, succeeded by a violent explosion. Much glass was broken, and other devastation of a minor kind wrought; above all, some hundreds of thousands of people were thoroughly frightened. The origin of the commotion was visible in a white cloud rushing across the sky, and leaving behind a dusty train. Of this débris, scattered from a height of fifteen miles, some fragments were picked up and analysed. They were composed of silicates of magnesia and iron, with very small quantities of aluminium, nickel, and calcium. These specimens were strictly "aerolites," a term used to designate any solid meteoritic matter that reaches the earth.

Equally conspicuous apparitions of the sort are not always equally clamorous. There are silent, as well as detonating fire-balls. The cause of the difference cannot certainly be assigned. It resides, perhaps, in the diverse constitution of the exploding bodies; it is, beyond doubt, unconnected with their height in the atmosphere. Thus, a remarkable meteor was seen, but not heard, by Dr. Rambaud, the astronomer-royal for Ireland, at Dunsink, February 8, 1894. The object, he says, "suddenly burst into view with an intense brilliance, and shone out against the cloudless blue sky with a greenish metallic lustre. It fell in a vertical direction until it disappeared behind some trees. In shape it resembled a very elongated pear, like most fire-balls of the sort. It emitted no visible sparks, and disappeared quite noiselessly." When first observed, it was at a height of about 87 miles above the Irish Channel; then crossing Lancashire, it descended so rapidly on its way, probably, to engulfment in the North Sea, that, when last noticed, it was scarcely, if at all, higher above the earth's surface than the Madrid meteorite at the moment of its formidable disruption. Astonished rustic beholders at Kingswood and Dudley averred that it burst "in the next field"; but this is a common illusion. Professor Langley relates that some witnesses of a marvellously swift meteor at a presumable elevation of some fifty miles, sallied out of their houses next day to make sure that it had not struck their chimneys.

Such phenomena are tolerably frequent, and have been recorded from the remotest antiquity. Homer lends a meteoric aspect to Athene, when she descends from Olympus to take the war-path by the shore of Scamander. Chronicles abound with accounts substantially identical with the telegrams supplied by Reuter's Agency on February 10, 1896. The fall of the "Crema meteorite" has a special interest as having been depicted by Raphael in his "Madonna di Foligno."[1] A multitude of stones were discharged by it on the banks of the Adda, six of which weighed each one hundred pounds and upwards; the sulphurous smell characteristic of fresh-fallen aerolites is mentioned in contemporary accounts of the event, which occurred September 4, 1511; and it is further said that

[1] Holden: "Publ. Astr. Pac. Society," vol. ii., p. 19. H. A. Newton: *Ibid.*, vol. iii., p. 91.

"sheep were killed in the fields, birds in the air, and fishes in the streams." No specimen of this sky-volley is known to exist. In elder times, objects of this class were worshipped; and Professor Newton[1] has collected many curious facts about the meteoric cult traceable in classical history. To this day, indeed, the central sanctuary of Mahometanism—the Kaaba—owes its sacredness to the embedment in its masonry of a blackened aerolite.

Until the beginning of the present century, only the ignorant believed it possible that stones could come from heaven; philosophers regarded them as generated in the clouds. They were at last convinced that the popular view was correct by Biot's investigation of the meteoric tempest which broke over L'Aigle, in the department of the Orne, April 26, 1803. He estimated at two thousand the number of fragments scattered over an area six by two and a half miles, one of which, weighing five pounds, is now in the South Kensington Museum. And at Pultulsk, January 30, 1869, one hundred thousand stones were reported to have been showered upon the earth. It is not often, indeed, that largesse from space is so lavishly made. Yet all meteors (with the rarest exceptions) rendered luminous by the resistance of its atmosphere, become, in one way or another, incorporated with its mass. Their materials are no doubt often reduced to fine dust and gas; yet six or seven hundred solid masses per annum are computed to reach the surface of sea or land, for the most part "unrecked-of and in vain." Of late, the scientific demand for them has grown keen, and their enhanced value has raised the legal question of their ownership. The decision of the American courts is that aerolites are not "wild game," but "real estate," and, as such, belong to the owner of the land upon which they fall.

No wonder they should be at a premium, those blackened and wasted samples of immeasurably distant globes. The velocities with which they entered our atmosphere alone suffice to prove their cosmical origin. Had it not trapped

[1] "Report Brit. Ass.," 1891, p. 805.

them, many, circuiting the sun in a hyperbolic curve, would have escaped for ever from our system. Their primitive disconnexion from it is implied by their swift motions, which considerably exceed, on an average, those of comets, and point to interstellar space as their proper habitat. The earth's orbital pacing has, however, to be added or subtracted as the case may be; so that the actual rate of encounter varies from ten to forty-five miles a second. Most of this is spent before the earth's surface is reached. Only considerable masses travelling at express speed bring any sensible proportion of it with them to the ground. But what is lost as motion reappears in other forms of energy, as light, heat, and sound. In front of the rushing body, the air—despite its inconceivable tenuity at elevations of fully one hundred miles—is suddenly compressed and raised to an exceedingly high temperature, while a corresponding vacuum behind gives rise to violent reactive currents. Professor Dewar calculated, by way of example, in 1887, that a body, three feet in diameter, moving eighteen miles a second at an altitude of twenty-three miles, where barometric pressure is reduced to one-fifth of an inch, would compress the air in its path 5,600 times, the resistance offered to its passage thus equalling that of thirty-seven atmospheres. The abrupt increase of heat accompanying compressions of this order, amounts to thousands of degrees, and tends to rend in pieces a body arriving from frigid abysses where matter can only exist in a stark and, so to speak, lifeless state. Explosions of occluded gases ensue; vaporised and incandescent particles are blown behind in a luminous train; and, at the most, some shattered solid remnants tumble to our continents, or plunge into our oceans. The few that are rescued for examination look much the worse for their final adventure. The signs of the furnace and the hurricane (both self-created), are visible in their jetty and fused surfaces, "thumb-marked," probably through the continual and irregular changes in the pressure exerted upon them. The crust is, however, a mere varnish, the interior, which is usually of a greyish hue, being entirely unaffected by heat. It remains, on the contrary, sunk in the depths of cold. Agassiz compared the aerolite which fell at Dhurmsala in India, in 1860, to the Chinese *chef d'œuvre*, a "fried ice";[1] and a large fragment of it, which fell in moist earth, was found coated with ice.[2]

Aerolites, or meteorites, as they may equally well be called, are roughly divided into "stones" and "irons"; the former being composed of various and peculiar minerals, the latter of iron, with a considerable percentage of nickel.[3] All show a more or less distinctive crystalline structure. Meteoric chemistry includes about thirty of the seventy or so terrestrial elements. The chief of them are: iron, nickel, carbon, oxygen, silicon, magnesium, sulphur, aluminium, phosphorus, with smaller quantities of chromium, cobalt, tin, copper, titanium, manganese, antimony, arsenic, lithium, hydrogen, nitrogen, argon, and helium. Argon and helium were expelled by heat from a piece of meteoric iron picked up in Augusta County, Virginia, the former coming off nearly a hundred times more plentifully than the latter. As the light of argon makes no show in the spectrum of any heavenly body, the proof of its cosmical diffusion thus obtained by Professor Ramsay is of great value. Besides argon and helium, hydrogen, carbonic acid, and carbonic oxide gases are found included in meteorites. They seem, as it were, to hybernate in the stony or metallic enclosures from which they can only be *boiled out*.

Although these wind-falls from space contain no strange elements, the manner of their composition is special to themselves. Their study constitutes a separate branch of mineralogy. They are certainly of igneous origin. They show no sign of water-action, and but little of oxidation. The nearest affinities of the minerals aggregated in them are with volcanic products from great depths. Thus meteorites seem broken-up fragments of the interior parts of globes like our own. A few among them contain solid carbon, either amorphous,

1 S. Meunier: "Encycl. Chimique," t. ii., p. 461.
2 Young: "Gen. Astr.," p. 435.
3 Cornish: *Knowledge*, vol. vi., p. 163.

or in the shape of graphite, or even crystallised into minute diamonds. In the Cañon Diablo siderite, or meteoric iron, all three varieties occurred together, some of the translucent particles proving, when put to the test of actual combustion, to be indeed "gems of purest ray serene," dwelling incognito in a strange environment!

The thin streaks of light called "shooting stars" differ in several respects from explosive meteorites. In the first place, they—probably without exception—form systems. Innumerable multitudes of them travel in the same paths round the sun. Moreover, those paths resemble cometary orbits; they are very elongated ellipses, inclined at all angles to the plane of the ecliptic, and traversed indifferently in either direction. Their velocities are thus sensibly parabolic, while fire-balls commonly attain hyperbolic speed. Finally, they are soundless. They slide by in ghostly silence. Most of them are probably not larger than a pea, yet were the shield of its atmosphere withdrawn, the earth would be rendered well-nigh uninhabitable by their pelting. Incredible numbers of them are encountered. They come by the million daily to be burnt, visibly to the naked eye, in the thin upper air. Kleiber's allowance is eleven, Newton's twenty millions; and these figures should be multiplied a score of times to include telescopic fire-specks. Now, the combined mass of all these particles goes to reinforce the mass of the earth; but it is relatively so small that ages must elapse before the contribution can become sensible. Our defeated meteoric assailants surrender to us also the heat of their arrested motion; which is, however, only as a spark added to the furnace of our supply from the sun.

Shooting stars, as we have seen, move in closed orbits. They are, then, a periodical phenomenon. Not that we ever see the same individual twice; its visibility implies its dissolution, but its companions are as the sands of the sea-shore. Their association is recognised by their agreement in direction and date. Unless their orbits intersected that of the earth, nothing could be known of them terrestrially; they come to our notice only through actual encounters, and encounters are possible only at the time of year when our planet is passing through the node. This is the given rendezvous, different, speaking generally, for each system; although, speaking particularly, many meteoric streams are so wide that the earth takes days, even weeks, to cut its way through them, and so may be overtaken by fresh onsets before the original one is exhausted. Each community is distinguished by the lie of its orbit—that is, by the point in the sky from which the flying arrows of light seem to diverge. This is known as the "radiant-point" of the system, and is its special characteristic.

The August meteors are a familiar example of such an association. Their annual recurrence is no new discovery. Long ago, in mediæval times, they were called the "tears of Saint Lawrence," because never looked for vainly on the 10th of August. But they are so far from being limited to that particular night, that Mr. Denning has picked up skirmishers and stragglers from the main body all the way from July 8 to August 22. They are distributed with tolerable evenness along an immensely long ellipse, traversed in 120 years; and, because they radiate from near the star η Persei, are known to science as the "Perseids."

The scattering of the November meteors—or "Leonids," since their point of emanation is marked by ζ Leonis—is on the same plan, with a difference: the Perseids might be compared to a plain gold ring; the Leonids, to a ring with a gem on it. They send us some shots every year on the 13th and 14th of November; but three times in a century they open fire for a regular bombardment. An early Leonid display took place in 902 A.D., noted in old chronicles as "the year of the stars." All night long on October 19—the node advances 14½ degrees in a thousand years—while the tyrant Ibrahim lay dying "by the judgment of God" before Cosenza, beholders far and near viewed with consternation the stars precipitating themselves from the sky. Recurrences of the phenomenon every thirty-three years received curiously little attention until Humboldt described, and insisted on the periodic nature of the meteoric tempest witnessed by him at Cumana on the morning of November 12, 1799. One scarcely less violent broke over Europe and Asia in 1832, and the American continent in 1833. From the Gulf of Mexico to Halifax the stars were seen to fall as silently as snow-flakes, and almost as thickly, yet after a less undirected fashion. Rather they darted and swooped, like falcons, with a purpose; and it was noticed that the lines of their flight could, with essential invariability, be traced back to one point, or small area in the heavens. This remark gave the clue to their nature. They were perceived to be necessarily cosmical bodies. For since the focus of the meteors remained unaffected by the earth's rotation, they showed themselves plainly extraneous to its domestic arrangements. "A new planetary world," exclaimed Arago, "has been disclosed to us!"

The anticipated repetition, in 1866, of the November shower of 1833, came off with *éclat.* Many still remember the amazing spectacle presented by the heavens in the early morning of November 14, in that year. In 1867, when the earth came round again to the same point of its orbit, the star-rain was still falling heavily; and even in 1868 it amounted to a fair sprinkle. Thus the swarm was, thirty years ago, already so extended that it spent three years in sweeping past the node, at the rate of twenty-seven miles a second. "The meteors themselves," according to Dr. Johnstone Stoney,[1] "are probably little pebbles, the larger about an ounce, or perhaps two ounces, in weight, and spaced in the densest part of the swarm at intervals of one or two miles asunder every way. The thickness of the stream is about 100,000 miles, which, however, is a mere nothing compared with its enormous length. The width is such that the earth, when it passes obliquely through the stream, is exposed to the downpour of meteors for about five hours." Each "pebble" revolves round the sun, and suffers planetary perturbation, in

[1] *Journal Brit. Astr. Ass.*, vol. vi., p. 432.

complete independence of its fellows, their orbits being only alike, not identical. The next full encounter with them will take place November 14, 1899; but avant-couriers may be looked for at the critical dates in 1897 and 1898, as well as a strong rear-guard in 1900.

The orbit of the November meteors is roughly bounded by the orbits of the earth and of Uranus. They pass perihelion very near our meeting-place with them; and since they run counter to the earth's motion, the velocity of collision is nearly equal to the sum of the two orbital velocities, or forty-four miles a second. They are almost the swiftest shooting-stars of our acquaintance.

The successful calculation of meteoric orbits by Adams, Schiaparelli, and Leverrier, promptly led to a discovery as important as it was unexpected. Late in 1866, Schiaparelli announced that the August meteors follow precisely the same track with a bright comet (1862, III.) discovered in 1862 by Tuttle, an American astronomer; and the reality of this singular relationship was, in the following year, verified by the detection of three similar examples. The Leonids, with a period of 33¼ years, proved to be close associates of Tempel's comet (1866, I.); a meteoric stream flowing down upon the earth annually on April 20, from the direction of the constellation Lyra, was perceived to move in the vast ellipse traced out in 415 years by the comet 1861, I.; finally a star-drift, first noticed December 6, 1798, was rightfully claimed as an appurtenance of Biela's comet.

Thus the fact of a close connexion between comets and meteors was at once rendered patent; and as to the nature of the connexion, the history of Biela's comet is particularly instructive. Since its disappearance, the meteor-swarm sharing its orbit has received a notable accession. The comet seems to have broken up into meteors. And this, we can scarcely doubt, is what has really occurred. Hence, when the earth passes moderately near where the comet *would* have been, had it survived in cometary shape (a conjuncture happening once in thirteen years), a vehement outburst of shooting stars is observed. On November 27, 1872, the "Bielids," or "Andromedes," came in tens of thousands from near γ Andromedæ, the very point whence the track of the disaggregated comet intersects the earth's orbit at an angle of twelve degrees. Their movements were leisurely; for they came up with our globe, instead of, like the Leonids, rushing to meet it. They seemed to sail, rather than shoot, across the sky. The calculated position of the originating body was, at this date, two hundred millions of miles *in advance* of the node, and it was three hundreds of miles *behind* the same point when the display was renewed in 1885. It is then certain[1] that at least five hundred millions of miles of Biela's route are densely strewn with meteoric fragments. The entire multitude, moreover, necessarily separated from the comet subsequently to an episode of disturbance by Jupiter in 1841. This is plainly shown by the fact that the members of the associated company pursue the modified track. The perturbation of 1841 was exerted upon them no less than upon the comet, with which, accordingly, they must then have formed one mass.

Biela's comet has thus taught us that such bodies meet their end by getting pulverised into meteoric particles; and further, that the particles disperse with extraordinary rapidity along the length of their orbits. Solar and planetary *differential* action produce this kind of effect, although they hardly explain its amount. Subordinate swarms are also created by disturbance. Such an one met the earth November 23, 1892, when Professor Young estimated that at least 30,000 Andromedes furrowed the sky at Princeton. Heavy star-showers, however, are perishable phenomena. They thin out with comparative rapidity into a continuous drizzle. At each recurrence, diffusion is seen to have made progress, until at last the "gem on the ring" has vanished. With the Perseids this is already the case. The stream flows without material interruption over a bed a hundred times wider than that of the Leonids. These meteors, too, will no

[1] H. A. Newton: "Proc. Amer. Phil. Society," vol. xxxii.

doubt eventually reach a similar condition. In the course of a couple of centuries, their thirty-three year period will be completely effaced. In 1799, the main body of them crossed the node in less than a year; at the close of the present century, the earth will probably make her annual round at least four times, before the march-past comes to an end. Obviously, it is about to become perennial. Leverrier concluded from his researches that the Leonid comet and the Leonid meteors, which then made part of its substance, were "captured" by Uranus in 126 A.D., and so introduced into the solar domain. The truth of the supposition may still be tested; should it be established, this remarkable system affords yet another example of the rapidity with which cometary materials become disintegrated and scattered.

The number of meteoric radiants now distinctly known is estimated by Mr. Denning at about three thousand; and we need not hesitate to ascribe to all these streams a cometary origin. It is true that the three thousand generating comets have, all but three, "gone over to the majority." But we have witnessed the obsequies of Biela, and it seems only logical to infer that those of its 2996 congeners were, in old times, celebrated after the same fashion, and are still kept in mind by the annual blaze, in their honour, of a few representative sky-rockets.

No component of a star-burst has so far *undoubtedly* come to the ground. The fire-works shown are of the most innocuous kind. Two *possible* exceptions are, however, on record. On April 4, 1095, a shower of Lyraids was visible in Western Europe. The stars, according to the Saxon Chronicle,[1] crowded "so thickly that no man could count them." And in France, one of the throng fell so accessibly that a bystander, having noted the spot, "cast water upon it, which was raised in steam with a great noise of boiling." But, unless the aerolite came from the same radiant as the stars, their simultaneous arrival was an unmeaning coincidence. It implied no connexion, physical or dynamical,

[1] Quoted by Sir F. Palgrave: "Phil. Trans.," vol. cxxx., p. 173.

between them. The same coincidence was renewed during the Andromede shower of November 27, 1885. Just before it began, a "ball of fire" struck the ground at Mazapil in Mexico, and proved to be a substantial piece of iron containing nodules of graphite. It weighed eight pounds. Yet here again that essential circumstance, the direction of its fall, remained unknown. We must then, for the present, suspend our judgment as to whether aerolites may be regarded, like shooting stars, as actual cometary débris.

Mr. Denning's patient watch of thirty years has led him to the singular discovery of "stationary radiants." The direction in which meteors appear to approach the earth is determined by the combination of theirs with the earth's movements. The effect is strictly analogous to the aberration of light. Meteoric radiants ought accordingly to shift on the sphere just as the heavenly bodies change their apparent places by the prescribed measure of aberration. And most do in this respect conform to theory, the Perseid radiant notably. On the other hand, certain well-known radiants continue fixed night after night in seeming independence of the earth's orbital advance; and there are a good many points in the sky whence shooting stars continue to *dribble* without sensible interruption during many months of each year. The fact is undeniable, although inexplicable.

The future progress of meteoric astronomy depends largely upon the introduction of the photographic mode of observation. Only by its aid can the precise determination of radiant-points be effected; and this is the chief desideratum. Its realisation before the close of the century may safely be predicted. Dr. Elkin, director of Yale College Observatory, had a "meteorograph" constructed for the purpose in 1894, and hopes to use it for the registration of the Leonids now hastening to meet us. Hitherto, only casual fire-balls have printed their tracks on sensitive plates. Success in obtaining permanent records of shooting stars diverging from a radiant will mark a turning-point in meteoric investigations.

ASTRONOMY

SECTION IV.—THE SIDEREAL HEAVENS

By J. E. GORE, F.R.A.S.

NEBULA IN ANDROMEDA, 31 MESSIER.
(*From a Photograph by Dr. Roberts.*)

SECTION IV.—THE SIDEREAL HEAVENS.

BY J. E. GORE, F.R.A.S.

CHAPTER I.

THE STARS AND CONSTELLATIONS.

THE study of the sidereal heavens is one of surpassing interest, and tends to raise our minds above the sordid things of time and the petty affairs of the little planet on which we dwell,—a globe absolutely large, of course, when compared with objects around us, but relatively very small in comparison with the vast stellar universe which surrounds us on all sides, a universe so vast that even the largest telescopes can only partially fathom its immeasurable depths.

For the study of the sidereal heavens, as revealed to us by the giant telescopes of modern times, it will be advisable to begin by a consideration of the starry sky as seen by the naked eye, without optical assistance of any kind. On a clear and moonless night, when the vault of heaven is spangled over with shining points of light, some bright, others fainter, and many more barely perceptible to the unaided vision, we are inclined to imagine that the stars visible to the naked eye are innumerable, and that any attempt to count them would be a hopeless task. This idea, however, is quite a mistake, and, indeed, merely an optical illusion, due partly to the scintillation or twinkling of the brighter stars, and stars near the limit of vision, and partly to their irregular distribution over the surface of the heavens. As a matter of fact, the stars visible to the naked eye can be easily counted; and they have been counted and catalogued. As every book in the catalogue of a large library can be identified, so every star visible to the unaided vision—and thousands even fainter, and only visible in telescopes—have been mapped, and their exact positions are as well known to astronomers as those of every town and village in Great Britain are known to geographers. The number of stars which can be seen with ordinary eyesight is, in fact, very limited, and does not exceed the number of inhabitants in a small town. Some years ago, a German astronomer, Heis, who was gifted with excellent eyesight, carefully mapped down all the stars visible to his eye without optical aid, and found the total number visible in the middle of Europe to be only 3,903. A similar work was undertaken for the Southern Hemisphere by Behrmann, another German astronomer, and the total number distinctly seen by both astronomers in both hemispheres of the star sphere is 7,249. Of course, at any given time and place only one half the star sphere is visible, the other half being below the horizon. It follows, therefore, that about 3,600 stars are visible at one time from any point on the earth's surface. As, however, everyone does not possess the keen vision of the astronomers referred to above, we may safely say that not more than 3,000 stars are, on the average, visible at a time to ordinary eyesight. On the other hand, persons gifted with exceptionally keen vision may possibly see even more than Heis and Behrmann did; but even to such eyes, the total number distinctly visible on a clear night without a moon would probably not exceed 5,000. We may easily satisfy ourselves as to the truth of this statement by taking a portion of the sky, and counting the number of stars which can be steadily seen. Everybody knows the Great Bear, sometimes called the "Plough," or "Charles' Wain." Four of the well-known stars in this remarkable group form a four-sided figure. Well, let the reader look carefully at this figure, and see how many stars can be detected within the space formed by imaginary lines joining the bright stars. Probably surprise will be felt at the small number which can be distinctly seen. Heis, with his keen vision, only shows eight on his map, and of these, four are very faint, and near the limit of even good eyesight. Probably very few eyes will see more than eight, and perhaps most persons will fail to see so many. As the whole hemisphere is roughly

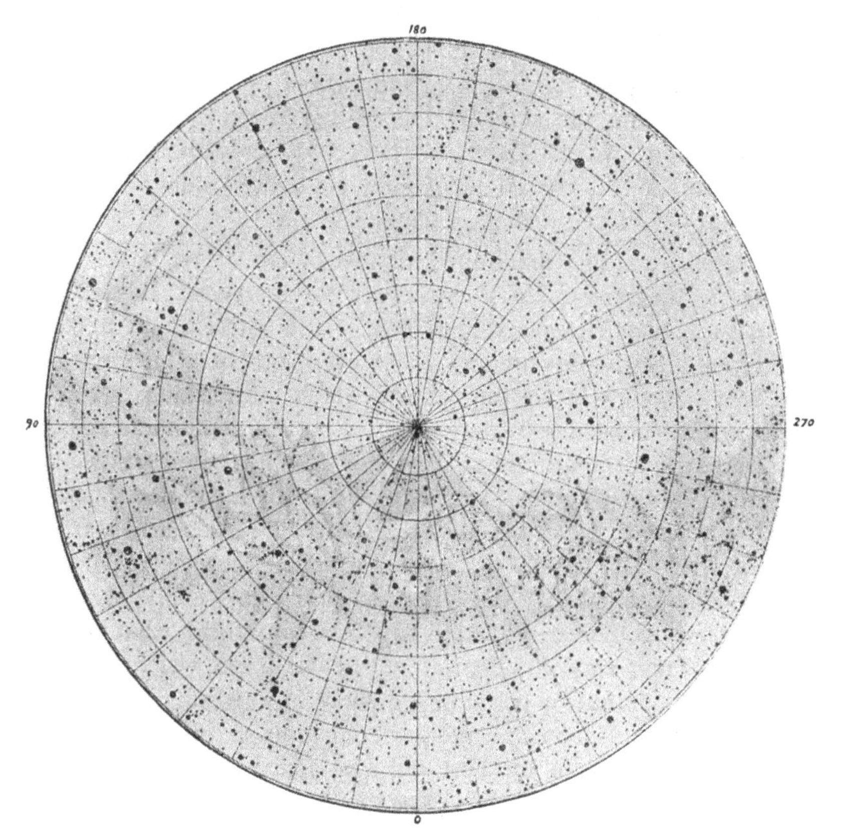

FIG. 1.—*Stars visible in the Northern Hemisphere.*
(From "Visible Universe.")

five hundred times larger than this spot, the number seen by Heis in the quadrilateral of the Plough would give a total of 4,000 stars visible at one time. Of course, some portions of the sky are much richer in stars than the spot selected; but, on the other hand, others are much poorer, so that perhaps this may be taken as a spot of average richness. From this single example it will be seen that the idea of countless multitudes of stars visible to the naked eye is a mistake. Probably the effect of a great number is partly due to our catching glimpses by "averted vision" of still fainter stars, which cannot, however, be seen steadily when the eye is turned directly towards them.

In speaking of stars visible to the naked eye, we do not, of course, include the stars in the Milky Way, that arch of cloudy light which spans the heavens; for although this wonderful zone is composed of faint stars, these stars are not individually visible without a telescope.

Notwithstanding the limited number of the visible, or lucid, stars, as they are called, the aspect of the starry sky still presents a spectacle of marvellous beauty and interest, and may be viewed with pleasure and profit even without a telescope. There are many interesting objects which may be seen without optical assistance of any kind. Look at the middle star of the three forming the "tail" of the Great Bear, or "handle" of the Plough. This star was called Mizar by the old Arabian astronomers. Close to it, good eyesight will see a small star, known as Alcor. This little star was called by the Arabians Alsuha, which means "the neglected small star." The name Alcor means the "test," and is supposed to indicate that the old astronomers considered it a test for keen vision; but the Arabians had a proverb, "I show him Alsuha, and he shows me the moon," a saying which seems to imply that it could be easily seen by these old astronomers. The faintest star of the seven, the one at the root of the tail, was called Megrez by the Arabian astronomers. This star is supposed to have diminished in brightness since ancient times, as it was rated of the third magnitude by Ptolemy, and of the second by Tycho Brahé, while at present it is not much above the fourth magnitude. It may possibly be variable in its light, like many other stars in the heavens.

Here it may be mentioned that the stars were divided into magnitudes or classes according to their brightness by the ancient astronomers, all the brightest stars being placed in the first magnitude, those considerably fainter being called second magnitude, those fainter still third magnitude, and so on to the sixth magnitude, or those just visible to ordinary eyesight. This classification has been practically retained by modern astronomers, but, of course, there are stars of all degrees of brightness from Sirius down to the faintest stars visible in the

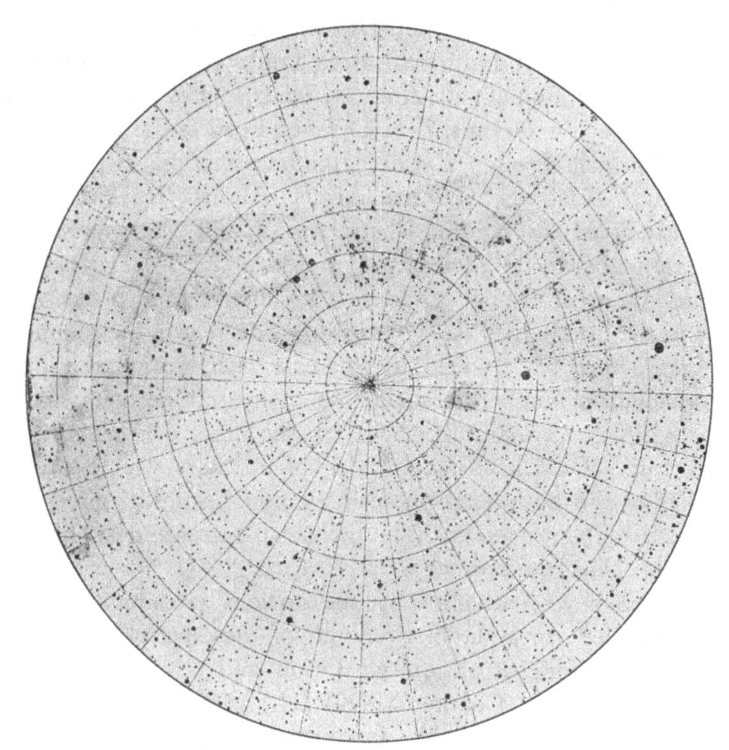

FIG. 2.—*Stars visible in the Southern Hemisphere.*
(From "Visible Universe.")

largest telescopes. Sirius is the brightest star in the heavens, and is equal to about six average stars of the first magnitude, such as Altair or Aldebaran. According to the Harvard photometric measures, the following are the brightest stars in the heavens in order of magnitude :—(1) Sirius, (2) Canopus, (3) Arcturus, (4) Capella, (5) Vega, (6) Alpha Centauri, (7) Rigel, (8) Procyon, (9) Achernar, (10) Beta Centauri, (11) Betelgeuse (slightly variable), (12) Altair, and (13) Aldebaran. Of these Canopus, Alpha, and Beta Centauri, and Achernar, do not rise above the horizon of London. Of those brighter than the second magnitude, the following are north of the Equator: Alpha Cygni, Pollux, Castor, Eta Ursæ Majoris, Gamma Orionis, Beta Tauri, Epsilon Ursæ Majoris, Alpha Ursæ Majoris, Alpha Persei, and Beta Aurigæ; and south of the Equator: Alpha Crucis, Fomalhaut, Antares, Spica, Beta Crucis, Gamma Crucis, Epsilon Orionis, Zeta Orionis, Epsilon Canis Majoris, Beta Carinæ, Epsilon Carinæ, Lambda Scorpii, Alpha Triangulum Australis, Gamma Argûs, Alpha Gruis, Epsilon Sagittarii, Alpha Hydræ, Theta Scorpii, and Delta Velorum. Of those below the second magnitude, and brighter than the third, there are about 34 in the Northern Hemisphere, and 61 in the Southern. As the brightness decreases, the numbers increase rapidly. Indeed, the increase is in geometrical progression, the number in each class of magnitude being about three times as many as those in the class one magnitude brighter. The exact magnitudes of all stars visible to the naked eye in both hemispheres have now been determined by the aid of photometers. These instruments are described in Section II. of the present work, Chapter XVII.

The stars were divided by the ancient astronomers into groups called constellations. Some of these were formed in the earliest ages of antiquity. Orion and the Pleiades are mentioned in Job (Chapter XXXVIII.), which is believed to be one of the oldest books in existence. Josephus ascribes the division of the stars into constellations to the family of Seth, the son of Adam; and according to the Book of Enoch the constellations were already known and named in the time of that patriarch. The brightest stars of each constellation are designated by the letters of the Greek alphabet, which were assigned to them by Bayer in the year 1603, Alpha generally denoting the brightest star, Beta the next in lustre, and so on. This is not, however, invariably the case, and

Bayer seems in many cases to have followed the outline of the imaginary figure from which the constellation derives its name, rather than the relative brightness of the stars composing the constellation. For example, the seven stars in the Plough are known as Alpha, Beta, Gamma, Delta (the faint one), Epsilon, Zeta, and Eta, beginning with the northern of the two in the square farthest from the tail, thus evidently following the shape of the figure, and not the order of relative brightness. When the letters of the Greek alphabet are exhausted, recourse is had to numbers, those in Flamsteed's catalogue being usually employed. Those only visible in telescopes are known by their numbers in various catalogues. The exact positions of the stars are fixed by determining their right ascensions and declinations, terms which on the celestial sphere correspond to longitude and latitude on the earth.

The stars Alpha and Beta of the Plough are called "the pointers," because a line drawn from Beta through Alpha points nearly to a star of the second magnitude, called the Pole Star, which lies near the pole of the celestial sphere, or the point round which the whole star sphere seems to rotate, owing to the rotation of the earth on its axis, in twenty-four hours. The distance from Alpha to the Pole Star is about five times the distance between Alpha and Beta.

If we draw an imaginary line from the star Epsilon through the Pole Star, and produce it to about the same distance on the opposite side of the Pole, it will pass through a well-known group called Cassiopeia's Chair. This consists of five fairly bright stars arranged in the form of an irregular W. A sixth star, much fainter than the others, forms with three of them a quadrilateral figure. It was near this faint star—known to astronomers as Kappa—that the famous "new," or temporary, star of Tycho Brahé, sometimes called the "Pilgrim Star," suddenly appeared in November, 1572, of which more hereafter.

If we continue the curve formed by the three stars in the tail of the Great Bear, it will pass near a very bright star of an orange colour. This is Arcturus, one of the brightest stars in the sky. If we can rely on the measures of distance which have been made of this brilliant star, it must be one of the largest bodies in the universe, much larger than our sun, which, placed at the distance assigned to Arcturus, would only shine as a small star, quite invisible indeed to the naked eye.

Returning again to the Great Bear, if we draw a line from Gamma to Beta and produce it, it will pass near a bright star of a yellow colour. This is Capella. It was called by the Arabian astronomers the "Guardian of the Pleiades." It is the brightest star of the constellation Auriga or "the Charioteer," referred to by Tennyson in the lines:

> "And the shining daffodil dies, and the Charioteer
> And starry Gemini hang like glorious crowns
> Over Orion's grave low down in the West,"

evidently referring to the disappearance of Orion below the western horizon in the evening sky of April. "Starry Gemini" is marked by two bright stars, Castor and Pollux, which may be found by drawing a line from Delta to Beta of the Great Bear, and producing it. Another line drawn from Delta to Gamma, and produced towards the south, will pass near a bright star called Regulus, the brightest star in the well-known "Sickle" in Leo or the Lion. Again, a line drawn from Regulus to Gamma in the Great Bear, and produced, will pass near another bright star, Vega in the Lyre. This is one of the brightest stars in the Northern Hemisphere, the three, Arcturus, Capella, and Vega, being nearly equal in brightness. The name Vega seems to be a corruption of the Arabic name *vaki*, or *al-nasr al-vaki*, "the falling eagle," the wings of the bird being represented by the stars Epsilon and Zeta Lyræ, which form, with Vega, a small triangle, called by the Arabians *al-alsafi*, the trivet. But what relation exists between a "falling eagle" and the musical instrument known as the Lyre (Persian *al-lûra*) is not very obvious. Possibly, however, as suggested by Schjellerup, the Arabic word, *al-schalzâk* a goose,—also applied to the constellation—refers to the resemblance in shape between a plucked goose and a

Greek lyre. The Greeks called the constellation χέλυς, a tortoise, which also somewhat resembles a lyre in shape.

Of the two stars which form a triangle with Vega, the northern, Epsilon, is a double star, which is said to have been seen double with the naked eye by several astronomers, but, probably, most people would fail to see it as anything but a single star, as the component stars are very close. An opera-glass will, however, show it distinctly. Each of the components is again double, so that the object forms a most interesting quadruple star when viewed with a good telescope.

To the east of Vega lies Cygnus, or the Swan, one of the finest of the constellations. It may be distinguished by the long cross formed by the principal stars which are known to astronomers as Alpha, Beta, Gamma, Delta, and Epsilon; Alpha, or Deneb, being the brightest and most northern of the five, and Beta the most southern and faintest. The name Deneb is derived from the Arabic word *dzanab al-dadjâdja*, or "the tail of the hen," referring to its position in the ancient figure, which represents a hen or swan flying towards the south.

To the south-east of Cassiopeia's Chair, we find the well-known festoon of stars which marks the constellation Perseus. Its brightest star is sometimes called Mirfak, a name derived from the Arabic word *marfik*, the elbow, referring, perhaps, to its position in the curved line of stars. South of Perseus, and the nearest bright star to Mirfak in that direction, is Algol, the famous variable star. Further south, we come to the constellation of Taurus, or the Bull, with the well-known groups of the Pleiades and Hyades. The Pleiades form a remarkable cluster, and when once recognised can never be mistaken. To ordinary eyesight six stars are visible, but those having keener vision can see more. A little south of the Pleiades is a V-shaped figure, the Hyades, with a bright star of a reddish colour. This is Aldebaran, a name derived from the Arabic *al-dabarân*, the attendant or follower, because it appears to follow the Pleiades in the diurnal motion. It was also called *aïn al-tsaur*, "the eye of the bull," and by several other names, such as *al-fanîk*, "the great camel," the other smaller stars forming the Hyades being called *al-kilas*, "the young camels!"

South of Taurus and Gemini comes the magnificent constellation of Orion, perhaps the most splendid collection of stars in the sky. This brilliant asterism contains many fine objects. Looking at it when it is visible in the winter sky, we notice a large quadrilateral figure formed by four conspicuous stars. The upper one to the left is called Betelgeuse, and is decidedly reddish in colour—very much resembling Aldebaran both in tint and brightness. Its name is derived from our Arabic word meaning the shoulder, because it is situated on the right shoulder of the giant Orion on the old celestial globes. The upper one to the right is called Bellatrix, or the female warrior! The real significance of some of those old names is sometimes difficult to understand. Of the lower stars, the one on the right is a fine white star of the first magnitude known as Rigel. It is situated on the left foot of the ancient figure of Orion, and its name is derived from the first part of the compound Arabic name *ridjl-al-djauzâ*, "the leg of the giant." The lower star on the left is known to astronomers by the Greek letter Kappa.

In the middle of the four-sided figure referred to above are three stars of the second magnitude, nearly in a straight line, forming "Orion's Belt." The upper one of the three is slightly fainter than the others, and has been suspected of being slightly variable in its light, but the variability is doubtful. South of these three conspicuous stars are three fainter stars, forming a nearly vertical line. This is "the Sword of Orion." The middle star of the three marks the position of "the great nebula in Orion," one of the finest objects in the heavens, of which more hereafter. To some eyes a nebulous glow is visible round this star. Even in a small telescope the nebula is an interesting object. On a very clear night the southern star of the three may be seen double with good eyesight. The stars forming Orion's Belt were called by the Arabian astronomers *mintakat al-djauza*, "the Belt of the Giant"; and the stars forming the "sword," *al-lakat*, the "gleaned ears of corn," and also *saif-al-djabbâr*, "the Sword of the Giant." Perhaps the latter word is the origin of the name Algebar, formerly applied to Rigel.

The three bright stars in Orion's Belt nearly point (to the south-east) to Sirius, the brightest star in the heavens. This is a splendid white star, and is so much brighter than any other fixed star that its identity cannot be mistaken.

If we draw a line from the star Gamma in the Plough to the Pole Star, and produce it, it will pass through a somewhat similar four-sided figure, but of much larger size, and the stars rather fainter. This is known as "the Square of Pegasus." The upper stars are known as Beta Pegasi (the one to the right) and Alpha Andromeda. To the east of Alpha Andromedæ is a star of the third magnitude, Delta, and to the east of Delta, a star of the second magnitude called Beta Andromedæ. A little north of Beta are two small stars, Mu and Nu, nearly in a line with Beta, and to the north of Nu is the famous "nebula in Andromeda," "the queen of the nebulæ," as it has been termed. It is just visible to the naked eye as a hazy spot of light, and it may be well seen in a good opera-glass or binocular. Even in a small telescope it is a really splendid object. The reader should fix its exact position carefully, as it has been frequently mistaken for a comet by observers whose knowledge of the heavens is not very accurate.

The following alignments may be found useful by beginners in the study of the starry sky:—

Castor and Pollux, already mentioned, nearly point south to the star Alpha Hydræ, an isolated reddish star of the second magnitude. It is also called Alphard, from the Arabic *al-fard*, "the solitary one," because there is no other bright star near it. It is described by Al-Sûfi, the Persian astronomer, as red in the tenth century. In the Chinese annals it is called "the Red Bird."

An isosceles triangle is formed by Castor (at the vertex), Alphard and Sirius. Procyon is nearly in the centre of this triangle. Two other roughly isosceles triangles are formed, having Aldebaran at the vertex of each, namely: Aldebaran, Castor, and Procyon, and Aldebaran, Procyon, and Sirius.

Castor, Alpha, Delta, and Beta Orionis are nearly in a straight line; also Beta Pegasi, Alpha Pegasi and Fomalhaut. A right angled triangle is formed by Arcturus, Spica, and Regulus, Spica being at the right angle.

In the Southern Hemisphere, the most remarkable group of stars is the well-known Southern Cross. It consists of four stars, known as Alpha, Beta, Gamma and Delta—Gamma being at the top of the cross, and Alpha at the bottom. These stars are popularly supposed to be of great brilliancy, but this is a mistake; their magnitudes, according to recent photometric measures, being Alpha, first magnitudes, Beta 1½, Gamma, second magnitude, and Delta, third magnitude. A little south of Delta is Epsilon, a star of the fourth magnitude, which rather spoils the symmetry of the cross-shaped figure. A little to the east of the Southern Cross are Alpha and Beta Centauri, two of the brightest stars in the sky. Another fine group of stars is Scorpio, or the Scorpion, of which the brightest star is Antares, a reddish star of about magnitude 1½, which is visible near the southern horizon in the months of June and July in England.

When the positions of the principal stars are known, it will be easy to find any other required object by means of star maps.

CHAPTER II.

DOUBLE, MULTIPLE, AND COLOURED STARS.

MANY of the stars when examined with a good telescope are seen to be double, some triple, and a few quadruple, and even multiple. These when viewed with the naked eye, or even a powerful binocular, seem to be single, and show no sign of consisting of two components. These telescopic double stars should be carefully distinguished from those which appear very close together with the naked eye, and which in opera-glasses or telescopes of small power might be mistaken for wide double stars by the inexperienced observer. These latter stars, such as Mizar—the middle star in the tail of the Great Bear, and its small companion, Alcor, referred to in the last chapter—have been called "naked eye doubles," but they are not, properly speaking, double stars at all. Telescopic double stars are far closer, and even the widest of them could not possibly be seen double without optical aid, even by those who are gifted with the keenest vision. Of these so-called "naked eye doubles," we may mention Alpha Capricorni, which on a very clear night may be seen with the naked eye to consist of two stars. On a very fine night two stars may be seen in Iota Orionis, the most southern star in Orion's Sword. The star Zeta Ceti has near it a fifth magnitude star, Chi, which may be easily seen with the unaided vision. The star Epsilon Lyræ (near Vega), is, as mentioned in the last chapter, a severe test for naked eye vision. Bessel, the famous German astronomer, is said to have seen it when thirteen years of age. Omicron Cygni (north of Alpha and Delta Cygni) forms another naked eye double, and other objects of this class may be noticed by a sharp-eyed observer.

The star Mizar, already referred to, is itself a wide telescopic double, and it seems to have been the first double star discovered with the telescope (by Riccioli in 1650). It consists of two components, of which one is considerably brighter than the other. It will give an idea of the closeness of even a "wide" telescopic double when we say that the apparent distance between Mizar and Alcor is nearly forty times the distance which separates the close components of the bright star. From this it will be seen that even a powerful binocular field-glass would fail to show Mizar as anything but a single star. The components may, however, be well seen with a 3-inch telescope, or even with a good 2-inch. The colours of the two stars are pale green and white. Between Mizar and Alcor is a star of the eighth magnitude, and others fainter. Mizar was the first double star photographed by Bond.

The Pole Star has a small companion at a little greater distance than that which separates the components of Mizar, but owing to the faintness of this small star, the object is not so easy as Mizar. A telescope with a good 3-inch aperture should, however, show it readily. Dawes saw it with a small telescope of $1\frac{3}{10}$-inch aperture, and Ward, who has wonderful vision, with only $1\frac{1}{4}$-inch.

The star Beta Cygni is composed of a large and small star, of which the colours are described as "golden-yellow and smalt-blue." This is a very wide double, and may be seen with quite a small telescope. Another fine double star is that known to astronomers as Gamma Andromedæ. The magnitudes of the components are about the same as those of Mizar, but a little closer. Their colours are beautiful ("gold and blue"). This is one of the prettiest double stars in the heavens. It is really a triple star, the fainter of the pair being a very close double star; but this is beyond the reach of all but the largest telescopes. The star Gamma Delphini is another beautiful object, the components being a little more unequal in magnitude, but the distance between them about the same as in Gamma Andromedæ. I have noted the colours with a 3-inch telescope as "reddish-yellow and greyish-lilac." Gamma Arietis, the faintest of the three well-known stars in the head of Aries, is another fine double star, a little closer than Gamma Delphini. This is an interesting object, from the fact that it was one of the first double stars discovered with the telescope—by Hooke, in 1664, when following the comet of that year. He says:—"I took notice that it consisted of two small stars very near together, a like instance of which I have not else met with in all the heaven." Eight years previous to this, however, in 1656, Huygens is said to have seen three stars in Theta Orionis, the well-known multiple star in the Orion nebula; and in 1650, Riccioli, at

Bologne, saw Zeta Ursæ Majoris (Mizar) double, as already stated.

Another beautiful double star is Eta Cassiopeiæ, the components being about equal in brightness to those of Gamma Delphini, but the distance less than one half, so that a higher magnifying power will be required to see them well. The colours are, according to Webb, yellow and purple; but other observers have found the smaller star garnet or red. This is a very interesting object, the components revolving round each other, and forming what is called a binary star.

Another fine double star is Castor, which is composed of two nearly equal stars separated by a distance about half that between the components of Gamma Andromedæ. This is also a binary or revolving double star, but the period is long. Gamma Virginis is another fine double star, with components at about the same distance as those of Castor, and the colours very similar. It is also a remarkable binary star, and further details respecting it will be given when we come to speak of the binary stars.

Among double stars of which the components are closer than those mentioned above, but which are within the reach of a good 3-inch telescope—a common size with amateur observers—the following may be noticed:—Alpha Herculis, colours, orange or emerald green; the light of this star is slightly variable. Gamma Leonis, another binary star with a long period; colours, pale yellow and purple. Epsilon Böotis, a lovely double star, the colours of which Secchi described as "most beautiful yellow, superb blue." This has been well seen with a 2¼-inch achromatic.

For observers in the Southern Hemisphere, the following fine double stars may be seen with a 3-inch telescope:—Alpha Centauri; this famous star, the nearest of all the fixed stars to the earth, is also a remarkable binary; its period, as recently computed by Dr. See, is 81 years, and the component stars are now at nearly their greatest distance apart, the distance being greater than that between the components of Mizar, so that any small telescope will show them. Theta

Eridani is a splendid pair, but closer than Alpha Centauri. It is, however, an easy object with a 3-inch telescope, and with a telescope of this size I noted the colours in India as light yellow and dusky yellow. The star known as *f* Eridani is a very similar double to Theta, but the components are fainter. I noted the colours in India as yellowish-white and very light green. There are, of course, many other double stars in both hemispheres within the reach of small telescopes; but those described above are perhaps the finest examples.

In addition to these comparatively wide double stars, there are many of which the components are so close that they are quite beyond the reach of a 3-inch or even a 4-inch telescope. Some, indeed, are so excessively close as to tax the highest powers of the largest telescopes yet constructed.

Of triple, quadruple, and multiple stars, there are several which may be well seen with a small telescope. Of these may be mentioned Iota Orionis, the lowest star in the Sword of Orion, which consists of a bright star accompanied by two small companions. In Theta Orionis, the middle star of the sword, four stars may be seen forming a quadrilateral figure, known to observers as the "trapezium." I have seen these in India—where the star is higher in the sky than in this country—with a 3-inch refractor reduced by a "stop" over the object glass to 1½ inch. There are two fainter stars in this curious object, which lie in the midst of the Orion nebula, but a somewhat larger telescope is required to see them. Within the trapezium are two very faint stars, which are only visible in the largest telescopes. In Sigma Orionis —a star closely south of Zeta, the lowest star in Orion's Belt—six stars may be seen with a 3-inch telescope. Indeed, Ward has seen ten with a slightly smaller telescope. Epsilon Lyræ may be seen double with a low power, and each star of the pair again double with a high power; but this is more difficult than the other close stars mentioned above.

When carefully examined, many of the stars show differences in colour. Among the brightest stars it will be noticed that Sirius, Rigel, and Vega, shine with a white or bluish-white light; Capella is distinctly yellowish; Arcturus yellow or orange; and Aldebaran and Betelgeuse have a well-marked reddish hue. There are no stars of a decided blue colour visible to the naked eye, at least in the Northern Hemisphere. The third magnitude star, Beta Lyræ, is said to be greenish, but its colour is not conspicuous. Betelgeuse is perhaps the ruddiest of the brighter stars, and its reddish tint contrasts strongly with the white light of Rigel, in the same constellation. Aldebaran, which lies not far from Betelgeuse, is of nearly the same hue. But the reddest star visible to the naked eye in the Northern Hemisphere is the fourth magnitude star, Mu Cephei. It is not, however, sufficiently bright to enable its colour to be well seen without optical aid, but with an opera-glass its reddish hue is beautiful and striking when compared with other stars in its immediate vicinity. It was called by Sir William Herschel the "garnet star," and its colour is certainly remarkable. Like so many of the red stars, it is variable in light, but numerous observations by the present writer seem to show that there is no regular period, and its light often remains for many weeks with little or no perceptible change.

Among other stars visible to the naked eye, the reddish colour is also conspicuous in Antares, Alphard, Eta, and Mu Geminorum, Mu and Nu Ursæ Majoris, Beta Ophiuchi, Gamma Aquilæ, and others in the Southern Hemisphere. Alphard was noted as red by the Persian astronomer, Al-Sûfi, in the tenth century, and it was called "the Red Bird," by the old Chinese observers.

Ptolemy, in his catalogue, calls the following stars "fiery red': Arcturus, Aldebaran, Pollux, Antares, Betelgeuse, and, curious to say, Sirius, which is now white. There is some little doubt as to the reality of this change of hue in Sirius, but Al-Sûfi distinctly describes the variable star, Algol, as red, whereas it is now white, or only slightly yellowish.

The finest examples of red stars are, however, found among those only visible with a telescope. Of these may be mentioned the star numbered 713 in Espin's edition of Birmingham's "Catalogue of Red Stars," which Franks describes as "orange vermilion," and the star Birmingham 248, which Espin notes as "magnificent blood red." Another very fine red star is the variable R Crateris, which Sir John Herschel described as "scarlet, almost blood colour," Birmingham "crimson," and Webb "very intense ruby." Observing it in India with a 3-inch telescope, I noted it as "full scarlet." It has near it a star of the ninth magnitude of a pale bluish tint. No. 4 of Birmingham's "Catalogue" is described by Espin as of an "intense red colour, most wonderful." The variable star U Cygni is very red, and is described by Webb as showing "one of the loveliest hues in the sky." Another red star is the remarkable, variable R Leonis, whose fluctuations in light will be described in the chapter on Variable Stars. Hind says: "It is one of the most fiery-looking variables on our list—fiery in every stage from maximum to minimum, and is really a fine telescopic object in a dark sky about the time of greatest brilliancy, when its colour forms a striking contrast with the steady white light of the sixth magnitude, a little to the north."

In the Southern Hemisphere there are some fine red stars. Epsilon Crucis, one of the stars of the Southern Cross, is said to be very red, and so are Mu Muscæ and Delta Gruis, the southern star of a naked eye double. Pi Gruis is also a wide double star, and Dr. Gould describes one of the pair as "deep crimson," while the other is "conspicuously white." The variable R Sculptoris is another fine red star, which Gould describes as "intense scarlet," and Miss Clerke says it "glows like a live coal in the field," a good description of these telescopic red stars. With reference to a small star in the field of view with Beta Crucis, one of the brightest stars in the Southern Cross, Sir John Herschel says: "The fullest and deepest maroon-red, the most intense blood-red of any star I have seen. It is like a drop of blood when contrasted with the whiteness of Beta Crucis."

Among the double stars there are numerous examples of coloured suns. Of these may be mentioned Alpha Herculis, the components of which are orange and emerald, or bluish-green, and described by Smith as "a lovely object, one of the finest in the heavens"; Epsilon Boötis, of which the colours are described by Secchi as "most beautiful yellow, superb blue"; Beta Cygni, "golden-yellow and smalt-blue"; Beta Cephei, "yellow and violet"; Delta Cephei, "yellow and blue"; Gamma Andromedæ, "gold and blue"; and Beta Piscis Australis, of which the colours were noted by the present writer in India as white and reddish-lilac.

It has been found that the red stars are most numerous in or near the Milky Way, and one portion of the Galaxy—between Aquila, Lyra, and Cygnus—was called by Birmingham "the red region in Cygnus." Yellow and orange stars seem to be most abundant in the constellations, Cetus, Pisces, Hydra, and Virgo, and the white stars in Orion, Cassiopeia, and Lyra.

THE DISTANCES

CHAPTER III.

THE DISTANCES AND MOTIONS OF THE STARS.

THE determination of the distances of the stars from the earth has always formed a subject of great interest to astronomers. The earlier observers appear to have thought that the problem was an insoluble one. The famous Kepler, judging from what he called the "harmony of relations," came to the conclusion that the distance of the fixed stars should be about 2,000 times the distance of Saturn from the sun. Saturn was then the outermost planet of the solar system. The distance of even the nearest star, as now known, is about 14 times greater than that supposed by Kepler. Huygens thought the determination of stellar distance by observation to be impossible, but made an attempt at a solution of the problem by a photometric comparison between Sirius and the sun. By this method, he found that Sirius is probably about 28,000 times the sun's distance from the earth, but modern measures show that this estimate is far too small, the distance of Sirius being probably over 500,000 times the sun's distance, or about 18 times greater than Huygens made it.

When the Copernican theory of the earth's motion round the sun was first advanced, it was objected that, if the earth moved in a large orbit, its real change of place should produce an *apparent* change of position in the stars nearest to the earth, causing them to shift their relative position with reference to more distant stars. Copernicus replied to this objection—and we now know that his reply was correct—by saying that the distance of even the nearest stars was so great that the earth's motion would have no perceptible effect in changing their apparent position in the heavens; in other words, the diameter of the earth's orbit round the sun would be almost a vanishing point if viewed from the distance of the nearest stars. This explanation of Copernicus was at first ridiculed, and even the famous astronomer, Tycho Brahé, could not accept such a startling conclusion. This celebrated observer failed indeed to detect by his own observations any annual change of place in the stars, but he fancied that the brightest stars showed a perceptible disc, like the planets, a fact which, if true, would imply that, if the distance of the stars was so great as Copernicus supposed, their real diameter must be enormous. The invention of the telescope, however, dispelled this delusion of Tycho Brahé, and showed that even the brightest stars showed no perceptible disc. This was proved by Horrocks and Crabtree, who noticed that, in occultations of stars by the moon, the stars disappeared instantaneously, a fact which proved that the apparent diameter of the stars must be a very small fraction of a second of arc.

Galileo suggested that possibly the distance of the nearer stars might be determined by careful measures of double stars, on the assumption that the brighter star of the pair—if the difference in brillancy is considerable—is nearer the earth than the fainter star. He says (in his "*Opere di Galileo Galilei*"), "I do not believe that all the stars are scattered over a spherical superficies *at equal distances from a common centre*, but I am of opinion that their distances from us are so various that some of them may be two or three times as remote as others, so that when some minute star is discovered by the telescope close to one of the larger, and yet the former is highest, it may be that some sensible change might take place among them." Acting on this idea, Sir William Herschel, at the close of the eighteenth century, made a careful series of measures of certain double stars. He did not, however, succeed in his attempt, as his instruments were not sufficiently accurate for such an investigation, but his labours were rewarded by the great discovery of binary or revolving double stars, most interesting objects, which will be considered in the next chapter.

Numerous but unsuccessful attempts were made by Hooke, Flamsteed, Cassini, Molyneux, and Bradley, to find the distance of some of the stars. Hooke, in the year 1669, thought he had detected a parallax of 27 to 30 seconds arc in the star Gamma Draconis, but we now know that no star in the heavens has anything like so large a parallax. It must be here explained that to find the distance of any star from the earth, we must first measure its "parallax," which is the apparent change in its place due to the earth's motion round the sun. As the earth makes half a revolution in six months, and as the earth's mean distance from the sun—or the radius of the earth's orbit—is about 93 millions of miles, the earth is, at any given time, about 186 millions of miles distant from the point in its orbit which it occupied six months previously. The apparent change of position in a star's place, known as parallax, is *one-half* the total displacement of the star as seen from opposite points of the earth's orbit. In other words, it is the angle subtended at the star by the sun's mean distance from the earth. The measured parallax of a star may be either "absolute" or "relative." An "absolute parallax" is the actual parallax. A "relative parallax" is the parallax with reference to a faint star situated near a brighter star, the faint star being assumed to lie, as suggested by Galileo, at a much greater distance from the earth. As, however, the faint star may have a small parallax of its own, the "relative parallax" is the difference between the parallaxes of the two stars. Indeed, in some cases a "negative parallax" has been found, which, if not due to errors of observation, would imply that the faint star is actually the nearer of the two. From the observed parallax, the star's distance in miles may be found by simply multiplying 93 millions of miles by 206,265 and dividing the result by the parallax. To find the time that light would take to reach us from the star—the light journey as it is called—it is only necessary to divide the number 3·258 by the parallax.

In attempting to verify the result found by Hooke for the parallax of Gamma Draconis, Molyneux and Bradley found an apparent parallax of about 20 seconds of arc, thus apparently confirming Hooke's result, but observations of other stars showing a similar result, Bradley came to the conclusion that the apparent change of position was not really due to parallax, but was caused by a phenomenon now known as the "aberration of light," an apparent displacement in the positions of the stars, due to the effect of the earth's motion in its orbit round the sun combined with the progressive motion of light. The result is that "a star is displaced by aberration along a great circle, joining its true place to the point on the celestial sphere towards which the earth is moving." The amount of aberration is a maximum for stars lying in a direction at right angles to that of the earth's motion. The existence of aberration is an absolute proof that the earth does revolve round the sun, for were the earth at rest—as some paradoxes contend—there would be no aberration of the stars. This effect of aberration must, of course, be carefully allowed for in all measures of stellar parallax. To show that "aberration" could not possibly be due to "parallax," it may be stated that aberration shifts the apparent place of a star in one direction, while parallax shifts it in the opposite direction.

From photometric comparisons, the Rev. John Mitchell, in the year 1767, concluded that the parallax of Sirius is less than a second of arc; a result which has been fully confirmed by modern measures. He considered that stars of the sixth magnitude are probably 20 to 30 times the distance of Sirius, and judging from their relative brilliancy alone, this result would also be nearly correct. But recent measures have shown that some of the fainter stars are actually nearer to us than some of the brighter, and that the brightness of a star is no criterion of its distance.

The first stars on which observations seem to have been made with a view to a determination of their distance seem to have been Aldebaran and Sirius. From observations made in the years 1792 to 1804 with a vertical circle and telescope of 3 inches aperture, Piazzi found for Aldebaran an "absolute" parallax of about 1½ seconds of arc. O. Struve and Shdanow, in 1857, using a refractor of 15 inches aperture, found a "relative" parallax of about half a second. This was further reduced by Hall with the 26-inch refractor of the Washington Observatory to about one-tenth of a second, and Elkin, with a heliometer of 6 inches aperture, finds a relative parallax of 0″·116, or about 30 years' journey for light. For Sirius, Piazzi found, in 1792-1804, an absolute parallax of four seconds, but this was certainly much too large. All subsequent observers find a much smaller parallax, recent measures giving a relative parallax of 0·370″ by Gill, and 0·407″ by Elkin. In the years 1802-1804, Piazzi and Cacciatori found an absolute parallax of 1″·31 for the Pole Star; but this has been much reduced by other observers. Pritchard, by means of photography, found a relative parallax of only 0·073″, which agrees closely with some other previous results, and indicates a "light journey" of about 44 years!

For the bright star Procyon, Piazzi found a parallax of about three seconds, but this is also much too large, a recent determination by Elkin giving 0·266″, a figure in fair agreement with results found by Auwers and Wagner. For the bright star Vega, Calandrelli, in the years 1805-6, found an absolute parallax of nearly four seconds, but this has also been much reduced by modern measures; Elkin, from observations in the years 1887-88, finding a relative parallax of only 0·034″. Brinkley found a parallax of over one second for Arcturus, but Elkin's result is only 0·018″. If this minute parallax can be relied on, Arcturus must be a sun of vast size.

Owing to the large "proper motion" of the star known as 61 Cygni, its comparative proximity to the earth was suspected, and in 1812, Arago and Mathieu found, from measures made with a repeating circle, a parallax of over half a second. Various measures of its parallax have since been made, ranging from about 0·27″ to 0·566″. Sir Robert Ball, at Dunsink, Ireland, found 0·468″, and Pritchard, by means of photography with a 13-inch reflector, found 0·437″. We may, therefore, safely assume that the parallax of 61 Cygni is about 0·45″. This implies a distance of 458,366 times the sun's distance from the earth, or about 42 billions of miles, and a "light journey" of about 7¼ years.

It is usually stated that 61 Cygni is the nearest star to the earth in the Northern Hemisphere, but for the star known as Lalande 21,185, Winnecke found 0·511″, and afterwards 0·501″. This has, however, been reduced by Kapteyn (1885-1887) to 0·434″; and recently a parallax of 0·465″ has been found by the photographic method for the binary star, Eta Cassiopeiæ. 61 Cygni is a wide double star, but it seems doubtful whether the components are physically connected, although several orbits have been provisionally computed.

Nearer to us than 61 Cygni is the bright southern star Alpha Centauri, which, so far as is known at present, is the nearest of all the fixed stars to the earth. The first attempt to find its distance was made by Henderson in the years 1832-33, using a mural circle of 4 inches aperture and a transit of 5 inches. He found an "absolute" parallax of about one second of arc, which subsequent measures have shown to be rather too large. Measures in recent years range from 0·512″ to 0·976″, but probably the most reliable are those made with a heliometer of 4½ inches aperture by Dr. Gill (1881-82), who found a "relative" parallax of 0·76″, and by Dr. Elkin, using the same instrument, 0·671″. Gill's result would place the star at a distance of 271,400 times the sun's distance from the earth, or about 25 billions of miles, a distance which light, with its great velocity of 186,300 miles a second, would take over 4¼ years to traverse.

It will be understood that the parallaxes found for even the nearest fixed stars are so small that their exact determination taxes the powers of the most perfect instruments and the skill of the most experienced observers. One thing, however, seems certain, that the brightest stars are not necessarily the nearest, and that comparatively faint stars may be actually nearer to the earth than some of the brightest gems which deck our midnight sky. Indeed, from a discussion of the observed parallaxes and "proper motions" of 11 stars, Gylden finds a mean parallax of only 0·083″ for stars of the first magnitude. This agrees closely with the value 0·089″ found by Dr. Elkin.

In old times the stars were supposed to be absolutely fixed in the celestial vault, that is to say, that their relative positions did not change. This was a very natural conclusion, for before the invention of the telescope it would have been impossible to detect any "proper motion"—as it is called—by naked eye observations. Hence the term "fixed stars," used to distinguish the stars from the planets, which are always shifting their positions in the heavens. The existence of proper motion, in some at least of the stars, seems to have been discovered by Halley, who found from his observations in 1715 that the bright stars, Sirius, Arcturus, and Aldebaran, had apparently shifted their positions since the date of the earliest observations. This discovery was confirmed by James Cassini in 1738. He found that Arcturus had apparently moved through some five minutes of arc in 152 years, or about two seconds a year, a result which agrees fairly well with more exact modern measures.

This interesting discovery of stellar motion has been fully confirmed by modern observations, and we now know that, far from the stars being "fixed," most of them have an apparent motion on the celestial vault. These motions are, however, very slow, and can only be detected by accurate measurements and a careful comparison of their positions after the lapse of a number of years. The largest proper motion hitherto detected is that of a star known as 1830 of Groombridge's catalogue, a small star of about 6½ magnitude, which lies in the constellation Ursa Major. This star has an apparent motion of seven seconds per annum, which, though relatively large, is of course absolutely small, as the observed motion would only suffice to carry it through a space equal to the moon's apparent diameter in about 266 years. Assuming a parallax of about one-sixth of a second found by Kapteyn, this apparent motion would indicate a real motion of about 128 miles a second at right angles to the line of sight. As, however, there may be also motion *in* the line of sight, the above velocity would be a minimum—if the parallax can be relied upon—and the actual motion may be considerably more. From its rapidity, 1830 Groombridge has been called by Prof. Newcomb "the runaway star."

Next in order of rapidity of motion comes the southern star known as Lacaille 9352, which lies in the constellation Piscis Australis, a little south of Fomalhaut. This seventh magnitude star has an apparent motion of 6·9 seconds, which, with a parallax of 0·285″ found by Gill, indicates a velocity of 71 miles per second. Next comes 61 Cygni, with a velocity of 30 miles, and Epsilon Indi—another southern star—with a velocity of nearly 68 miles a second. These velocities are, however, exceeded by other stars if the measured parallaxes are correct. Thus the star Mu Cassiopeiæ, with a proper motion of 3·7 seconds, has, according to Pritchard's photographic measures, a parallax of only 0·036″, which would indicate a velocity of no less than 302 miles a second! and the small parallax found by Elkin for Arcturus would imply the startling velocity of 376 miles a second!

It is a remarkable fact that the eight stars with the largest proper motions are all below the fourth magnitude in brightness, and as a large proper motion probably indicates proximity to the earth, the conclusion seems evident that the brightest stars are not as a rule the nearest. Of twenty-five stars, with proper motions greater than two seconds of arc, there are only two—Arcturus and Alpha Centauri—whose magnitude exceeds the third. Indeed, more than half the stars with motions greater than one second are invisible to the naked eye!

Many stars have proper motions of less than a second of arc per annum. Very small proper motions have also been detected, which only reveal themselves after the lapse of a great number of years, and it seems probable that there are no really "fixed stars" in the heavens. For stars of the sixth magnitude, M. Ludwig Struve finds an average motion of only eight seconds in a hundred years, or about one-twelfth of a second per annum. If we assume that stars of the sixth magnitude are, on the average, of the same size and brightness as stars of the first magnitude, their distance from the earth would be ten times greater. Consequently, stars of the first magnitude should have an average proper motion of about eighty seconds in one hundred years. This, however, is not the case. The twenty brightest stars show an average motion of only sixty seconds in a hundred years. And the motion of stars of the second magnitude is relatively still slower. Instead of an average motion of fifty seconds in a hundred years—which they should have if the brightness were inversely proportional to the distance—it has been found that twenty-two stars of the second magnitude show an average motion of only seventeen seconds. This result seems to show that the brighter stars are not so near us as their brilliancy would lead us to suppose, a conclusion which has been already proved by actual measures of their distance.

From a consideration of the results found for stellar

parallax, Mr. Thomas Lewis, F.R.A.S., of the Greenwich Observatory, comes to the following conclusions[1]:—

"(1) Leaving out a few of the brightest stars, the parallaxes are constant down to 2·70 magnitude.

"(2) After 2·70 mag. is reached, the parallaxes are doubled, and remain practically constant to 8·40 mag.

"(3) Up to the 3rd mag. the velocities are very small, averaging about 9 miles per second, while after the 3rd mag. the velocity is 38 miles per second.

"Hence we may fairly deduce—

"(1) That there are a few stars (about 8) of exceptional brilliancy in our immediate neighbourhood, and scattered about amongst these a number of small stars (at present about 40 are known).

"(2) Stars of mag. 1·0 to 3·0 are, as a class, far outside this inner space, and have very small velocities.

"(3) The small stars here dealt with have apparently large velocities across the line of sight.

"These results show that the generally received idea that parallaxes are to be sought for in stars with large proper motion is correct, and we may add that this holds good, no matter what may be the star's magnitude."

The "proper motion" of a star only indicates its motion at right angles to the line of sight—that is, its motion on the surface of the celestial vault—and gives us no information as to whether the star is approaching to or receding from the earth. This motion "in the line of sight" cannot be detected by micrometrical measures with an ordinary telescope, and would probably have remained for ever unknown had the spectroscope not been invented. Dr. Huggins was the first to show that motions in the line of sight could be determined by measuring the displacement of the spectral lines caused by the approach or recession of the source of light, the lines being slightly shifted towards the blue end of the spectrum

[1] *Observatory*, April, 1895.

when the star is approaching the earth, and towards the red end when it is receding from us. The effect would, of course, be exactly the same if the star were at rest and the earth in motion. By carefully measuring this observed displacement of the spectral lines, the velocity in the line of sight can be easily computed. Dr. Huggins' observations were fully confirmed by Dr. Vogel.

The earlier determinations of motion in the line of sight were made by eye measurements with a micrometer, and owing to the difficulty and delicacy of these measures, the results were very discordant. The method has recently been much improved by photographing the spectra and measuring the positions of the lines on the photograph. Both methods agree in showing that the following stars, among others, are certainly *approaching* the earth: Arcturus, Vega, Procyon, Pollux, Altair, Spica, Alpha Cephei, Alpha Persei, Alpha Arietis, 61 Cygni, and the Pole Star; and the following are certainly *receding:* Capella, Rigel, Betelgeuse, Aldebaran, and Regulus.

Measures of photographic stellar spectra have yielded much more accurate results than the old method. Some of the velocities found in this way by Dr. Vogel—who has given especial attention to this subject—are very considerable. For the bright star Rigel he finds a velocity of recession of about 39 miles a second, for Aldebaran 30 miles, and for Capella 15 miles. He finds that the Pole Star is approaching the earth at the rate of 16 miles a second, and Procyon about 7 miles.

Dr. Bélopolsky has recently investigated the *absolute* velocity in space of the brighter component of 61 Cygni—that is, the motion across the line of sight combined with the motion *in* the line of sight. Assuming a parallax of half a second and a proper motion of 5·2 seconds, he finds that the motion across the line of sight, corrected for the sun's motion in space, is about 22½ miles per second. The motion *in* the line of sight, also corrected for the sun's motion, he finds, from photographs taken at Pulkova, to be about 27 miles a second towards the earth. Combining these motions, he finds the absolute velocity of the star in space to be about 35 miles a second, or nearly double the velocity of the earth in its orbit.

This method of measuring velocities in the line of sight has also been applied to the nebulæ. Mr. Keeler has observed and measured a displacement of the line known as the chief nebular line in several planetary nebulæ, and finds considerable motion in the line of sight. For example, in the nebula numbered 6790 in the "New General Catalogue," he finds a motion of recession of about 38 miles a second. Some of these motions may possibly be due, in part at least, to the sun's motion in space, carrying the earth with it, a motion which will now be considered. The method has also led to the discovery of the so-called "spectroscopic binary stars," a most interesting class of objects, which will be considered in the next chapter.

The proper motions of the stars long since suggested the idea that possibly the observed motion may be—to some extent, at least—merely apparent, and due to the real motion of the sun and solar system through space. The first investigation of this interesting question was made by Sir William Herschel in 1783, and he came to the conclusion that the sun is moving towards a point near Lambda Herculis, a result not differing widely from modern determinations. The reality of Herschel's result has been fully confirmed by subsequent investigations, and Argelander placed it beyond doubt by a comparison of the positions of a large number of stars determined at Abo with those found by Bradley in 1752. The accuracy of Argelander's result was confirmed by Otto Struve. According to the elder Struve, the results arrived at by Argelander, O. Struve, and Peters, is to place the point towards which the sun is moving, between the stars Pi and Mu Herculis, "at a quarter of the apparent distance of these stars from Pi Herculis," and they estimated the annual motion at about 33½ million miles geographical. The general accuracy of this conclusion has been verified by modern researches, although the results found by different astronomers vary to some extent. The accompanying diagram shows some of

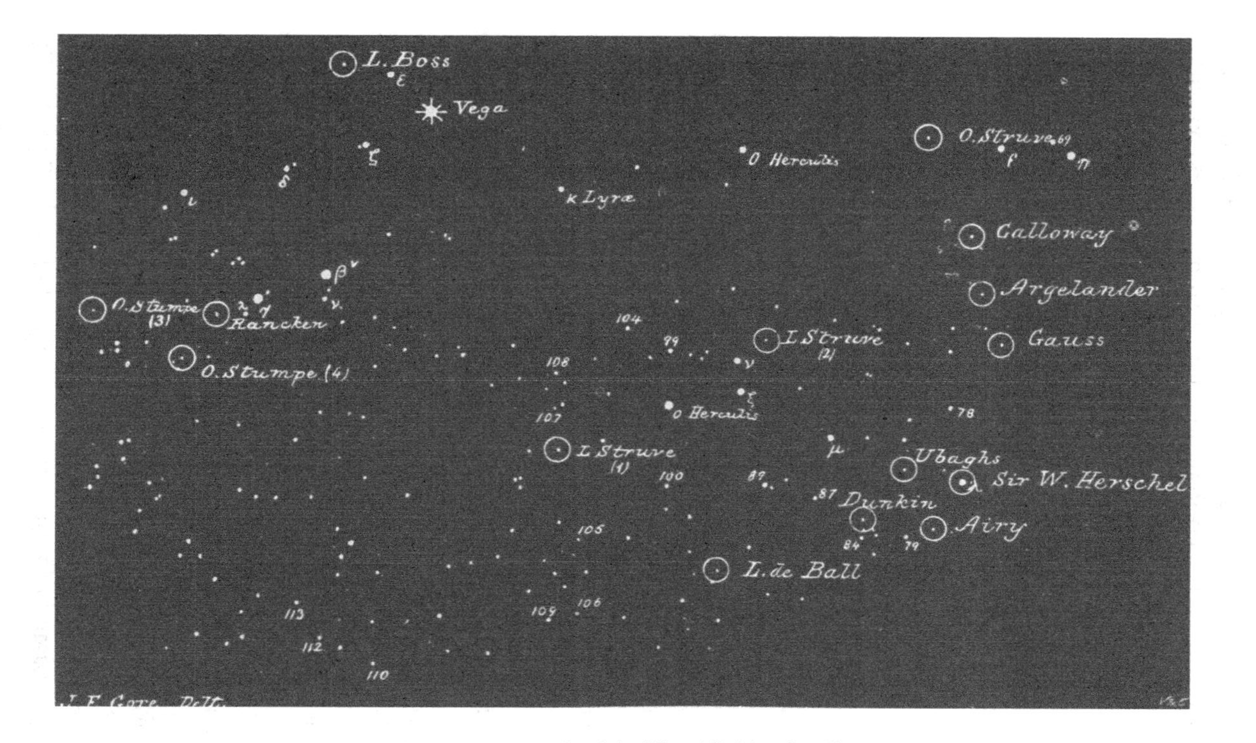

FIG. 3.—*Diagram showing "Solar Apex," and the different Positions found by various Computors.*
(From "Visible Universe.")

the different positions found by various computers. The later determinations seem to place the "apex of the solar motion," as it is termed, not far from the bright star Vega, or further to the east than Herschel placed it. The velocity of the sun's motion in space has not been so well determined as its direction. L. Struve's computations would indicate a velocity of about 14 miles a second; but other results give a much smaller velocity.

From a recent investigation of the nature of the sun's motion in space by Mr. G. C. Bompas,[1] he considers that the various positions of the sun's "apex" show a tendency to a drift along the edge of the Milky Way, and that this drift "seems to point to a plane of motion of the sun nearly coinciding with the plane of the Milky Way, or, perhaps, more nearly with the plane of that great circle of bright stars first described by Sir Wm. Herschel as inclined about 20° to the galaxy, and which passes through Lyra, in or near which constellation the solar apex lies," and he concludes, from the motion of the nearer stars, "that the sun moves in a retrograde orbit from east to west, and in a plane inclined a few degrees to that of the Milky Way." With reference to this very interesting conclusion, which may, perhaps, be confirmed by further observations, Mr. Bompas quotes the following passage from "The Visible Universe," p. 197, by the present writer :—"With reference to a possible motion of the stars in some general system, M. Rancken has found, from an examination of 106 stars, a tendency to drift along the course of the Milky Way from Aquila towards Cygnus and Cassiopeia, and past Capella through Orion to Argo. The *larger* motions, shown in Proctor's map of 'proper motions,' exhibit this tendency in a marked degree between Cygnus and Capella, and less clearly on the Sirius, but the smaller motions not so well," and Mr. Bompas points out that this apparent drift of the stars in the Milky Way, from west to east, "is just such as would be occasioned by a real motion of the sun in that plane, in a contrary direction from east to west."

[1] *Observatory*, Jan., 1896.

CHAPTER IV.

BINARY STARS.

DOUBLE and multiple stars may be either optical or real. Optical double stars are those in which the component stars are merely apparently close together, owing to their being seen in nearly the same direction in space. Two stars may *seem* to be close together, while, in reality, one of them may be placed at an immense distance behind the other. Just as two lighthouses at sea may, on a dark night, appear close together when viewed from a certain point, whereas they may be really miles apart. In the case of double stars it is, of course, always difficult to determine whether the apparent closeness of the stars is real or merely optical. But when, from a long series of observations of their relative position, we find that one is apparently moving round the other, we know that the stars must be comparatively close, and linked together by some physical bond of union. These most interesting objects are known to astronomers as binary or revolving double stars. The probable existence of such objects was predicted from abstract reasoning by Mitchell in the eighteenth century; but the discovery of their actual existence was made by Sir William Herschel, while engaged on an attempt to determine the distance of some of the double stars from the earth. "Instead of finding, as he expected, that annual fluctuation to and fro of one component of a double star with respect to the other—that alternate increase and decrease of their distance and angle of position, which the parallax of the earth's annual motion would produce—he observed, in many cases, a regular progressive change; in some cases bearing chiefly on their distance, in others on their position, and advancing steadily in one direction, so as clearly to indicate a real motion of the stars themselves," and measurements made during the subsequent 25 years fully proved the truth of the illustrious astronomer's discovery. It was found that in many double stars an orbital motion round each other was evident after a number of years of careful observation of their relative positions. Unlike the planetary orbits, which are nearly circular, at least those of the larger planets of the solar system, it was found that the orbits of these double stars differ, in many cases, widely from the circular form, in some cases, indeed, approaching in shape more the orbit of a comet than a planet.

The binary stars are among the most interesting objects in the heavens. The number now known probably amounts to nearly one thousand. In most of them, however, the motion is very slow, and in only about seventy cases has the change of position, since their discovery, been sufficient to enable an orbit to be computed. In most cases the plane of the real orbit, or ellipse, described by the companion round the principal star, is inclined to the line of sight. We therefore see the orbit foreshortened into a more elongated ellipse.

The relation of the apparent ellipse—or the ellipse we see described by one star round the other—to the real ellipse will be easily understood by the following illustration. Suppose a cylinder or rod of an elliptical, not circular, section to be cut across obliquely to its axis. This oblique section will represent the *real* orbit of a binary star, and the section at right angles to the axis, the *apparent* orbit. The angle between these two sections will represent the inclination of the real orbit to the plane of projection, or background of the sky. In the apparent orbit, the primary star, which is assumed to be situated in one of the foci of the real ellipse, does not lie in the focus of the apparent ellipse, and from its observed position in this latter ellipse we can deduce, mathematically, the particular angle at which the oblique section must be made to agree with the observed place of the primary star, and other details respecting the real ellipse.

Savary, in 1830, was the first astronomer who attempted to compute the orbit of a binary star, namely, the star Xi Ursæ Majoris. This remarkable pair was discovered by Sir

William Herschel in 1780, and as the period of revolution is about 61 years, a considerable portion of the ellipse had been described in 1830, when it was attacked by Savary. Since that year, orbits have been computed for a number of binary stars by several computers, among whom may be mentioned Sir R. Ball, Behrmann, Casey, Celoria, Doberck, Dunér, Elkin, Fritsche, Glasenapp, Sir J. Herschel, Hind, Jacob, Mädler, Mann, Schur, See, Thiele, Villarceau, and the present writer. The computation of a double star orbit is a matter of considerable trouble and difficulty, and cannot be described here. An account of the principal results arrived at by astronomers in this interesting branch of sidereal astronomy may, however, prove of interest to the general reader.

We will first consider the binary stars with short periods of revolution, which are, of course, the most interesting, and those whose orbits can be computed with greater accuracy than binaries having periods of considerable length. The binary star with the shortest period known at present seems to be the fourth magnitude star Kappa Pegasi. It was discovered as a wide double star by Sir William Herschel in 1786, the companion star being of the ninth magnitude. In August, 1880, Mr. Burnham, the famous American double star observer, examining the star with the 18½ inch refractor of the Dearborn Observatory, found the brighter star to be a very close double, with a distance between the components of only a quarter of a second of arc. A few years' observations showed that this pair were in rapid motion round each other, and from measures up to the year 1892, Burnham finds a period of 11·37 years. A later determination by Dr. See makes the period 11·42 years, so that we may conclude that the orbit is now pretty accurately determined. The plane of the orbit is highly inclined to the line of sight. Dr. See makes the inclination 81°.

Another binary star, with a period of about the same length, is Delta Equulei, which was discovered to be a close double by Otto Struve in 1851. As in the case of Kappa Pegasi, the orbit is highly inclined to the line of sight. In the year 1887,

Wrublewsky, the Russian computer, found a period of about 11½ years, with an orbit nearly circular. A new orbit was published in 1895 by Dr. See, who finds a period of 11·45 years, and an orbit agreeing fairly well with that of Wrublewsky, the orbit differing little from the circular form, and inclined to the line of sight at the high angle of 79 degrees. Burnham found only a "slight elongation" in the star with the great 36-inch telescope of the Lick Observatory in July, 1889. The distance between the components does not at any time exceed half a second of arc, so that it is always beyond the reach of all but the largest telescopes.

Next in order of shortness of period comes the southern binary star Zeta Sagittarii, for which an orbit was first computed in the year 1886 by the present writer, who found a period of 18·69 years. The orbit was re-computed in 1893, with the aid of recent measures by Mr. J. W. Froley, who finds a period of 17·71 years. The orbit of this star will, I think, require still further revision, but the period of about 18 years is probably not far from the truth.

Another remarkably rapid binary star is 85 Pegasi, for which Schaeberle computed a period of 22·3 years, but a later orbit by Prof. Glasenapp makes the period 17½ years, and Burnham thinks it will certainly be less than 20 years. Dr. See, however, finds a period of 24 years. The primary star is about the sixth magnitude, and the companion only the eleventh, a difference of five magnitudes, which implies that the larger star is one hundred times brighter than the companion.

Next in order of rapidity of motion we have the southern binary star 9 Argûs. For this pair, Burnham finds a period of 23·3 years, and Dr. See 22 years, the other elements of the orbit being also in close agreement. In this case also the orbit plane is highly inclined to the line of sight.

The star 42 Comæ Berenices has a period of about 25¾ years, according to Otto Struve. The orbit is remarkable from the fact that its plane passes through or nearly through the earth, and is, therefore, projected into a straight line, the companion star oscillating backwards and forwards on each side of its primary. I find that the plane of the orbit is at right angles to the general plane of the Milky Way.

The star Beta Delphini—the most southern of the four stars in the "Dolphin's Rhomb"—is also a fast-moving binary, discovered by Burnham in 1873, for which periods have been computed of 22·97 years by Glasenapp, 26·07 years by Dubjago, 27·66 years by Dr. See, and 30·91 years by the present writer. Burnham thinks the period will prove to be about 28 years. The spectrum of the light of Beta Delphini is similar to that of our sun, so that the two bodies should be comparable in intrinsic brilliancy. From my orbit of the pair, the "hypothetical parallax" is 0·052″—that is, this is the parallax the star would have on the supposition that the combined mass of its components is equal to the mass of the sun. Now, assuming the value of the sun's stellar magnitude which I have recently computed (*Knowledge*, June, 1895)—namely, 27·15—I find that the sun, if placed at the distance indicated for Beta Delphini, would be reduced to a star of 5·84 magnitude. As the star was measured 3·74 at Harvard, we have a difference of 2·1 magnitude, denoting that the binary—if of the same mass as the sun—must be about seven times brighter. As the spectrum is of the same type, this seems improbable, and we must conclude that the star's parallax is more than 0·052″.

Another remarkable binary star with a comparatively short period is Zeta Herculis. This pair have now performed three complete revolutions since their discovery in 1782 by Sir William Herschel. Several orbits have been computed, but Dr. See's period of 35 years is probably the best. The companion is now not far from its maximum distance (1½ seconds) from the primary star, and is within the reach of moderate-sized telescopes. The companion is, however, rather faint, being only 6½ magnitude, while the primary star is of the third. When at their nearest, some observers have spoken of an "occultation" of one star by the other, but no real occultation ever takes place, the components never approaching

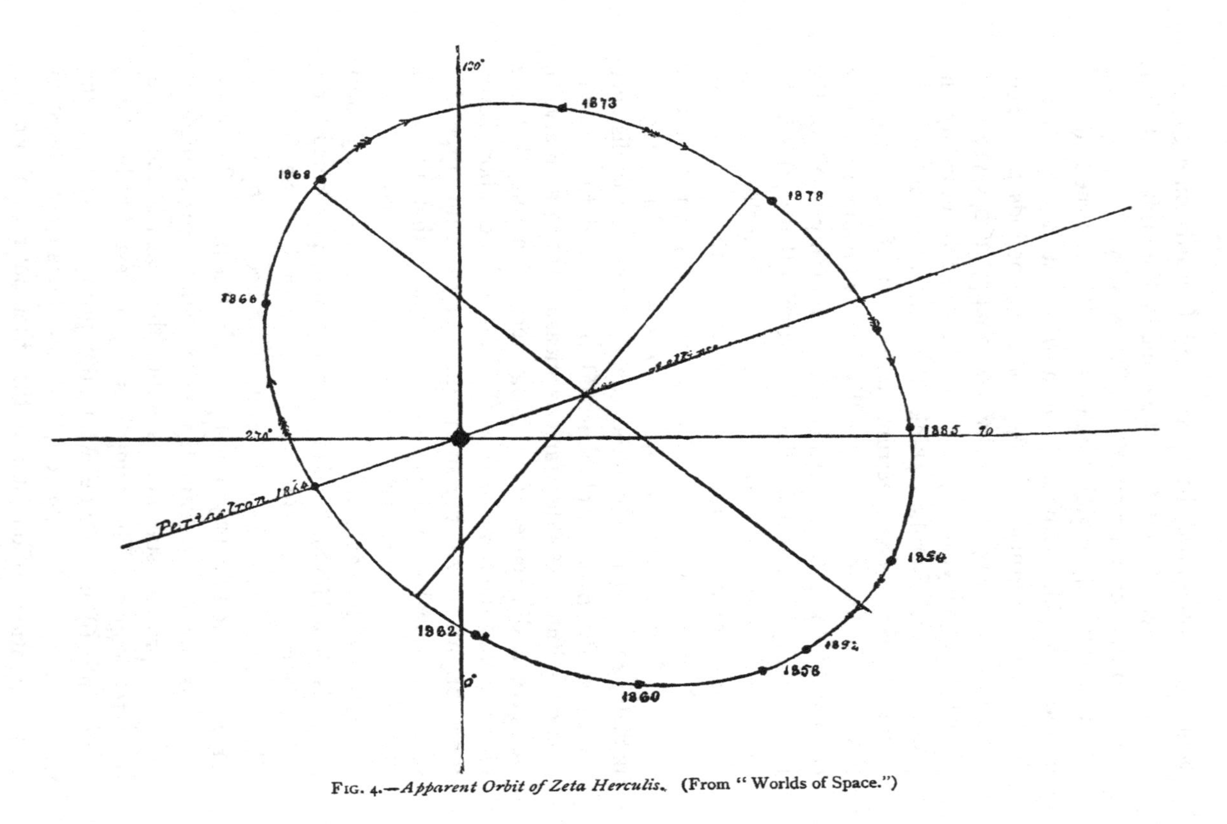

FIG. 4.—*Apparent Orbit of Zeta Herculis.* (From "Worlds of Space.")

within half a second of arc. The companion merely disappears owing to its faintness in telescopes of moderate power. An occultation of one component of a binary star by the other cannot take place except—as in the case of 42 Comæ—when the plane of the orbit passes through the earth.

In the case of the binary star, Eta Coronæ Borealis, it was some forty years ago, uncertain whether its period was 43 or 66 years, but now that two complete revolutions have been performed since its discovery by Sir William Herschel in 1781, the question has been finally decided in favour of the shorter period. Numerous orbits have been computed, but these by Dr. Doberck and Dr. Dunér are probably the best. Those give a period of about 41½ years. The components are nearly equal in brightness, but at their present distance are not within the reach of small telescopes.

The brilliant star Sirius is also an interesting binary star. The companion, which is relatively very faint—about tenth magnitude,—was discovered by Alvan Clark in 1862. The existence of some such disturbing body was previously suspected by astronomers, owing to observed irregularities in the proper motion of Sirius. Several orbits, giving periods of about 50 years, have been computed. Some measures in recent years, however, seemed to show that this period was somewhat too short, but a period of about 58½ years, computed by the present writer in 1889, will probably prove too long. Some few years ago, Burnham found the companion an easy object with the 36-inch refractor of the Lick Observatory, but towards the end of the year 1890 it passed beyond the power of even this giant telescope. It will probably, however, emerge very soon now from the rays of its brilliant primary.[1] Burnham finds a period of about 52 years, but the German astronomer, Auwers, who has carefully investigated the observed irregularities in the proper motion of Sirius, adheres to a period of about 49½ years. The great brilliancy of Sirius, the brightest star in the heavens, naturally suggests a sun of great size. Recent investigations, however, do not

[1] It has been recently seen again in America.

favour this idea. Assuming a parallax of 0″·39 (about a mean of the results found by Elkin and Gill), Auwers finds the mass of the system to be about three times the mass of the sun, the mass of the companion being about equal to the sun's mass. Placed at the distance of Sirius, the sun would, I find, be reduced to a star of about 1½ magnitude. As Sirius is about 1 magnitude brighter than the zero magnitude—that is, about 2 magnitudes brighter than a standard star of the first magnitude—it follows that it is about 2½ magnitudes, or about ten times brighter than the sun would be in the same position. Its spectrum is, however, of the first type, and the star is therefore not comparable with the sun in brilliancy. The above result would indicate that stars of the first or Sirian type are intrinsically brighter than our sun.

Sirius is about 11 magnitudes brighter than its faint companion. This makes the light of Sirius about 25,000 times the light of the small star. If, therefore, the two bodies were of the same intrinsic brilliancy, their diameters would be in the ratio of 158·5 to 1, and if of the same density, the mass of Sirius would be nearly five million times the mass of the companion! But, according to Auwers' calculations, the companion's mass is about one-half that of its primary. The two bodies must, therefore, be differently constituted, and, indeed, the companion must be nearly a dark body. It has been suggested that the companion may possibly shine by reflected light from Sirius; but this I have shown elsewhere to be quite impossible.[1] Even with a diameter equal to that of the sun, I find that with reflected light only it would be quite invisible in all parts of its orbit, even with the great Lick telescope. It must, therefore, shine with inhererent light of its own, and it seems probable that it is a large body, cooling down and approaching the complete extinction of its light. If Sirius has any planets revolving round it—like those of our solar system—they must for ever remain invisible in our largest telescopes. This remark, of course, applies to all the fixed stars, single and double. They may possibly have attendant families of

[1] *Journal of the British Astronomical Association*, March, 1891.

planets, like our sun, but if so, the fact can never be ascertained by direct observation. I find that the plane of the orbit of Sirius is at right angles to the general plane of the Milky Way.

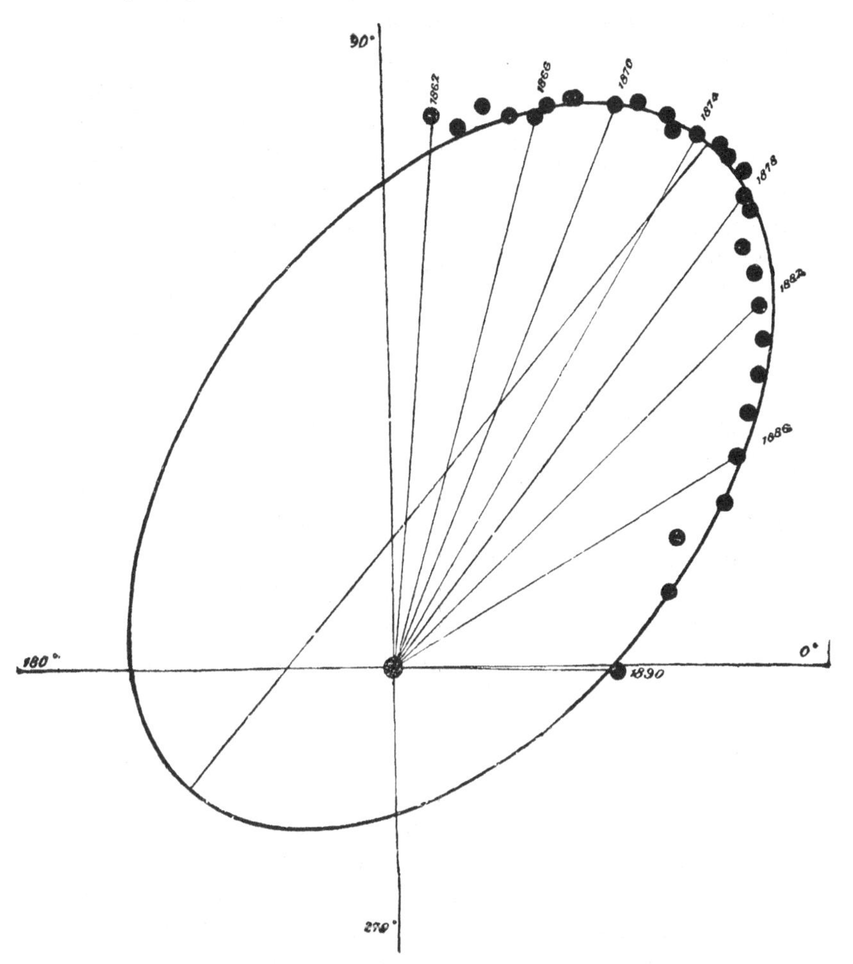

FIG. 5.—*Apparent Orbit of the Companion of Sirius.*
(From "Old and New Astronomy.")

The star Zeta Cancri is a well-known triple star, the close pair revolving in a period of about 60 years. Nearly two revolutions have now been completed since its discovery by

Sir William Herschel in 1781. All three stars probably form a connected system, but the motion of the third star round the binary pair is very slow and irregular. The motion of this interesting system has recently been investigated by Professor Seeliger, and he comes to the conclusion that, to make the observations agree with calculation, it is necessary to assume that the third star is in reality a very close double, the components of which revolve round their centre of gravity in about 17½ years, and both round the known binary pair. If this be so, we have here a remarkable quadruple pair; but it must be added that all efforts with large telescopes to see the companion star double have failed, and that the existence of the fourth star rests only on theory. Burnham, in 1889, using a power of 1500, failed to see any other component.

Another interesting binary star is Xi Ursæ Majoris. As already stated, this was the first pair for which an orbit was computed. More than a complete revolution has now been performed since its discovery by Sir William Herschel in 1780. The period has, therefore, been well determined, and seems to be about 60 years. Although the components are not near their maximum distance at present, they are still within the reach of moderate telescopes, the distance being about 1¾ seconds, and the magnitudes of the components, not very unequal, about 4 and 5.

The bright southern star, Alpha Centauri, the nearest of all the fixed stars to the earth, so far as is known at present, is also a remarkable binary star. It seems to have been first noticed as a double star by Richaud in 1690. Several orbits have been computed, ranging from about 75 to 88½ years, but recent calculations by Mr. A. W. Roberts and Dr. See make the period about 81 years, which agrees closely with Dr. Elkin's period of 80⅓ years. Combining Dr. Gill's parallax of 0″·76 with Elkin's elements, I find the sum of the masses nearly twice the mass of our sun, and the mean distance between the components about 23 times the earth's distance from the sun, or somewhat greater than the distance between the sun and Uranus. Dr. Doberck finds a period of about 79 years, and assuming a parallax of 0″·75, he finds the mean distance between the components 24·6 times the earth's distance from the sun; and he points out that if we suppose that their diameter does not differ much from that of our sun, each component "would appear from the other as a mere star to unaided vision, the distance being too great to show a disc."[1] From a recent investigation of the proper motion and position of Alpha Centauri, Mr. A. W. Roberts finds that the masses of the components are nearly equal, and the combined mass equal to twice the mass of our sun, a conclusion in close agreement with the result found above from the orbit. According to Dr. Gill, the difference in brightness of the two components is 1·25 magnitude, and Professor Bailey makes their photometric magnitudes 0·50 and 1·75. As this difference would make the brighter component over three times brighter than the companion, it follows that its surface must be much brighter, and Mr. Roberts concludes that the companion has proceeded "some distance on the down track from a sun to an ordinary planet." Assuming my value of the sun's stellar magnitude (about 27), I find that the sun, if placed at the distance of Alpha Centauri, would appear of about the same brightness as the star does to us. As, according to Professor Pickering, the spectrum of Alpha Centauri is of the second or solar type, it would seem that in mass, brightness, and physical condition, the star closely resembles our sun.

We next come to another very interesting binary star, known to astronomers as 70 Ophiuchi. It is a very fine double star, the magnitudes of the components being about 4 and 6, and the colours yellow and orange. More than a complete revolution has now been described by the components since its discovery by Sir William Herschel in 1779. Numerous orbits have been computed with periods ranging from $73\frac{3}{4}$ to 98 years. An orbit computed by the present writer, in 1888, gave a period of 87·84 years, and this was confirmed in 1894 by Burnham, who found a period of 87·85 years. A subse-

[1] *Nature*, Feb. 13, 1896.

quent investigation by Schur gives a period of 88·356 years. My orbit, combined with Krüger's parallax of 0″·162, give for the combined mass of the components 2·777 times the mass of the sun, and the distance between them 27·777 times the earth's distance from the sun, or somewhat less than the distance of Neptune from the sun. Schur has, however, recently found a parallax of 0″·286, which would reduce the mass of the system, and also the distance between the components. Recent observations show that the companion is now in advance of the theoretical position indicated by Schur's orbit, and Dr. See thinks that the observed irregularities in the orbital motion of the pair indicate the existence of a third body, and that either the primary star or the companion, probably the latter, is a very close binary star. Careful search, however, for a third body, made with large telescopes, have failed to reveal its existence, and so the matter remains in suspense. Placed at the distance indicated by Krüger's parallax, I find that our sun would be reduced to a star of about magnitude 3½, which shows that the sun and star are of about equal brightness. The spectrum is of the solar type, according to Vogel. I find that the plane of the orbit is at right angles to the plane of the Milky Way.

The star Gamma, in Corona Borealis, is a close and difficult binary star. Dr. Doberck finds a period of 95½ years, and Celoria about 85¼. As in the case of 42 Comæ, the plane of the orbit nearly passes through the earth, and the apparent orbit is, consequently, nearly a straight line. I find that the plane of the orbit is at right angles to the plane of the Milky Way.

The star Xi Scorpii is a remarkable triple star, like Zeta Cancri, the magnitudes of the components being about 4½, 5, and 7½. The components of the close pair have described a complete revolution since their discovery by Sir William Herschel in 1780. Dr. Doberck finds a period of about 96 years, and Schorr 105 years. The real orbit is nearly circular, but owing to its high inclination, about 70°, the apparent orbit is a very elongated ellipse. All three stars have a common proper motion through space, and, probably, form one system, but the motion of the third star is very

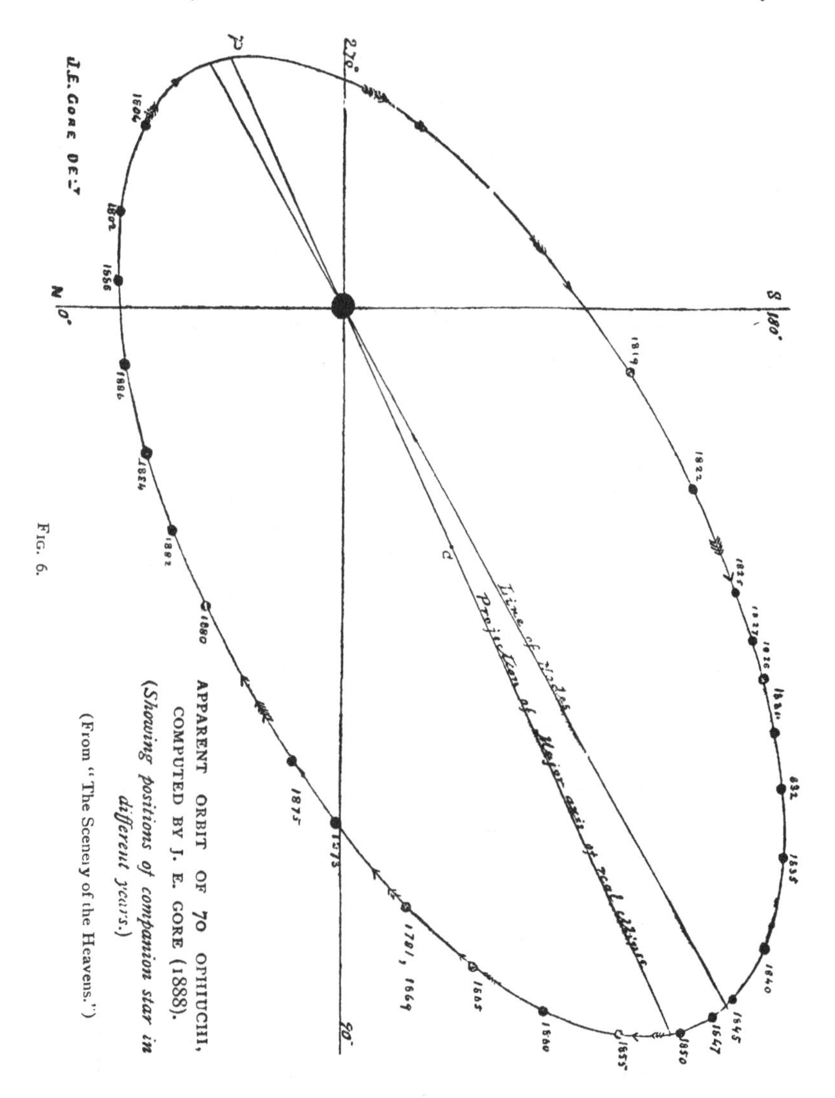

APPARENT ORBIT OF 70 OPHIUCHI, COMPUTED BY J. E. GORE (1888).
(*Showing positions of companion star in different years.*)
(From "The Scenery of the Heavens.")

Fig. 6.

slow, and its period of revolution must be several hundred years.

The star o^2, or 40 Eridani, is another interesting object. It is a star of about $4\frac{1}{2}$ magnitude, with a distant ninth magnitude companion, which is a double and binary star. It is sometimes stated that the bright star is the binary, but this is quite incorrect; the large star is single—at least, as far as is known at present. An orbit for the binary pair was computed, in 1886, by the present writer, who found a period of 139 years; but Burnham, using later observations, finds a period of 180 years. A physical connection may possibly exist between the binary pair and the bright star, as both have the same common motion through space, but the angular motion, if any, is very slow. Professor Asaph Hall found a parallax of about one-fifth of a second of arc, and this, combined with Burnham's orbit, gives the combined mass of the binary pair about two-thirds of the sun's mass, a result which seems remarkable, for the sun, placed at the distance indicated by Hall's parallax would, I find, shine as a star of about the third magnitude, or considerably brighter than the principal star of 40 Eridani. Owing to the faintness of the binary pair, the nature of its spectrum has not been determined. Computed by a well-known formula, its "relative brightness"—that is, its brightness compared with that of other binaries—is very small.

A very famous binary star is that known to astronomers as Gamma Virginis. Its history is a very interesting one. It lies close to the celestial equator, about one degree to the south and about fifteen degrees to the north-west of the bright star Spica (Alpha of the same constellation), with which it forms the stem of a Y-shaped figure, formed by the brightest stars of the constellation Virgo, or the Virgin, Gamma being at the junction of the two upper branches. The brightness of Gamma Virginis is a little greater than an average star of the third magnitude. Photometric measures made at Oxford and Harvard Observatories agree closely, and make its brightness about 2·7 magnitude—that is to say, rather nearer the third than the second magnitude. Variation of light has, however, been suspected in one or both components, and this question of light variation will be considered further on. The Persian astronomer, Al-Sûfi, in his description of the heavens, written in the tenth century, rates it of the third magnitude, and describes it as "the third of the stars of *al-auvâ*, which is a mansion of the moon," the first and second stars of this "mansion" being Beta and Eta Virginis, the fourth star Delta, and the fifth Epsilon, these five stars forming the two upper branches of the Y-shaped figure above referred to. Gamma was called *Zawiyah-al-auvâ*, "the corner of the barkers!" perhaps from its position in the figure, which formed the thirteenth Lunar Mansion of the old astrologers. It was also called *Porrima* and *Postvarta* in the old calendars. These ancient names of the stars are curious, and their origin doubtful.

The fact that Gamma Virginis really consists of two stars very close together seems to have been discovered by the famous astronomer, Bradley, in 1718. He recorded the position of the components by stating that the line joining them was then exactly parallel to a line joining Alpha and Delta of the same constellation. This was, of course, only a rough method of measurement, and the position thus found by Bradley being probably more or less erroneous, has given much trouble to computers of the orbit described by the component stars round each other, or, rather, round their common centre of gravity. Bradley does not give the apparent distance between the component stars; but we may conclude from the orbit, which is now well determined, that they were then at nearly their greatest possible distance apart. It is curious that between Bradley's time and 1794, the star was on several occasions occulted by the moon; but none of the observers refer to its duplicity. It was again measured by Cassini in 1720, by Tobias Mayer in 1756, and by Sir William Herschel in 1780. These measures showed that the distance between the components was steadily diminishing, and that the position angle of the two stars was also decreasing. This decrease in the position angle—measured from the north round by the east, south, and west, from 0 to 360°—shows that the apparent orbital motion is what is called retrograde, or in the direction of the hands of a clock, direct or "planetary motion" being in the opposite direction. The star was again measured by Sir John Herschel and South in the years 1822-38, by Struve in the same years, and by Dawes and other observers from 1831 to the present time. The recorded measures are very numerous, and have enabled computers to determine the orbit with considerable accuracy. The rapid decrease in the apparent distance from 1780-1834 indicated that the apparent orbit is very elongated, and that possibly the two stars might "close up" altogether, and appear as a single star even in telescopes of considerable power. This actually occurred in the year 1836, or, at least, the stars were then so close together that the most powerful telescopes of that day failed to show Gamma Virginis as anything but a single star. Of course, it would not have been beyond the reach of the giant telescopes of our day. From the year 1836 the pair began to open out again, and at present the distance is again approaching a maximum. It is now within the reach of small telescopes, and forms a fine telescopic object with a moderate-sized instrument.

The general character of the orbital motion may be described as follows:—In 1718, at the time of Bradley's observation, the companion star was to the north-west of the primary star; it then gradually moved towards the west and south, and in 1836, when at its minimum distance, it was to the south-east. From that date it again turned towards the north, and at present it is north-west of the primary star, and not far from the position found by Bradley in 1718.

The first to attempt a calculation of the orbit described by this remarkable pair of suns was Sir John Herschel, who in the year 1831 found a period of about 513 years. In 1833, he re-calculated the orbit, and found nearly 629 years. We now know that both these periods are much too long; but the data then available were insufficient for the calculation of an accurate orbit. From these results Herschel predicted that "the latter end of the year 1833, or the beginning of the year 1834, will witness one of the most striking phenomena which sidereal astronomy has yet afforded, *viz.*, the perihelion passage of one star round another, with the immense angular velocity of between 60° and 70° per annum, that is to say, of a degree in five days. As the two stars will then, however, be within little more than half a second of each other, and as they are both large and nearly equal, none but the very finest telescopes will have any chance of showing this magnificent phenomenon. The prospect, however, of witnessing a visible and measurable change in the state of an object so remote, in a time so short, may reasonably be expected to call into action the most powerful instrumental means which can be brought to bear on it." This prediction was not verified until the year 1836, when the pair "closed up out of all telescopic reach," except at the Dorpat Observatory, where a magnifying power of 848 still showed an elongation in the telescopic disc of the star. The orbit found by Sir John Herschel was a tolerably elongated ellipse, with its longer axis lying north-east and south-west. This was not quite correct, for we now know that this axis lies north-west and south-east, and that the apparent orbit is much more elongated than Sir John Herschel at first supposed. This was soon recognised by Herschel himself, and he came to the conclusion that he and other computers had been misled by Bradley's observation in 1718. He then rejected this early, and evidently faulty, observation, and using the measures up to 1845, he found a period of about 182 years, which we now know to be near the truth. The orbit was also computed by the famous German astronomer, Mädler, who found periods of 145, 157, and 169 years; by Hind, 141 years; by Henderson, 143 years; by Jacob, 133½, 157½, and 171 years; by Adams, 174 years; by Flammarion, 175 years; and by Admiral Smyth, 148 and 178 years. All these periods, we now know, are too small. Fletcher found 184½ years, and Thiele 185 years. Two orbits were computed by Dr. Doberck, in recent years, with periods of 180½ and 179½ years; but very recently (1895) the orbit has been re-computed by Dr. See, and he finds a period of 194 years. A comparison of the observed and computed positions shows, he thinks, that his elements are the most exact yet determined for any binary star.

The apparent orbit of the pair is a very elongated ellipse, and as Admiral Smyth said, "more like a comet's than a planet's." The real ellipse has a very high eccentricity, nearly 0·9—indeed, the greatest of all the known binary stars, and not much less than that of Halley's comet.

As I said above, the variability of the light of one or both components of Gamma Virginis has been strongly suspected. So far back as 1851 and 1852, O. Struve paid particular attention to this point. His observations in these years show that sometimes the component stars were exactly equal in brilliancy, and sometimes the southern star—the one generally taken as the primary—was from 0·2 to 0·7 magnitude brighter than the other. There seems to be little doubt that some variation really takes place in the relative brightness of the pair. This is clearly indicated by the measures of position angle. For example, in the year 1886, Professor Hall recorded the position as 154·9, evidently measuring from the northern star as the brightest of the two; while, in 1887, Schiaparelli gives 334°·2—or about 180° more—thus indicating that he considered the *southern* star as the primary, or brighter, of the pair. Burnham found 153°·4 in 1889, and Dr. See 332°·50 in 1891. This is also shown by earlier measures, for Otto Struve found the southern star half a magnitude brighter than the other on April 3, 1852, while on April 29 of the same year he found them "perfectly equal." He thought the variation was about 0·7 of a magnitude, but that the climate of Poulkova, where he observed, was not suitable for such observations. This variation is very interesting, and the question should be thoroughly investigated with a good telescope.

As the distance of Gamma Virginis from the earth has not been determined, it is not possible to calculate the actual dimensions of the orbit and the mass of the system. If we assume that the combined mass of the components is equal to the sun's mass, I find from Dr. See's orbit that the "hypothetical parallax" would be 0·119″, implying a distance of

1,733,319 times the sun's distance from the earth. If, however, we suppose that the mass of each of the components is equal to the sun's mass, or the mass of the system double that of the sun—perhaps a more probable supposition—I find that the parallax would be about one-tenth of a second, denoting a distance of 2,062,650 times the sun's distance from the earth. Placed at this last distance, the sun would, I find, be reduced to a star of about 4½ magnitude, or about 1¾ magnitudes fainter than Gamma Virginis appears to us. This difference implies that, supposing each of the component stars of the binary to have a mass equal to the sun's mass, their combined light is about five times greater than the sun would emit if placed at the same distance, and as the components are nearly equal in brightness, each of them would be 2½ times brighter than the sun. According to Vogel, the star's light gives a spectrum of the first or Sirian type, but according to the Draper "Catalogue of Stellar Spectra," the spectrum is of the solar type. If the spectrum is of the first type, its brilliancy is easily explained; for, as I have shown elsewhere, the Sirian stars, are intrinsically much brighter in proportion to their mass than those of the solar type. But if its spectrum is of the solar type, it is not so easy to explain its brilliancy. Computing by a well-known formula, I find its relative brightness is nearly five times greater than that of Xi Ursæ Majoris, the spectrum of which is of the solar type. If, to account for its brilliancy, we assume that the star is nearer to the earth than the parallax assumed above would imply, then the mass of the system must be less than the mass of our sun. As we have seen above, doubling the supposed mass increased the distance; so, on the other hand, if we diminish the distance, we must diminish the mass also. Thus, if we reduce the distance to one-half, we must reduce the mass to one-eighth of the sun's mass. A distance of one-third would give a mass of $\frac{1}{27}$th, and a distance of one-fourth would imply a mass only $\frac{1}{64}$th of the sun's mass. To reduce the sun to the same brightness as Gamma Virginis, it should be removed to a distance indicated by a parallax of one-tenth of a second multiplied by the square root of five, or 0·223″. If, however, the star's parallax were so much as this, it is probable that it would have been detected and measured long ago. In the case of the binary star Castor, I find from the orbit and a small parallax found by Johnson (about one-fifth of a second) that its mass is only $\frac{1}{19}$th of the sun's mass, but in this case the spectrum is of the Sirian type, and stars of this type are very bright in proportion to their mass. The colours of the components of Gamma Virginis, which are very similar to those of Castor—white or pale yellow—would suggest that they may belong to the same type.

Another interesting binary star is Eta Cassiopeiæ. The components are about 4 and $7\frac{1}{2}$ magnitude, and the pair have described a considerable portion of their orbit since its discovery in 1779 by Sir William Herschel, the distance diminishing from about 11 seconds to $4\frac{3}{4}$. Periods ranging from 149 to $222\frac{1}{2}$ years have been found by different computers. The most recent computation makes it about 196 years. Assuming a parallax of 0·154″ found by Struve, the mass of the system will be from $5\frac{3}{4}$ to $10\frac{3}{4}$ times the mass of the sun, according to the length of the period we assume. A much larger parallax of 0″·3743 was, however, found by Schweizer and Socoloff, which would considerably reduce the mass, and recently a still larger parallax of 0″·465 has been found by photography, which, with Grüber's elements of the orbit, would reduce the mass of the system to $\frac{1}{6}$th of that of the sun.

The bright star Gamma Leonis, situated in the well-known "Sickle in Leo," is also a binary star, but only a small portion of the orbit has been described since its discovery by Sir William Herschel in 1782. Dr. Doberck finds a period of 407 years. It is remarkable for its very high "relative brightness," which is curious, as its spectrum is of the solar type. This pair forms a fine object for a small telescope.

The star known as 12 Lyncis is a triple star, the components being 5, 6, and $7\frac{1}{2}$ magnitude. The close pair form a binary system, for which an orbit has been computed by the present writer, who finds a period of about 486 years. Sir

John Herschel predicted in 1823 that the angular motion of the pair would " bring the three stars into a straight line in 57 years." This prediction was fulfilled in 1887, when measures by Tarrant showed that the stars were then exactly in a straight line.

The bright star Castor is a famous double star, and has been known since the year 1718, when it was observed by Bradley and Pond. It was also observed by Maskelyne in 1759, and frequently by Sir William Herschel from 1799 to 1803. Numerous orbits have been computed, with periods ranging from 199 years by Mädler, and 1,001 years by Doberck. Wilson found a period of about 983 years, and Thiele about 997 years, so that the longest period would seem to be nearest the

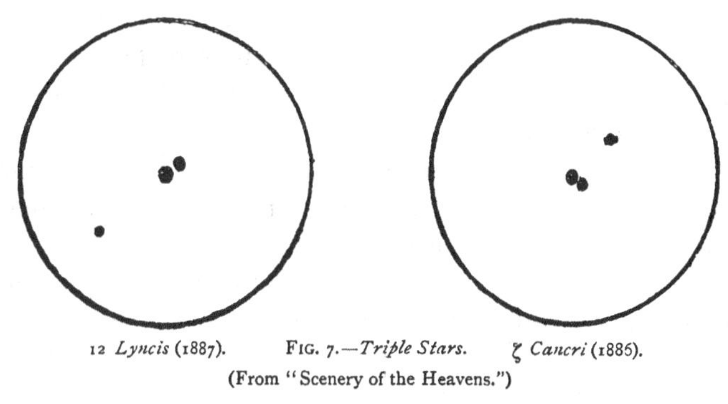

12 *Lyncis* (1887). FIG. 7.—*Triple Stars.* ζ *Cancri* (1885).
(From "Scenery of the Heavens.")

truth. According to a somewhat doubtful parallax found by Johnson, the distance of Castor from the earth is about double that of Sirius. With this distance, and Doberck's elements of the orbit, I find that the mass of the system of Castor is only $\frac{1}{19}$th of the sun's mass, a result which would imply that the components are masses of glowing gas! The spectrum of Sirius is of the first, or Sirian, type, another example of the great brilliancy of stars of this type. Quite recently (1896), Dr. Bélopolsky has found, with the spectroscope, that the brighter component is a close binary star with a dark companion, like Algol. The period of revolution is about 3 days, and the relative orbital velocity about $20\frac{3}{4}$ miles a second. Dr. Bélopolsky's observations show that the system is receding from the earth at the rate of about $4\frac{1}{2}$ miles per second. Assuming the bright and dark companion to be of equal mass, and hence the absolute orbital velocity of each one half the relative velocity found by Bélopolsky, I find that, if the orbit is circular, the distance between the components is about 85,400 miles, or slightly less than the sun's diameter, and their combined mass about $\frac{1}{87}$th of the sun's mass. This result would imply a still smaller mass for the whole system of Castor than that found from the orbit of the two bright components, but tends strongly to confirm the opinion already expressed, that the components of this remarkable system are merely masses of glowing gas. Assuming that all three components are of equal mass, the combined mass of the system would be $\frac{1}{58}$th of the sun's mass. From this result we can easily compute the stars' parallax, which, from Dr. Doberck's orbit, I find to be $0''\cdot2873$, a quantity which might be measured by the photographic method.

With reference to the colours of the components of binary stars, the following relation between colour and relative brightness has been established[1] :—

(1.) When the magnitudes of the components are equal, or approaching equality, the colours are generally the same, or similar.

(2.) When the magnitudes of the components differ considerably, there is also a considerable difference in colour.

A new class of binary stars has been discovered within the last few years by means of the spectroscope. These have been called "spectroscopic binaries," and the brighter component of Castor, referred to above, is an example of the class. They are supposed to consist of two component stars, so close together that the highest powers of the largest telescopes fail to show them as anything but single stars. Indeed, the velocities indicated by the spectroscope show that they must be so close that the components must for ever remain invisible by the most powerful telescopes which could ever be constructed by man. In some of these remarkable objects, the

[1] "Planetary and Stellar Studies," p. 257.

doubling of the spectral lines indicates that the components are both bright bodies, but in others, as in Algol, the lines are merely shifted from their normal position, not doubled, thus denoting that one of the components is a dark body. In either case, the motion in the line of sight can be measured by the spectroscope, and we can, therefore, calculate the actual dimensions of the system in miles, and thence its mass in terms of the sun's mass, although the star's distance from the earth remains unknown. Judging, however, from the brightness of the star, and the character of its spectrum, we can make an estimate of its probable distance from the earth.

Let us first take the case of Algol. This famous variable star has, according to the Draper catalogue, a spectrum of the Sirian type. It may, therefore, be comparable with that brilliant star in intrinsic brightness and density. Assuming the mass of Sirius at 2·20 times the mass of the sun, as found by Auwers, and that of the brighter component of Algol at four-ninths of the sun's mass, as given by Vogel,[1] I find that for the *same distance* Sirius would be about 2·8 times brighter than Algol. But photometric measures show that Sirius is about 22 times brighter than Algol, from which it follows—since light varies inversely as the square of the distance—that Algol is 2·77 times further from the earth. Assuming the parallax of Sirius at 0·39″, this would give for the parallax of Algol 0·14″, or a journey for light of about 23 years. From the dimensions of the system, as given by Vogel—about 3,230,000 miles from centre to centre of the components—this parallax would give an apparent distance between the components of less than $\frac{1}{200}$th of a second, a quantity much too small to be visible in our largest telescopes, or probably in any telescope which man can ever construct. From a consideration of irregularities in the proper motion of Algol and in the period of its light changes, Dr. Chandler infers the existence of a third dark body and a parallax of 0·07″. As this is exactly one-half the parallax found above, it implies a distance just double of what I have found, and would, of course,

[1] See Chapter V.

indicate that Algol is intrinsically four times brighter than Sirius. This greater brilliancy would suggest greater heat, and would agree with its small density, which, from its diameter, as given by Vogel—1,061,000 miles—I find to be only one-third of that of water.

Let us now consider the case of Beta Aurigæ, which spectroscopic observations show to be a close binary star with a period of about four days, and a distance between the components of about eight millions of miles. This period and distance imply that the mass of the system is about five times that of the sun. As in this case the spectral lines are doubled at regular intervals of two days, and not merely shifted, as in the case of Algol, we may conclude that both the components are bright bodies, and we may not be far wrong in supposing that they are of equal mass, each having $2\frac{1}{2}$ times the mass of the sun. As the spectrum of Beta Aurigæ is of the same type as Sirius, we may compare it with that star, as we did in the case of Algol. Assuming the same density and intrinsic brightness for both Beta Aurigæ and Sirius, I find that Beta Aurigæ should be about twice as bright as Sirius. Now, according to the Oxford photometric measures, Sirius is 2·89 magnitudes, or 14·32 times brighter than Beta Aurigæ. Hence it follows that the distance of Beta Aurigæ should be about $5\frac{1}{2}$ times greater than the distance of Sirius. Hence, assuming the parallax of Sirius at $0''·39$, that of Beta Aurigæ should be about $0''·061$. From actual measures of the parallax of Beta Aurigæ, made by the late Prof. Pritchard at Oxford, he found, from two companion stars, a mean parallax of $0''·062$, a result in remarkably close agreement with that computed above from a consideration of the star's mass and light compared with that of Sirius. As the actual distance between the components of Beta Aurigæ is equal to the sun's diameter divided by 11·625, we have the maximum angular separation between the components equal to $0''·062$ divided by 11·625, or about $\frac{1}{200}$th of a second, or nearly the same as in the case of Algol.

The bright star Spica has also been found by the spectroscope to be a close binary star. Vogel finds a period of four days with a distance between the components of about $6\frac{1}{4}$ millions of miles, and assuming that the components have equal mass and are moving in a circular orbit, he finds the mass of the system about 2·6 times the mass of our sun. This would give each of the components 1·3 times the mass of the sun, and it follows that the light of Spica—which gives a spectrum of the Sirian type—should, for equal distances, exceed that of Sirius about 1·4 times. Now, the photometric measures at Oxford show that Sirius is 1·91 magnitude, or 5·8 times brighter than Spica. Hence it follows that the distance of Spica should be 2·85 times the distance of Sirius. This would make the parallax of Spica about 0″·137. So far as I know, a measurable parallax has not yet been found for this star. Brioschi, in 1819-20, observing with a vertical circle of four inches aperture, found a negative parallax, which would imply that its parallax is too small to be measurable. Still, the above result would seem to indicate that its parallax might be measurable by the photographic method. The parallax found above would imply that the maximum distance between the components of Spica would not exceed $\frac{1}{10}$th of a second, a quantity much too small to be detected by the most powerful telescopes. In addition to its orbital motion, Vogel finds that Spica is approaching the sun at the rate of over 9 miles per second.

We now come to Zeta Ursæ Majoris (Mizar), which has also a spectrum of the Sirian type, and which the spectroscopic measures indicate is a close binary star with a period of about 104 days, and a combined mass equal to forty times the mass of the sun. Proceeding as before, we find that the light of Mizar should be about 8·7 times that of Sirius. But the photometric measures show that Sirius is about three magnitudes, or about sixteen times brighter than Mizar Hence the distance of Mizar should be nearly twelve times the distance of Sirius. This gives for the parallax of Mizar about 0″·033. Klinkerfues found a parallax of 0″·0429 to 0″·0477, which does not differ widely from the above result.

As the velocity of the orbital motion shown by the spectroscope indicates a distance between the components of about 143 millions of miles, or about the distance of Mars from the sun, it follows that the maximum distance between the components would be 0″·032, multiplied by 1½, or 0″·048, a quantity beyond the reach of our present telescopes.

The well-known variable star, Delta Cephei, has recently been added to the list of "spectroscopic binaries." From observations with the great 30-inch refractor of the Pulkowa Observatory in the summer of 1894, M. Bélopolsky finds that the star is probably a very close double, the companion being a nearly, or wholly, dark body, as in the case of Algol, and the orbit a very eccentric one. The observed variation of light indicates, however, that there is no eclipse, as occurs in Algol, so that the fluctuations in the light of Delta Cephei are probably due to some other cause. The spectrum of the star is of the solar type, so that in this respect it differs from the other spectroscopic binaries referred to above. The observations show that the system is approaching the sun at the rate of about 15 miles a second. Spectroscopic observations also suggest that the well-known variable star Beta Lyræ may also consist of two close companions. Further details respecting these observations will be given in the next chapter.

From a recent investigation of the proper motion of the star Tau Virginis, Dr. Fritz Cohen thinks it is probably a close binary, the companion star of which has not yet been detected.

It should be mentioned that in the case of Beta Aurigæ, Spica, Zeta Ursæ Majoris, and Castor, as there is no variation of light, as in Algol, the plane of the orbit is probably inclined to the line of sight. This would have the effect of increasing the computed mass of the system, and thus diminishing the calculated parallax. As the above calculations have been made on the assumption that the plane of the orbit passes through the earth, it follows that the computed parallax is a maximum, and that these remarkable objects may be really further from the earth than even the minute parallaxes found above would indicate. As the parallaxes of the nearest stars, such as Alpha Centauri, 61 Cygni, Sirius, and some other stars, are considerably greater than those found above, it would seem that our solar system is not situated in a region of binary stars, and that these wonderful objects lie beyond our immediate neighbourhood. It is also remarkable that, with the exception of Delta Cephei, they have all spectra of the Sirian type, including those Algol variables whose spectra have been examined.

By the aid of the parallaxes computed above, we can easily calculate the relative brightness of the sun compared with that of the spectroscopic binaries. Assuming that the sun is 27 magnitudes brighter than the Zero magnitude, or 28 magnitudes brighter than a standard star of the first magnitude, and taking the parallax of Algol as 0''·07, I find that the sun, placed at the distance indicated by this parallax, would be reduced to a star of 5·35 magnitude, or about three magnitudes fainter than Algol, which implies that Algol is about 15½ times brighter than our sun. In the case of Beta Aurigæ, if the sun were placed at the distance indicated by the parallax of 0''·061, it would be reduced to a star of 5·65 magnitude, or about 3·7 magnitudes fainter than Beta Aurigæ, which would imply that Beta Aurigæ is about thirty times brighter than the sun. In the case of Spica we have the sun reduced to a star of about the fourth magnitude, or about three magnitudes fainter than Spica, indicating that Spica is, like Algol, about 15½ times brighter than the sun, although the mass of Spica is only 2·6 times the mass of the sun. Finally, in the case of Mizar, we have the sun reduced to a star of about the seventh, or about five magnitudes fainter than Mizar, indicating that Mizar is no less than one hundred times brighter than our sun. These results show the great relative brilliancy of stars with a Sirian spectrum, when compared with that of the sun, a consideration which has already been arrived at from other considerations.

CHAPTER V.

VARIABLE AND TEMPORARY STARS.

TO ordinary observers, the light of the stars seems to be constant. Even to those who are familiar with the constellations, the stars appear to maintain their relative brilliancy unchanged. To a great extent this is, of course, true ; the great majority of the stars remaining of the same brightness from day to day, and from year to year. There are, however, numerous exceptions to this rule. Many of the stars, when carefully watched, are found to fluctuate in their light, being sometimes brighter, and sometimes fainter. These are known as "variable stars"—one of the most interesting class of objects in the heavens. Some of these have been known for a great number of years, and their variations having been carefully watched, the laws governing their light changes have been well determined.

We will first consider the variable stars with long periods of variation, as these generally show the largest fluctuations of light. Among these, the first star in which variation of light seems to have been noticed is the extraordinary object, Omicron Ceti, popularly known as Mira, or the "wonderful" star. It appears to have been first noticed by David Fabricius in the year 1596. He observed that the star now called Omicron, in the constellation Cetus, was of the third magnitude on April 13 of that year, and that in the following year it had disappeared. Bayer saw it again in 1603, when forming his maps of the constellations, and assigned to it the Greek letter Omicron, but does not seem to have noticed the fact that it was the same star which had been observed by Fabricius seven years previously. No further attention seems to have been paid to it until 1638 and 1639, when it was observed at Francker by Professor Phocylides Holwarda to be of the third magnitude in December, 1638, invisible in the following summer, and again visible in October, 1639. From 1648 to 1662 it was carefully observed by Hevelius, and in subsequent years by several observers. Its variations are now regularly followed from year to year, and it forms one of the most interesting objects of its kind in the heavens. Its light varies from about the second magnitude to the ninth, but its brightness at maximum is variable to a considerable extent. Heis found its *average* brightness at maximum in the years 1840-58 to be about the third magnitude, but on November 6, 1799, Sir William Herschel found it but little inferior to Aldebaran. On the other hand, at the maximum of 1868, November 7, Heis found it only of the fifth magnitude, and fainter than he had seen it for twenty-seven years. Sawyer also observed a maximum of about the fifth magnitude (4·9) on November 10, 1887. M. Dumenel finds (1896) that in the last twelve periods the magnitude at maximum varied from 2·5 to 4·7.[1]

It is stated in several books on astronomy, on the authority of Hevelius, that in the years 1672-76 Mira was invisible at the epoch of maximum. This is, however, quite a mistake, for it was long since (1837) pointed out by Bianchi that the supposed non-appearance of Mira in those years can be simply accounted for by the fact that the star was near the sun at the time of maxima, and could not be observed. If the star happens to be at a maximum in April or May, it will be too near the sun to be seen, and as the mean period is about 331 days, this occurs every ten years. For this reason the maxima seems to have passed unobserved in the years 1852, 1853, and 1854, and again in 1883. The star will be very favourably placed for observation in the year 1897, and some following years. It has also been stated that Mira wholly disappears at the maximum, but this is another error, for the star never becomes fainter than 9½ magnitude at any time, and always remains visible in a 3-inch telescope. The colour of the star is decidedly reddish, but this hue seems to be more marked at minimum than at maximum. The spectrum is a remarkable one of the third type, in which

[1] *Comptes Rendus*, March 30, 1896.

bright lines have been seen by Espin, Maunder, and Secchi. At the minimum of February, 1896, the spectrum was photographed by Professor Wilsing, and he found it very similar to a photograph taken by Professor Pickering some years previously. The recent photograph shows the lines of hydrogen broad and bright. There seems to be no other bright lines except those of hydrogen. The blue end of the spectrum is very similar to that of our sun, but towards the red end there are "dark flutings, fading towards the red." The bright hydrogen lines have only been seen at maximum, but the instruments used by Professor Wilsing were not sufficiently powerful to show whether they are also visible at minimum.[1] Professor Pickering thinks that "probably most of the stars of long period give a spectrum resembling that of o Ceti, and having the hydrogen lines G, *h*, α, β, γ, and δ, bright about the time of maximum. When the photographic spectrum is faint, only the brighter lines, G and *h*, are visible." Within the last few years, Mrs. Fleming, while examining the photographs of stellar spectra taken for the Henry Draper Memorial, has detected a number of variable stars of long period by the presence of bright lines in their spectra. These are mostly telescopic stars.

Although the average period of Mira is about 331 days, it is subject to marked irregularities, which Argelander has attempted to represent by an elaborate formula. In recent years, however, the epochs of maxima have deviated considerably from the dates computed from this formula, and at the maximum of February, 1896, the star did not reach its maximum light until nearly two months after the predicted time.

Perhaps the long period variable star next in order of interest—at least to observers in the Northern Hemisphere—is that known as Chi Cygni. It was discovered by Kirch in 1686. A mistake is often made about the identity of this remarkable object. It is sometimes confused with the neighbouring star, 17 Cygni of Flamsteed's catalogue. At the

[1] *Nature*, April 30, 1896.

time of Flamsteed's observation, the variable star—which is the true Chi Cygni of Bayer's map (made in 1603)—happened to be faint, and Flamsteed, not being able to find Bayer's star, affixed the Greek letter χ to his No. 17. It was proposed by Struve to call Flamsteed's star χ^1, and the variable χ^2; but there seems to be no necessity to perpetuate Flamsteed's error, which has been frequently pointed out. All authorities on the variable stars now give this variable its proper designation—χ Cygni. The star varies at maximum from 4 to 6½ magnitude, and at the minimum it sinks to below the thirteenth magnitude. At some maxima, therefore, it is easily visible to the naked eye, and at others it is just below the limit of ordinary vision. At the maximum of 1847, it was visible to the naked eye for a period of 97 days. The average period is about 406 days; but, according to Schönfeld—a well-known authority on the variables—observations indicate a small lengthening of the period. Observations in recent years show that the minimum occurs about 185 days before the maximum. This gives 221 days for the fall from maximum to minimum, and illustrates a feature common to many of the variable stars, namely, that the increase of light is more rapid than the decrease. This peculiarity is especially marked in the short period variables, which will be considered further on. Chi Cygni is said to be "strikingly variable in colour." Espin's observations in different years show it "sometimes quite red, at others only pale orange-red." In the spectroscope, its light shows a splendid spectrum of the third type (or banded spectrum, very characteristic of these long period variables), in which bright lines were observed by Espin in May, 1889. One of these bright lines seems to be identical with the coronal line D_3, the characteristic line of helium.

R Leonis is another remarkable variable star, which is sometimes visible to the naked eye at maximum. It lies closely south of the star known as 19 Leonis. It was discovered by Koch in 1782. At the maximum, its brightness varies from 5·2 to 7 magnitude, and at minimum it fades to about the tenth magnitude. The mean period is about 313 days; but this is subject to some irregularities, and Chandler finds "good evidence of cyclical variation of period, with a long term." The star is red in all phases of its light, and forms a fine telescopic object. Close to it are two small stars, which form, with the variable, an isosceles triangle. The spectrum is a fine one of the third type, a type very characteristic of these long period variables. Espin finds that the bright bands of the spectrum are brighter when the star is increasing in light, and fainter when decreasing. At the maximum of 1889, he found bright lines in its spectrum.

Another long period variable star which is visible to the naked eye at maximum is R Hydræ—the Upsilon Hydræ of Bayer—but it is rather too far south to be well observed in this country. Its variability was discerned by Maraldi in 1704; but the star was also observed by Hevelius in 1672. Its light at maximum varies from $3\frac{1}{2}$ to $5\frac{1}{2}$ magnitude, and at minimum it fades to nearly the tenth magnitude. The period has diminished considerably since the year 1708, when it was about 500 days. This had decreased to about 487 days in 1785, to 461 days in 1825, and to 437 days in 1870, and it seems to be still diminishing. Formulæ have been computed by Gould and Chandler, but do not agree. Schmidt found that the minimum occurs about 200 days before the maximum. The star is very reddish, and the spectrum is a fine one of the third type, which Dunér describes as of "extraordinary beauty," the typical bands of this type of spectrum being very large, and perfectly black. At the maximum of 1889, Espin observed a bright line in its spectrum, and finds—as in R Leonis—that the bright bands are brighter when the star is increasing in light, and fainter as it decreases.

There is a very remarkable variable star in the Southern Hemisphere known as Eta Argûs. It lies in the midst of the great nebula in Argo, and the history of its fluctuations in light is very interesting. Observed by Halley in 1677 as a star of the fourth magnitude, it was seen of the second magnitude by Lacaille in 1751. After this, it must have again faded, for Burchell found it of only the fourth magnitude from

1811 to 1815. From 1822 to 1826, it was again of the second magnitude, as observed by Fallows and Brisbane; but on Feb. 1, 1827, it was estimated of the first magnitude by Burchell. It then faded again, for on Feb. 29, 1828, Burchell found it of the second magnitude. From 1829 to 1833, Johnson and Taylor rated it of the second magnitude; and it was still of this magnitude, or a little brighter, when Sir John Herschel commenced his observations at the Cape of Good Hope in 1834. It does not seem to have varied much in brightness from that time until December, 1837, when Herschel was astonished to find its light "nearly tripled." He says:[1] "It very decidedly surpassed Procyon, which was about the same altitude, and was far superior to Aldebaran. It exceeded α Orionis, and the only star (Sirius and Canopus excepted) which could at all be compared with it was Rigel, which, as I have already stated, it somewhat surpassed."

From this time its light continued to increase. On the 28th December it was far superior to Rigel, and could only be compared with α Centauri, which it equalled, having the advantage of altitude, but fell somewhat short of it as the altitudes approached equality. The maximum of brightness seems to have been obtained about the 2nd January, 1838, on which night, both stars being high and the sky clear and pure, it was judged to be very nearly matched, indeed, with α Centauri, sometimes the one, sometimes the other, being judged brighter; but, on the whole, α was considered to have some little superiority. After this, the light began to fade. Already on the 7th and 15th January, α Centauri was unhesitatingly placed above, and Rigel as unhesitatingly below, it. On the 20th, it was "visibly diminished — now much less than α Centauri, and not *much* greater than Rigel. The change is palpable." And on the 22nd, Arcturus (the nearest star in light and colour to α Centauri which the heavens afford), when only 10° high, surpassed η, the latter being on the meridian; η was still, however, superior to β Centauri, α Crucis, and Spica, and continued so (and even superior to

[1] "Cape Observations," p. 34.

Rigel) during the whole of February, nor was it until the 14th April, 1838, that it had so far faded as to bear comparison with Aldebaran, though still somewhat brighter than that star. In 1843, it again increased in brightness, and in April of that year it was observed by Maclear to be brighter than Canopus, and nearly equal to Sirius! It then faded slightly, but seems to have remained nearly as bright as Canopus until February, 1850, since which time its brilliancy gradually decreased. It was still of the first magnitude in 1856, according to Abbott, but was rated a little below the second magnitude by Powell in 1858. Tebbutt found it of the third magnitude in 1860; Abbott a little below the fourth in 1861. Ellery rated it fifth magnitude in 1863, and Tebbutt sixth magnitude in 1867. In 1874 it was estimated 6·8 magnitude at Cordoba, and only 7·4 in November, 1878. Tebbutt's observations from 1877-86 show that it did not rise above the seventh magnitude in those years, and in March, 1886, it was rated 7·6 magnitude by Finlay at the Cape of Good Hope. This seems to have been the minimum of light, for in May, 1888, Tebbutt found that it "had increased fully half a magnitude" since April, 1887, and might "be rated as a star of 7·0 magnitude." From photometric measures made with the meridian photometer in Peru in the years 1889-91, Professor Bailey found its mean magnitude to be 6·32, so that probably the star is now slowly rising to another maximum. Bailey found the hydrogen lines Hβ, Hγ, and Hδ, bright in the spectrum of its light. Wolf suggested a period of 46 years, and Loomis, 67; but Schönfeld thought that a regular period is very improbable. The star is very reddish in colour.

There are many other variables of long period, but they are too numerous to be described in detail in a work of this character. Particulars respecting some of them will be found in "The Scenery of the Heavens," by the present writer.

We will now consider the variables of short period, which are particularly interesting objects, owing to the comparative rapidity of their light changes. The periods vary in length from about 17¼ days down to a few hours. Perhaps the most interesting of these short period variables, at least to the amateur observer, is the star Beta Lyræ, which is easily visible to the naked eye in all phases of its light. It can be readily identified, as it is the nearest bright star to the south of the brilliant Vega, and one of two stars of nearly the same magnitude, the second being Gamma Lyræ. The variability of Beta Lyræ was discovered by Goodricke in the year 1784. The period is about 12 days, 21 hours, 46 minutes, 58 seconds. At maximum the star is about 3·4 magnitude, and there are two minima, one of magnitude 3·9, and the other—the chief minima—of 4·5 magnitude. That is, the star has at maximum $2\frac{3}{4}$ times the light of the chief minimum, and 1·6 times the light of the secondary minimum. In other words, if we represent the light of the star at maximum by 27 candles, placed at a suitable distance from the eye, the secondary minimum will be represented by 17 candles, and the chief minimum by 10 candles. These fluctuations, although not very great, can be easily recognised with the naked eye by comparison with the neighbouring star Gamma Lyræ. Professor Pickering thought that this variation in the light of Beta might be explained by supposing that the star rotated on its axis in the period indicated by the variation, that the ratio of the axis of the rotating spheroid is as 5 to 3, and that there is a darker portion at one of the ends, which is "symmetrically situated as regards the longer axis." Recent observations with the spectroscope, however, render this explanation doubtful, and indicate rather that the star is a very close double or "spectroscopic binary," although it does not seem certain that an actual eclipse of one component by the other takes place, as in the case of Algol. Bright lines were detected in the star's spectrum by Secchi so far back as 1866. In 1883, M. Von Gothard noticed that the appearance of these bright lines varied in appearance, and from an examination of photographs taken at Harvard Observatory in 1891, Mrs. Fleming found displacements of bright and dark lines in a double spectrum, the period of which agreed fairly well with that of the star's light changes. Professor Pickering thence concluded that the star consists of two components, one stellar and the other gaseous, but this conclusion has been somewhat modified by subsequent investigations. M. Bélopolsky, from photographs taken with the great 30-inch telescope at the Pulkowa Observatory, confirms the periodical displacement in the bright spectral lines "in a period identical with that of the star's usual double fluctuation," but Keeler and Vogel agree that the observed displacements are incompatible with the supposed occurrence of eclipses. Vogel, however, is "convinced that Beta Lyræ represents a binary or multiple system, the fundamental revolutions of which, in 12 days 22 hours, in some way control the light change, while the spectral variations, although intimately associated with the star's phases, are subject, besides, to complicated disturbances running through a cycle perhaps measured by years."[1] The helium line, D_3, is visible in the spectrum.

Another interesting star of short period is Delta Cephei, which is one of three stars forming an isosceles triangle a little to the west of Cassiopeia's Chair, the variable being at the vertex of the triangle, and the nearest of the three to Cassiopeia. Its variability was also discovered by Goodricke in 1784. It varies from 3·7 to 4·9 magnitude, with a period of 5 days, 8 hours, 47 minutes, 40 seconds. The amount of the variation is, therefore, the same as in the case of Algol, the star's light at maximum being about three times its light at minimum. The period and light curve, however, show, according to Schönfeld, some irregularities, the computed times of maxima and minima being sometimes in error to the extent of over an hour. These are, however, small, and, on the whole, the star seems to be very uniform in its fluctuations. From seven years' observations, Argelander found no deviation from perfect uniformity. The curve representing the light variations is not, however, very smooth, particularly during the decrease of light, when a nearly stationary period seems to occur from 16 to 24 hours after the maximum. The rise from minimum to maximum occupies about one-third of the period, another example of the feature so characteristic of

[1] *Journal of the British Astronomical Association*, vol. iv., No. 11, p. 21.

variable stars, namely, that the increase of light is quicker than the decrease. As already stated (Chapter IV.), observations of the spectrum recently made by M. Bélopolsky, with the great Pulkowa telescope, show that, like Beta Lyræ, the star is probably a close binary, the period of the observed fluctuations in the positions of the spectral lines agreeing with that of the star's light changes. In this case, however, the lines are not doubled, as in Beta Lyræ, but merely displaced from their normal position, indicating that, as in the case of Algol, one of the components is a dark body. There are, however, no indications that any eclipse of the bright star by its dark companion takes place. Indeed, the nature of the light changes, which are continuous and not confined to a few hours, as in Algol, are inconsistent with the occurrence of an eclipse. We must, therefore, conclude that the fluctuations of light are caused in some way by physical disturbances produced by the approach and recession of the two component bodies in an elliptic orbit round their centre of gravity. The observations indicate that the component stars, when furthest apart in their orbital revolution, are separated by a distance three times as great as when at their point of nearest approach. The observations also show that Delta Cephei is approaching the earth at the rate of about 8¾ miles a second. Its spectrum is of the second or solar type, differing in this respect from the other spectroscopic binaries, which show a spectrum of the first or Sirian type. The colour of the star is yellow, and it has a distant bluish companion of about the fifth magnitude, which may possibly have some physical connexion with the brighter star, as both stars have a common proper motion through space.

Another remarkable star of short period is Eta Aquilæ, the variability of which was discovered by Pigott in 1784. It varies from magnitude 3·5 to 4·7, with a period of 7 days, 4 hours, 14 minutes, but Schönfeld found marked deviations from a uniform period. It will be seen that the amount of the light change, 1·2 magnitude, is the same as that of Delta Cephei. Its colour is yellow, and its spectrum, like that of

Delta Cephei, of the second or solar type. The minimum takes place about three days before the maximum.

Zeta Geminorum is another variable star with a comparatively short period. It varies from about 3·7 to 4·5 magnitude, with a period of 10 days, 3 hours, 41½ minutes. Here the variation of light is only 0·8 of a magnitude, or, in other words, the light at maximum is about double the light of minimum, as in the case of the Algol type variable, Lambda Tauri. Its light curve, unlike that of Delta Cephei and Eta Aquilæ, is nearly symmetrical; that is, the period occupied in the increase of light is about the same as that of the decrease. Prof. Pickering thinks that Zeta Geminorum is possibly a "surface of revolution," one side of the rotating star being about four-fifths of the brightness of the other; but Prof. Lockyer finds it to be a "spectroscopic binary," like Beta Lyræ and Delta Cephei.

Among variables with very short periods may be mentioned the southern star R Muscæ, which is close to Alpha Muscæ. It varies from 6·6 to 7·4, and goes through all its changes in the short period of 21 hours 20 minutes. The minimum takes place about nine hours before the maximum. It was discovered at the Cordoba Observatory, and Dr. Gould remarks that "its average brightness is so near the limit of ordinary visibility in a clear sky at Cordoba, that the small regular fluctuations of light place it every few hours alternately within or beyond this limit."

A remarkable variable star of short period was discovered in 1888 by Mr. Paul in the southern constellation Antlia. It varies from magnitude 6·7 to 7·3, with the wonderfully short period of 7 hours, 46 minutes, 48 seconds, all the light changes being gone through no less than three times in twenty-four hours! It was for some years believed that the variation was of the Algol type, but recent measures made at the Harvard College Observatory shows that it belongs to the same class as Delta Cephei and Eta Aquilæ.

A telescopic variable with a wonderfully short period was discovered by Chandler in 1894. It lies a little to the west of the star Gamma Pegasi, and has been designated U Pegasi. It varies from magnitude 8·9 to 9·7, and was first supposed to be of the Algol type with a period of about two days, but further observations showed that the period was much shorter, and only 5 hours, 31 minutes, 9 seconds. The light curve is quite different from the Algol type, and also from that of Delta Cephei and other short period variables, the times of increase and decrease of light being about equal, as in the case of Zeta Geminorum. This fact, combined with the remarkable rapidity of its light changes, which are gone through four times in less than twenty-four hours, makes this remarkable star a most interesting object. Possibly there may be other stars in the heavens with a similar rapidity of variation which have hitherto escaped detection.

Several southern variables of short period have been discovered in recent years by Mr. A. W. Roberts at Lovedale in South Africa.

Unlike the variable stars of long period which seem scattered indifferently over the surface of the heavens, the great majority of the short period variables are found in a zone which nearly coincides with the course of the Milky Way. The most notable exceptions to this rule are W Virginis with the comparatively long period of 17¼ days, and U Pegasi, above described, which has the shortest known period of all the variable stars. Another peculiarity is that most of them are situated in what may be called the following hemisphere, that is between 12 hours and 24 hours of right ascension. The most remarkable exception to this rule is Zeta Geminorum. The above rules do not apply to variables of the Algol type, which we will now proceed to consider.

Algol, or Beta Persei, is a famous variable star, and the typical star of the class to which it belongs. Its name, Algol, is derived from a Persian word, meaning the 'demon," which suggests that the ancient astronomers may have detected some peculiarity in its behaviour. The real discovery of its variation was, however, made by Montanari in 1667, and his observations were confirmed by Maraldi in 1692. Its fluctuations of light were also noticed by Kirch and Palitzsch, but the true character of its variations was first determined by the English astronomer, Goodricke, in 1782. Its fluctuations of light are very curious and interesting. Shining with a constant, or nearly constant, brightness for a period of about 59 hours as a star of a little less than the second magnitude, it suddenly begins to diminish in brightness, and in about 4½ hours it is reduced to a star of about magnitude 3½. In other words, its light is reduced to about one-third of its normal brightness. If we suppose three candles placed side by side at such a distance that their combined light is merged into one, and equal to the usual brightness of Algol, then if two of these candles are extinguished, the remaining candle will represent the light of Algol at its minimum brilliancy. It is stated in several books on astronomy that Algol varies to the extent of two magnitudes, but this is quite incorrect, as a change of two magnitudes would imply that the light at maximum is over six times the light at minimum, which is more than double the star's real variation. The star remains at its minimum, or faintest, for only about 15 minutes. It then begins to increase, and in about 5 hours recovers its normal brightness, all the light changes being gone through in a period of about 10 hours out of nearly 69 hours, which elapse between successive minima. These curious changes take place with great regularity, and the exact hour at which a minimum of light may be expected can be predicted with as much certainty as an eclipse of the sun.

Goodricke, comparing his own observations with one made by Flamsteed in the year 1696, found the period from minimum to minimum to be 2 days, 20 hours, 48 minutes 59½ seconds, and he came to the conclusion that the diminution in the light of the star is probably due to a partial eclipse by "a large body revolving round Algol." This hypothesis was fully confirmed in the years 1888-89 by Professor Vogel with the spectroscope. As no close companion to Algol is visible in the largest telescopes, we must conclude that either the satellite is a dark body, or else so close to the primary that no telescope could show it. As has been stated in Chapter III., the motion of a star in the line of sight can be ascertained by measuring displacements in the positions of the spectral lines. Now, if the diminution in Algol's light is due to a dark body revolving round it, and periodically coming between us and the bright star, it follows that both components will be in motion, and both will revolve round the common centre of gravity of the pair. A little before a minimum of light takes place, the dark companion should therefore be approaching the eye, and, consequently, the bright companion will be receding. During the minimum there will be no apparent motion in the line of sight, as the motion of both bodies will be at right angles to the visual ray. After the minimum is over, the motion of the two bodies will be reversed, the bright one approaching the eye, and the dark one receding. Now, this is exactly what Vogel found. Before the diminution in the light of Algol begins, the spectroscope showed that the star is receding from the earth, and after the minimum, that it is approaching the eye. That the companion is dark and not bright, like the primary, is evident from the fact that the spectral lines are merely shifted from their normal position and not doubled, as would be the case were both components bright, as in the case of some of the "spectroscopic binaries" —for example, Beta Aurigæ—which has been considered in the chapter on binary stars (Chapter IV.). Vogel found that before the minimum of light, Algol is receding from the earth with the velocity of 24½ miles a second, and after the minimum it is approaching at the rate of 28½ miles a second. The difference between the observed velocities indicates that the system is approaching the earth with a velocity of about 2 miles a second. Knowing, then, the orbital velocity, which is evidently about 26½ miles a second, and assuming the orbit to be circular, it is easy, with the observed period of revolution, or the period of light variation, to calculate the diameter of the orbit in miles, although the star's distance from the earth remains unknown. Further, comparing its period of revolution and the dimensions of the orbit with that of the earth round the sun, it is easy to calculate, by Kepler's third law of motion, the mass of the system in terms of the sun's mass, and the probable size of the component bodies. Calculating in this way, Vogel computes that the diameter of Algol is about 1,061,000 miles, and that of the dark companion 830,300 miles, with a distance between their centres of 3,230,000 miles, and a combined mass equal to two-thirds of the sun's mass, the mass of Algol being four-ninths, and that of the companion two-ninths, of the mass of the sun. Taking the diameter of the sun as 866,000 miles, and its density as 1·44 (water being unity), I find that the above dimensions give a mean density for the components of Algol of about one-third that of water, so that the components are probably gaseous bodies, as Hall has already concluded.

From the recorded observations of minima in past years, it has been found that the period of variation of Algol's light has been slowly diminishing since Goodricke's time, and Dr. Chandler finds the present period is about 2 days, 20 hours, 48 minutes, 51 seconds, or about 8½ seconds less than Goodricke made it. Chandler thinks that this variation in the length of the period is cyclical, and that it has now about reached its smallest value, and will soon begin to increase again. He believes that this variation is probably due to the orbital revolution of the pair round a third body in a period of about 130 years. M. Tisserand, however, explains the irregularities by supposing an elliptical orbit, and a slight flattening or polar compression in the primary star. Professor Boss is inclined to favour Chandler's hypothesis.

It is a curious fact that Al-Sûfi, the Persian astronomer, in his "Description of the Heavens," written in the tenth century, speaks distinctly of Algol as a red star (*étoile, brillant; d'un éclat, rouge*), while at present it is white, or at the most, of a yellow colour. A similar change of colour is supposed to have taken place in the case of Sirius, but the change in Algol seems more certain, as Al-Sûfi's descriptions are generally most accurate and reliable.

Stars of the Algol type of variable are very rare objects, only a dozen or so having been hitherto discovered in the whole heavens. Those visible to the naked eye, when at their normal brightness, are: Algol, Lambda Tauri, Delta Libræ, R Canis Majoris, and U Ophiuchi. The variation of Lambda Tauri was discovered by Baxendell in 1848. It varies from magnitude 3·4 to 4·2, and its period from minimum to minimum of light is about 3 days, 22 hours, 52 minutes, 12 seconds. Its fluctuations have not been so well studied as those of Algol, but it is known that the "period is subject to marked inequalities," sometimes amounting to 3 hours. The variation of light is less than that of Algol, the light at maximum being only twice the light at minimum. Two candles at a suitable distance would therefore represent the maximum light, and one candle the minimum brightness. All the light changes take place in a period of about 10 hours. The star is white like Algol.

The variability of Delta Libræ was discovered by Schmidt in 1859. It varies from magnitude 4·9 to 6·1, with a period of 2 days, 7 hours, 51 minutes, 22·8 seconds. The period is, however, according to Schönfeld, subject to some irregularities. The variation of light is about the same as that of Algol, the light at maximum being about three times the light at minimum. The variation takes about 12 hours, of which the decrease occupies 5½ hours. The star is white like Algol.

The variability of R Canis Majoris was detected by Sawyer in 1887. The variation is from 5·9 to 6·7 magnitude, or about equal in amount to that of Lambda Tauri, and the period 1 day, 3 hours, 15 minutes, 55 seconds.

U Ophiuchi was also discovered by Sawyer in 1881. Its variation is from magnitude 6·0 to 6·7, or slightly less than that of Lambda Tauri, and the period 20 hours, 7 minutes, 41·6 seconds, but subject to an apparent diminution. The maximum brightness lasts for about 16 hours, and all the fluctuations of light take place in the short period of 4 hours. Its colour is white, like most stars of the Algol type.

U Cephei is a very interesting variable of the Algol type, discovered by Ceraski in 1880. It varies from 7·1 to 9·5, with a period of 2 days, 11 hours, 49 minutes, 45 seconds. Here the variation of light is greater than that of Algol, the light at maximum being nearly seven times the light at minimum. Its rapidity of variation is very great, sometimes exceeding a magnitude in an hour. The light variations occupy about 6 hours, and the minimum lasts for about an hour and a half, Professor Pickering thinks that the variation of light is, as in the case of Algol, caused by an eclipsing satellite, but that in this case the eclipse may possibly be total, the light at minimum being that due to the satellite, which may have some inherent light of its own. Lord Crawford examined the star with the spectroscope, and found that at the minimum the blue end of the spectroscope faded, and the red was intensified, which seems to suggest that the light of the star in that phase shines through a gaseous medium, and that the eclipsing body may be surrounded with an atmosphere.

Another interesting Algol variable is that known as Y Cygni, which was discovered by Chandler in 1886, while using it as a comparison star for the short period variable X Cygni. It varies from 7·1 to 7·9 magnitude, or about the same amount as Lambda Tauri, with a period of 1 day, 11 hours, 56 minutes, 48 seconds. It has alternate bright and faint minima, which suggest, according to Dunér, that the star consists of two *bright* components, one of them being brighter than the other, and both revolving round their common centre of gravity in an elliptic orbit, with a period double that of the light variation. Yendell, who has carefully observed the star's fluctuations, fully concurs in Dunér's views, and says "the substantial corrections of his fundamental assumption appears to be proved beyond the possibility of a cavil."

The variability of the star known as S Cancri was discovered by Hind in 1848. It varies from 8·2 to 9·8, or it is said, at some minima, to 11·7, with the comparatively long period of 9 days, 11 hours, 37 minutes, 45 seconds. The variations of light occupy about 21½ hours. If the minimum of 11·7 is correct, we have a variation of no less than 3½ magnitudes, which implies that the normal light of the star is 25 times its light at a faint minimum. If this be so, the eclipse must be nearly total. Argelander found that after the minimum the light increases very rapidly, and he thinks that the descent from the maximum is even more rapid.

Some interesting examples of the Algol type of variable have been discovered in recent years. One detected by Chandler, in 1894, and now known as Z Herculis, varies from about the seventh to the eighth magnitude, and has a period of 3 days, 23 hours, 48½ minutes. Faint and very bright minima alternate in periods of 47 and 49 hours, the ratios of the light at maximum and minima being 3, 2, and 1. These Professor Dunér considers, indicate that the star consists of two revolving components of equal size, one of which is twice as bright as the other, and he computes that the components revolve round their common centre of gravity in an elliptic orbit, the plane of which is in the line of sight, and the semi-axis major about six times the diameter of the stars. If we assume that the diameter of each component is equal to the diameter of our sun, I find, from the above date, that the combined mass of the system is about 1½ times the mass of the sun.

Another remarkable example of the Algol type was discovered by Miss Wells in 1895. The star lies a little north of the "Dolphin's rhomb," and at its normal brightness is about magnitude 9½. The period of variation is about four days. The variation somewhat resembles that of U Cephei. Professor Pickering says: "For nearly two hours before and after the minimum it is fainter than the twelfth magnitude. It is impossible at present to say how much fainter it becomes, or whether it disappears entirely. It increases at first very rapidly, and then more slowly, attaining its full brightness, magnitude 9·5, about five hours after the minimum. One hundred and thirty photographs indicate that, during the four days between the successive minima, it does not vary more than a few hundredths of a magnitude. The variation may be explained by assuming that the star revolves round a comparatively dark body, and is totally eclipsed by it for two or three hours, the light at minimum, if any, being entirely that of the dark body."[1] This seems to be an unique object, and it should be carefully followed through its minimum with a large telescope.[2]

With reference to the Algol type of variable stars, Chandler finds that "the shorter the period of the star, the higher the ratio which the time of oscillation bears to the entire period." Thus, in U Ophiuchi, with a period of about 20 hours, the light changes occupy five hours, or one-fourth of the period, while in S Cancri, which has a period of 227½ hours, the fluctuations of light take up 21½ hours, or only about one-tenth of the period. In all cases in which the Algol type variables have been examined with the spectroscope, the spectrum has been found to be of the first or Sirian type, and they seem to be the only stars with spectra of the Sirian type whose light is variable. It should be noted, however, that, on the eclipse theory, the variation of light in these stars is due merely to an occultation of one star by another, and not to any physical change in the star itself. The bright star Spica, although shown by the spectroscope to be a close binary star, like Algol, is not variable, because, in this case, the plane of its orbit is inclined to the line of sight, and hence the comparison star does not transit the disc of its primary. Seen from some other point in space, it would probably be an Algol variable.

A remarkable peculiarity about the variable stars in general is that none of them have any considerable proper motion. As a large proper motion is generally considered to indicate proximity to the earth, we may conclude, with great probability, that the variable stars, as a rule, lie at a great distance from our system. In other words, it appears that the sun does not lie in a region of variable stars, and, with the excep-

[1] *Journal of the British Astronomical Association*, vol. vi., No. 6, p. 312.

[2] Recent observations show that the total variation is 2·71 magnitudes—largest variation known for an Algol star.

tion of Alpha Cassiopeiæ and Alpha Herculis, a measurable parallax has not yet been found, so far as I know, for any known variable star.

Plotting the known variables on star charts, I find a marked tendency to cluster into groups. Thus, in and near the constellation, Corona Borealis, there are five; near Cassiopeia's Chair, five. In Cancer there are four in a limited area. Near Eta Argûs there are several, and in a comparatively small region in the northern portion of Scorpio there are no less than fifteen variable stars.

We now come to the interesting and mysterious class of objects known as "new" or "temporary" stars. These phenomena are of very rare occurrence, and but few undoubted examples of the class are recorded in the annals of astronomy. Possibly in some cases they have been merely variable stars, of irregular period and fitful variability; but others may have been due to a real catastrophe, such as the collision of two dark bodies in space, or, possibly, the passage of a bright or dark body through a gaseous nebula.

The earliest temporary star of which we have any reliable information seems to be one which is recorded in the Chinese annals of Ma-tuan-lin, as having appeared in the year 134 B.C. in the constellation Scorpio. Its position seems to have been somewhere between the stars Beta and Rho of Scorpio. Pliny informs us that it was the sudden appearance of a new star which induced the famous astronomer Hipparchus to form his catalogue of stars, the first ever constructed. As the date of Hipparchus' catalogue is 125 B.C., it seems highly probable that the new star referred to by Pliny was the same as that recorded by the Chinese astronomer as having appeared nine years previously.

A new star is said to have appeared in the year 76 B.C. between the stars Alpha and Delta in the Plough, but the accounts are vague.

In 101 A.D., a small "yellowish-blue" star is said to have appeared in the "sickle" in Leo, but its exact position is not known. In 107 A.D., a new star is mentioned near Delta,

Epsilon and Eta in Canis Major, three bright stars south-east of Sirius. In 123 A.D., another new star is recorded by Ma-tuan-lin to have appeared between Alpha Herculis and Alpha Ophiuchi.

The Chinese annals record that on Dec. 10, 173 A.D., a brilliant star appeared between Alpha and Beta Centauri in the Southern Hemisphere. It remained visible for eight months, and is described as resembling "a large bamboo mat!"—a curious description. There is at present close to the spot indicated, a known variable star—R Centauri—of which the period seems to be long and the variation of light irregular. Possibly an unusually bright maximum of this variable star formed the star of the Chinese annals, or perhaps the variable star is the remnant of the outburst which took place in the first century. The variable is a very reddish star, and at present varies from about the sixth to the tenth magnitude.

A new star is recorded in the year 386 A.D. as having appeared between Lambda and Phi Sagittarii. Near the position indicated, Flamsteed observed a star, No. 65 of his catalogue, which is now missing; and it has been conjectured that the star seen by Flamsteed may possibly have been a return of the star mentioned in the Chinese annals.

Cuspianus relates that a star as bright as Venus appeared near Altair in 389 A.D., during the reign of the Emperor Honorius, and that he had himself seen it. There is some doubt, however, about the exact date, as other accounts give the year 388 or 398. The star seems to have disappeared in about three weeks.

In the year 393 A.D., another strange star is recorded in the tail of Scorpio. An extraordinary star is said to have been seen near Alpha Crateris in 561 A.D. Here again a known variable and red star—R Crateris—is close to the position indicated by the ancient records.

The Chinese annals record a new star in 829 A.D., somewhere in the vicinity of the bright star Procyon, and in this locality there are several known variable stars.

The Bohemian astronomer, Cyprianus Leoviticus, mentions the appearance of new stars in Cassiopeia in the years 945 A.D. and 1264, and it has been conjectured that perhaps these were apparitions of Tycho Brahé's famous star of 1572 (to be presently described), forming a variable star with a period of over 300 years. Lynn and Sadler, however, have shown that the supposed stars of 945 and 1264 were, in all probability, comets

Extraordinary stars are recorded near Zeta Sagittarii in 1011 A.D., near Mu Scorpii in 1203, and near Pi Scorpii on July 1, 1584. It is remarkable how many of these objects seem to have appeared in this portion of the heavens.

A very brilliant star is mentioned by Hepidannus as having appeared in Aries in May, 1012. He describes it as "dazzling the eye." Other temporary stars are mentioned in 1054 A.D., near Zeta Tauri, and in 1139, near Kappa Virginis; but the accounts of these are very vague, and it seems by no means certain that they were really new stars.

No possible doubt, however, can be entertained with reference to the appearance of the object which suddenly blazed out in Cassiopeia's Chair in November, 1572. It was called the "Pilgrim Star," and was observed by the famous astronomer, Tycho Brahé, who has left us a very elaborate account of its appearance, position, etc. Although usually spoken of as Tycho Brahé's star, it seems to have been really discovered by Cornelius Gemma on the evening of November 9. That its appearance was very sudden may be inferred from Cornelius Gemma's statement, that it was not visible on the preceding night in a clear sky. Tycho Brahé's attention was first attracted to it on November 11. His description of the new star is as follows—as quoted by Humboldt:[1] —"On my return to the Danish islands from my travels in Germany, I resided for some time with my uncle, Steno Bille, in the old and pleasantly situated monastery of Herritzwadt, and here I made it a practice not to leave my chemical laboratory until the evening. Raising my eyes, as usual,

[1] "Cosmos," Bohn's edition, vol. iii., p. 205.

during one of my walks, to the well-known vault of heaven, I observed with indescribable astonishment, near the zenith in Cassiopeia, a radiant fixed star of a magnitude never before seen. In my amazement, I doubted the evidence of my senses. However, to convince myself that it was no illusion, and to have the testimony of others, I summoned my assistants from the laboratory, and inquired of them, and of all the country people that passed by, if they also observed the star that had thus suddenly burst forth. I subsequently heard that in Germany, waggoners and other common people first called the attention of astronomers to this great phenomenon in the heavens—a circumstance which, as in the case of non-predicted comets, furnished fresh occasion for the usual raillery at the expense of the learned. This new star I found to be without a tail, not surrounded by any nebula, and perfectly like all other fixed stars, with the exception that it scintillated more strongly than stars of the first magnitude. Its brightness was greater than that of Sirius, α Lyræ, or Jupiter. For splendour, it was only comparable to Venus when nearest to the earth (that is, when only a quarter of her disc is illuminated). Those gifted with keen sight could, when the air was clear, discern the new star in the daytime, and even at noon. At night, when the sky was overcast, so that all other stars were hidden, it was often visible through the clouds, if they were not very dense (*nubes non admodum densas*). Its distances from the nearest stars of Cassiopeia, which throughout the whole of the following year I measured with great care, convinced me of its perfect immobility. Already, in December, 1572, its brilliancy began to diminish, and the star gradually resembled Jupiter, but by January, 1573, it had become less bright than that planet. Successive photometric estimates gave the following results: for February and March, equality with stars of the first magnitude (*stellarum affixarum primi honoris*—for Tycho Brahé seems to have disliked Manilius' expression of *stellæ fixæ*); for April and May, with stars of the second magnitude; for July and August, with those of the third; for October and November, those of the fourth magnitude. Towards the month of November, the new star was not brighter than the eleventh in the lower part of Cassiopeia's Chair. The transition to the fifth and sixth magnitude took place between December, 1573, and February, 1574. In the following month the new star disappeared, and, after having shone seventeen months, was no longer discernible to the naked eye." (The telescope was not invented until thirty-seven

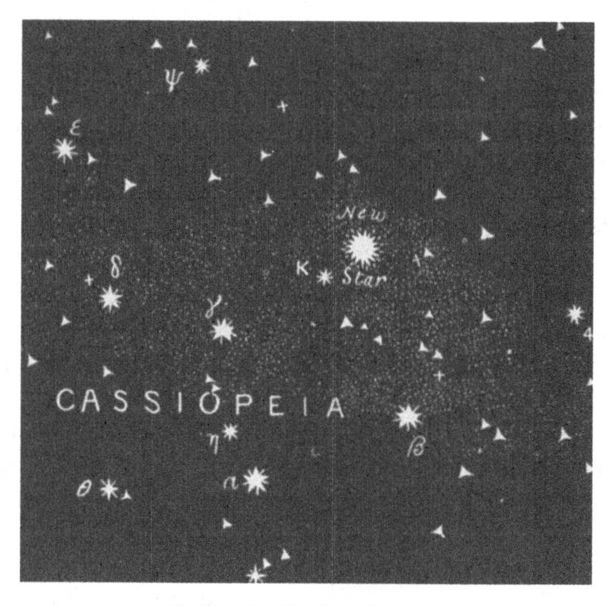

FIG. 8.—*The Temporary Star of* 1572.
(From "Planetary and Stellar Studies.")

years afterwards.) Humboldt adds:—"At its first appearance, as long as it had the brilliancy of Venus and Jupiter, it was for two months white, and then passed through yellow into red. In the spring of 1573, Tycho Brahé compared it to Mars; afterwards he thought it nearly resembled Betelgeuse, the star in the right shoulder of Orion. The colour for the most part was like the red tint of Aldebaran. In the spring of 1573, and especially in May, its white colour returned (*albedinam quandam sublividam induebat, qualis Saturni stellæ subesse videtur*). So it remained in January, 1574; being, up to the time of its entire disappearance in the month of March, 1574, of the fifth magnitude, and white, but of a duller whiteness, and exhibiting a remarkably strong scintillation in proportion to its faintness."

According to a sketch of the position given in Tycho Brahé's work, referred to above, the star was situated a little to the north of Kappa Cassiopeiæ, the faintest star in the Chair. This position is confirmed by Argelander's examination of Tycho Brahé's observations. The spot is a rather blank one to the naked eye, and even with an opera glass, only a few faint stars are visible. Quite close to the place fixed by Argelander, d'Arrest observed in 1865 a star of the eleventh magnitude, which seems to have escaped Argelander's notice. Hind and Plummer observed this small star in 1873, and thought they could detect fluctuations in its light to the extent of about one magnitude. Espin has also observed it, and the region has been photographed by Dr. Roberts. Some have thought that Tycho Brahé's star might possibly be identical with the Star of Bethlehem, and this idea has been supported by Cardanus, Chladni, and Klinkerfues, but Lynn and Sadler have shown that the theory is quite untenable, and it has now been rejected by all astronomers.

Ma-tuan-lin speaks of a star in 1578 "as large as the sun" (!) but does not state its position.

The star known as P (34) Cygni is sometimes spoken of as a "Nova," or new star; but it is still visible to the naked eye as a star of the fifth magnitude. It was observed of the third magnitude by Jansen in 1600 and by Kepler in 1602. After the year 1619, it appears to have diminished in brightness, and is said to have vanished in 1621; but it may merely have become too faint to be seen with the naked eye. It was again observed of the third magnitude by Dominique Cassini in 1655, and it afterwards disappeared. It was again seen by Hevelius in November, 1665. In 1667, 1682, and 1715, it is recorded as of the sixth magnitude, and there is no further record of any marked increase in its light. A period of about 18 years was assumed by Pigott; but this is now disproved, and it seems probable that the star is a variable of irregular period and fitful variability, and not, properly speaking, a temporary star. Its present colour is yellow, and bright lines have been seen in its spectrum.

Another remarkable object of the temporary class was observed by Kepler in 1604 in Ophiuchus, and is described by him in his work, "De Stella Nova in pede Serpentarii." He and his assistants were observing the planets Mars, Jupiter, and Saturn, which were then near each other in this region of the heavens, a few degrees to the south-east of the star Eta Ophiuchi, and on the evening of October 10, Brunowski, a pupil of Kepler's, noticed that a new and very brilliant star was added to the group.[1] When first seen, it was white, and exceeded in brightness Mars and Jupiter, but seems not to have quite equalled Venus in brilliancy. It slowly diminished, and in January, 1605, it was brighter than Antares but less than Arcturus. At the end of March, 1605, it had faded to the third magnitude. Its proximity to the sun then prevented further observations for several months. In March, 1605, it had disappeared to the naked eye. It was also observed by Galileo and by David Fabricius, whose observations place it about midway between the fifth magnitude star Xi and 58 Ophiuchi. Its exact position, however, does not seem to be known with such accuracy as that of Tycho Brahé's star, nor is there any known star very close to the spot indicated by Schönfeld from an examination of Fabricius' observations. It seems possible that Kepler's star may have been seen previously by Ptolemy, for in his catalogue he gives a star of the fourth magnitude close to the position of Kepler's star; but there is some doubt about the exact position indicated by Ptolemy. The Chinese annals mention a "ball-like" star as having appeared near Pi Scorpii on September 30, 1604, and remaining visible until March, 1606, which may possibly be identical with Kepler's star.

[1] It was, however, asserted by Herlicius that he had seen it on Sept. 27.

A new star of the third magnitude was observed near Beta Cygni by the Carthusian monk Anthelmus in 1670. It remained visible for about two years, and is said to have increased and diminished several times before its final disappearance. Schönfeld computed its exact position from observations made by Hevelius and Picard. Quite close to the spot indicated, a star of the eleventh magnitude has been observed at the Greenwich Observatory, and fluctuations of light were suspected in this small star by Hind and others. Hind says that, to his eye, "there is a hazy, ill-defined appearance about it which is not perceptible in other stars in the same field of view. Mr. Talmage received the same impression; and I may add that Mr. Baxendell, who has examined it with Mr. Worthington's reflector, observed that no adjustment of focus would bring the star up to a sharp focus." This hazy appearance is very suggestive, as it indicates that the "Nova" may possibly have faded into a small planetary nebula, as in the case of the new star in Cygnus, observed by Schmidt in 1876, and the new star in Auriga, found by Dr. Anderson in 1892. Near the position of Anthelm's new star is a known variable star, S Vulpeculæ, discovered by Hind in 1861, which might be suspected to be identical with Anthelm's star; but Hind has shown that the variable has no proper motion which would account for the difference of position since 1670, and he concludes that, "from the fixity of its position during eight years, it may be inferred that the variable is distinct from Anthelm's." It has been supposed that the star 11 Vulpeculæ in Flamsteed's catalogue is identical with Anthelm's star; but Baily could not find any evidence to show that Flamsteed's star ever really existed, and he says: "Under the presumption, however, that it may be a variable and not a *lost* star, I have preserved its recorded position with a view of inducing astronomers to look out for it from time to time."

On the evening of April 28, 1848, Hind, observing at Mr. Bishop's private observatory, in Regent's Park, London, noticed a new star of about the fifth magnitude, between

Zeta and Eta Ophiuchi. Its colour was reddish-yellow, and it seems to have subsequently increased in brightness to nearly the fourth magnitude, but it soon faded to the tenth or eleventh magnitude. This curious object has become very faint in recent years. In 1866, it was of the twelfth magnitude, and in 1874 and 1875, not above the thirteenth.

On May 28, 1860, Pogson discovered a new star in the globular cluster, 80 Messier, which lies between Antares and Beta Scorpii. When first noticed, it was about the seventh magnitude, and its brightness was sufficient to obscure the cluster. In other words, the cluster was apparently replaced by a star. On June 10, the star had nearly disappeared, and the cluster again shone with great brilliancy, and with a condensed centre. The observations of Auwers and Luther confirm those of Pogson. Pogson states that he examined the cluster on May 9, but noticed nothing peculiar; and, according to Schönfeld, the cluster presented its usual appearance on May 18, when examined at the Königsberg Observatory. The apparition of the temporary star was, therefore, probably sudden, as in the case of other "new" stars. The phenomenon was possibly caused by a collision between two of the stars composing the cluster, which is, at least, apparently very condensed.

A very remarkable star, sometimes called the "Blaze Star," suddenly appeared in Corona Borealis, in May, 1866. It was first seen by the late Mr. Birmingham, at Tuam, Ireland, about midnight, on the evening of May 12, when it was of the second magnitude, and equal to Alphecca, "the gem of the coronet." Its appearance must have been very sudden, for Schmidt, the Director of the Athens Observatory, stated that he was observing the constellation on the same evening, about 2½ hours previous to Birmingham's discovery, and observed nothing unusual. He was certain that no star, of even the fifth magnitude, could possibly have escaped his notice. On the following night it was seen by several observers in different parts of the world. M. Faye, the French astronomer, in his work—"L'Origine du Monde"—attributes the discovery to M. Courbebaisse, a French engineer, and does not mention Mr. Birmingham! He says M. Courbebaisse first saw it on the evening of May 13. This may be true; he was not the only observer who saw it on that evening; but it was, undoubtedly, *first* seen by Mr. Birmingham on the *preceding* night, and to Mr. Birmingham alone is certainly due the credit of the discovery. The star rapidly diminished in brightness, and on May 24 of the same year, had faded to 8½ magnitude. It afterwards increased to about 7·8 magnitude, but soon diminished again. Soon after its discovery it was found that the star was not really a new one, as it had been previously observed at Bonn by Schönfeld, in May, 1855, and March, 1856, while making the observations for Argelander's *Durchmusterung*, in which it appears as No. 2765, in degree 26. On both occasions it was rated as 9½ magnitude, and no suspicion of variable light seems to have arisen. When viewed with the naked eye at the time of its greatest brilliancy, it was remarked by some observers that it twinkled decidedly more than other stars in the vicinity, and that this peculiarity made it very difficult to form a correct estimation of its relative brilliancy. During the years 1866 to 1876, fluctuations in its light were observed by Schmidt, and he deduced a probable period of about 94 days, with a variation from the seventh to the ninth magnitude. This conclusion was confirmed by Schönfeld, and the star would therefore seem to be an irregular variable, and not a true temporary star.

A very remarkable and interesting variable star was discovered by Schmidt at Athens, near Rho Cygni, on the evening of November 24, 1876, when it was about the third magnitude, and somewhat brighter than Eta Pegasi. Schmidt stated that he had observed the vicinity on several occasions between November 1 and 20, and was certain that no star of even the fifth magnitude could possibly have escaped his notice, so that the star probably blazed out very suddenly, as most of these extraordinary objects have done. Between November 20 and 24, the sky was overcast, so the exact time of its appearance is unknown. The star would seem to be quite new, as there is no star in any of the catalogues in the position of the "Nova," the nearest being one of the ninth magnitude, which occurs in the Bonn observations. The new star rapidly faded, and on November 30 had descended to the fifth magnitude. On the night of its discovery it was remarked that its brightness was such as to render its near neighbour, 75 Cygni (a sixth magnitude star), invisible; while on December 14 and 15, 75 Cygni, in its turn, nearly obliterated the light of the stranger. In the 48 hours following the night of November 27, the star diminished in light to the extent of nearly $1\frac{1}{2}$ magnitude! It afterwards faded very regularly to August, 1877, and showed no oscillations of brightness as have been observed in other temporary stars. On the evening of its discovery, Schmidt considered the star to be of a strong golden-yellow, and that it afterwards remained of a deep golden-yellow, but at no time was it as ruddy as 75 Cygni. I could see no trace of colour in the star with a 3-inch telescope in the Punjab on January 12, 1877, but it had then faded to the eighth magnitude. On February 7, 1877, I estimated it ninth magnitude. A few days after its discovery, it was examined with the spectroscope, and its spectrum showed bright lines similar to the "Blaze Star" in Corona, which appeared in May, 1866. One of the bright lines was thought to be identical with the line numbered 1474 by Kirchoff, visible in the spectrum of the solar Corona during total eclipses of the sun. The other bright lines were identified by M. Cornu of the Paris Observatory with some of the lines of hydrogen, sodium, and magnesium. In September, 1877, the star was examined with a 15-inch refractor by Lord Lindsay (now Lord Crawford), who found "the light coming from it almost entirely monochromatic, that is, of only one colour, the star appearing exactly the same as when looked at without the spectroscope, the direct prism having no effect on it," and he considers that "there is little doubt that the star has changed into a planetary nebula of small angular diameter!" On September 3, the star's magnitude was $10\frac{1}{2}$; "faint blue, near another star of same size rather red" Lord Crawford remarks that no observer, discovering the object in its present state, would, after viewing it through a prism, hesitate to pronounce as to its nebulous character,[1] but no disc was detected with powers ranging up to 1000 diameters. Ward found the star only sixteenth magnitude in October, 1881, and it was estimated fifteenth magnitude at Mr. Wigglesworth's Observatory in September, 1885. At Lord Crawford's Observatory the exact position of the star, with reference to above fifty closely adjacent stars, was carefully determined with the micrometer. The vicinity was photographed by Dr. Roberts on September 27, 1891, with an exposure of two hours, and "the *Nova* appears as a star of about the thirteenth magnitude." Observations in 1894 and 1895, made its magnitude about 14·8, with an apparently continuous spectrum.[2]

In August, 1885, a star of about the seventh magnitude made its appearance close to the nucleus of the Great Nebula in Andromeda (Messier 31), a remarkable nebula, which will be described in the next chapter. The new star was independently discovered by several observers towards the end of August. It was not visible to Tempel at the Florence Observatory on August 15 and 16, but is said to have been seen by M. Ludovic Gully on August 17. It was, however, certainly seen by Mr. I. W. Ward at Belfast on August 19, at 11 p.m., when he estimated it 9½ magnitude, and it was independently detected by the Baroness Podmaniczky on August 22, by M. Lajoye on August 30, by Dr. Hartwig, at Dorpat, on August 31, and by Mr. G. T. Davis, at Theale, near Reading, on September 1. On September 3, the star was estimated 7½ magnitude by Lord Crawford and Dr. Copeland, and its spectrum was found to be "fairly continuous." On September 4, Mr. Maunder, at the Greenwich Observatory, found the spectrum "of precisely the same character as that of the nebula, *i.e.*, it was perfectly continuous, no lines, either bright or dark, being visible, and the red end was wanting." Dr. Huggins, however, on September 9, thought he could see

[1] The spectrum, however, seems to have since become continuous.

[2] *Astronomical Journal*, No. 100.

a few bright lines in its spectrum, a continuous spectrum being visible from the line D to F. The star gradually faded away. On December 10, 1885, it was estimated of the fourteenth magnitude at the Radcliffe Observatory, Oxford, and on February 7, 1886, it was rated only sixteenth magnitude with the 26-inch refractor of the Washington Observatory. A series of measures by Professor Hall, from September 29, 1885, to February 9, 1886, showed "no certain indications of any parallax," so that the star and the nebula, in which it probably lies, are evidently situated at a vast distance from the earth. Seeliger has investigated the decrease in the light of the star on the hypothesis that it was a cooling body, which had been suddenly raised to an intense heat by the shock of a collision, and finds a fair agreement between theory and observation. Auwers points out the similarity between this outburst and the new star of 1860, in the cluster 80 Messier (already described), and thinks it probable that both phenomena were caused by physical changes in the nebulæ in which they occurred. Proctor considered that the evidence of the spectroscope shows that the new star was situated *in* the nebula, and in this opinion I fully concur.

Several temporary stars have been detected in recent years by Mrs. Fleming, from an examination of photographs of stellar spectra, taken at the Harvard Observatory, for the Draper Memorial. Plates of the constellation Perseus show the existence of a star in 1887, the spectrum of which shows the bright lines of hydrogen, and it was on this account assumed to be a long period variable. During the following eight years, however, 81 photographs of the same region show no trace of the star, and it has been frequently looked for with a telescope, but without success. It would, therefore, seem probable that the star was a temporary one. Its magnitude was about the ninth.

A remarkable and very interesting temporary star was discovered in 1892 in the constellation Auriga. On February 1, of that year, an anonymous post-card was received by Dr.

Copeland at the Royal Observatory, Edinburgh, with the following announcement:

"Nova in Auriga. In Milky Way, about two degrees south of χ Aurigæ, preceding 26 Aurigæ. Fifth magnitude, slightly brighter than χ."

Such an announcement evidently required immediate attention, and on that evening, Dr. Copeland and his assistants looked for the new star, and easily found it with an opera-glass at 6 hours 8 minutes. They estimated it of the sixth magnitude, and equal to 26 Aurigæ. It was of a yellow colour. When examined with a prism placed before the eye-piece of a 24-inch reflector, its spectrum was seen to resemble the "Blaze Star" of 1866 in Corona. "The C line was intensely bright, a yellow line about D fairly visible; four bright lines, or bands, were conspicuous in the green; and, lastly, a bright line in the violet (probably Hγ) was easily seen." Notice of the discovery was at once telegraphed to Greenwich and Keil Observatories, and the star was photographed at Greenwich on the same night. It is not in the Bonn star charts, which show stars to nearly the tenth magnitude. In *Nature* of February 18, 1892, a letter appeared, signed Thomas D. Anderson, in which the writer stated that the post-card was sent by him, and he gives the following details respecting the discovery:

"Prof. Copeland has suggested to me that as I am the writer of the anonymous post-card mentioned by you a fortnight ago (p. 325), I should tell your readers what I know about the Nova.

"It was visible as a star of the fifth magnitude certainly for two or three days, very probably even for a week, before Prof. Copeland received my post-card. I am almost certain that at two o'clock on the morning of Sunday, the 24th ult., I saw a fifth magnitude star making a very large obtuse angle with β Tauri and χ Aurigæ, and I am positive that I saw it, at least, twice subsequently during that week. Unfortunately, I mistook it on each occasion for 26 Aurigæ, merely remarking to myself that 26 was a much brighter star than I used to think it. It was only on the morning of Sunday, the 31st ult., that I satisfied myself that it was a strange body. On each occasion of my seeing it, it was slightly brighter than χ. How long before the 24th ult. it was visible to the naked eye I cannot tell, as it was many months since I had looked minutely at that region of the heavens.

"You might also allow me to state, for the benefit of your readers, that my case is one that can afford encouragement to even the humblest of amateurs. My knowledge of the technicalities of astronomy is, unfortunately, of the most meagre description; and all the means at my disposal on the morning of the 31st ult., when I made sure that a strange body was present in the sky, were Klein's 'Star Atlas,' and a small pocket-telescope, which magnifies ten times."

Soon after the discovery of the new star, an examination was made by Professor Pickering of photographs taken of the region at Harvard Observatory, previous to Dr. Anderson's discovery. It was found that on eighteen photographs taken between the dates November 3, 1885, and November 2, 1891, there is no trace of the new star; but in those taken from December 16, 1891, to January 31, 1892, a star of the fifth magnitude is shown in the position of the new star. "In another series of plates taken with the transit photometer, no record of the new star up to December 1, 1891, was obtained, although χ Aurigæ (magnitude 5·0) was always visible, but the plates taken on the nights of December 10, 1891, and ending January 20, 1892, indicated clearly the position of the new star." Professor Pickering says: "It appears that the star was fainter than the eleventh magnitude on November 2, 1891, than the sixth magnitude on December 1, and that it was increasing rapidly on December 10. A graphical construction indicates that it had probably attained the seventh magnitude within a day or two of December 2, and the sixth magnitude on December 7. The brightness increased rapidly until December 18, attaining its maximum about December 20, when its magnitude was 4·4. It then began to decrease slowly, with slight fluctuations, until January 20, when it was slightly below the fifth magnitude. All these changes took place before its discovery, so that it escaped observation nearly two months. During half of this time it was probably brighter than the fifth magnitude."

It would seem from the above remarks that the star did not —like some other temporary stars—attain its full brilliancy at once, but increased gradually in brightness. After the decrease of light in January, 1892, it seems to have again risen to another maximum, for photographs taken at the Greenwich Observatory after its discovery show that the star rose to a magnitude of 3·5 (photographic) on February 3, and then began to fade again slowly during February, but rapidly during the month of March. Owing to cloudy weather in the west of Ireland, I could not observe the new star until February 14. The following are my observations, made with a binocular field-glass, the comparison stars being Chi Aurigæ, 26 Aurigæ, and D M + 30°, 898:—February 14, 4·55 magnitude; February 15, 5·56; February 16, 5·84; February 18, 5·51; February 21, 5·56; February 24, 5·66; February 28, 5·44; March 1, 5·68; March 5, 5·66; March 10, 7·3; March 11, 7¾; March 16, 8½, or fainter; March 18, 9 magnitude, or less, "only *very* faint stars seem near the place of the Nova; clear sky, no moon." The general accuracy of the above observations were confirmed by the photographic estimates of the star's light made at Greenwich,[1] and also by Schaeberle's observations of its brightness.

After March 18, the light of the star steadily and rapidly decreased, and on April 1, it had faded to nearly the fifteenth magnitude, and afterwards to about the sixteenth. In August, 1892, it brightened again, as it was found by Corder of about the ninth magnitude on August 21. Dr. J. Holetschek of the Vienna Observatory observed it from August 24 to September 2, 1892, and estimated it about 9½ magnitude. In October, 1892, most observers rated it between 10 and 10½ magnitude. Observations by Mr. C. E. Peck, "from October 3, 1893, to May 4, 1894, only vary from 10·1 to 11·0 magnitude, and observations up to the end of 1894 give the

[1] *Journal of the British Astronomical Association*, March, 1892.

same results."[1] In 1895 Professor Barnard found that it "is still visible as a small star, and has not changed in physical appearance since the autumn of 1892. It remains perfectly fixed with reference to the comparison stars."[2]

Examined with the spectroscope soon after its discovery, many bright lines were seen in its spectrum, and it was found that "the bright lines in the spectrum of the new star were accompanied by dark ones on their more refrangible sides," that is, the dark lines were on the blue side of the bright ones. This suggested the idea that the outburst was probably due to a collision between two bodies, one of which, having a spectrum of dark lines, was rushing towards the earth, and the other, with a bright-line spectrum, was receding. Lockyer supposed the outburst to be due to a collision between two swarms of meteorites. Dr. Huggins advanced the view that the phenomenon was due to the near approach of two gaseous bodies. "But," he says, "a casual near approach of two bodies of great size would be a greatly less improbable event than an actual collision. The phenomena of the new star scarcely permits us to suppose even a partial collision, though, if the bodies were diffused enough, or the approach close enough, there may have been, possibly, some interpenetration and mingling of the rare gases near the boundaries." But Maunder and Seeliger consider this hypothesis to be untenable. Mr. Monck suggested that a star or swarm of meteorites rushing through a gaseous nebula might explain the phenomena. Seeliger advocates a similar theory. Maunder also favours a collision theory.

A photograph of the spectrum taken by Maunder on February 22, 1892 (when the photographic magnitude was 4·78, and visual magnitude about 5·7), showed a displacement of the dark lines, which implied a relative motion of the two supposed colliding bodies of about 820 miles a second! Vogel found that the bright lines showed a double maxima, and he thought that these were due to "two different bodies moving

[1] *Journal of the British Astronomical Association*, February, 1895, vol. v. No. 4.
[2] *Ibid.*, April, 1895, p. 328.

with different velocities, so that the spectrum of the Nova consists of, at least, three spectra superposed. The measurement of the photograph gives the body showing the dark line spectrum as approaching the earth with a speed of nearly 420 miles per second, one of the two bright line bodies as approaching with a speed of 22 miles, whilst the other is receding with a speed of 300 miles a second."[1]

At the time of its increase of brightness, in August, 1892, Professor Barnard, observing it with the great 36-inch Lick telescope, says, the "Nova appeared as a small, bright nebula, with a star-like nucleus of the tenth magnitude. The nebulosity was pretty bright and dense, and was 3″ in diameter. Surrounding this was a fainter glow, perhaps half a minute in diameter." At this time, Professor Campbell of the Lick Observatory found that its spectrum showed the characteristic nebular lines. This observation was confirmed by Dr. Copeland on August 25 and 26, and by Herr Gothard, who photographed the spectra of a number of nebulæ, and compared them with his photograph of the spectrum of the new star. He says, "Each new photograph increased the probability, which may be considered as a proved fact, that the *spectrum not only resembles, but that the aspect and position of the lines show it to be identical with the spectra of the planetary nebulæ*. In other words, the new star has changed into a planetary nebula."

A nebulous spectrum was also found by Espin. From observations of the spectrum in November, 1894, Professor Campbell finds that "the spectrum is not only nebular, but it is approaching the average type of nebular spectrum," and he adds, "We may say that only five 'new stars' have been discovered since the application of the spectroscope to astronomical investigations, and that three of these had substantially identical spectroscopic histories." Espin found the star distinctly nebulous on December 9, 1895, and its magnitude about 10½.

Another new star was discovered by Mrs. Fleming by the

[1] *Journal of the British Astronomical Association*, February, 1892.

photographic method in the southern constellation, Norma, in the year 1893. When at its brightest, it seems to have been about the seventh magnitude. It was situated in the Milky Way, a little to the east of the pair of stars known as Gamma one and Gamma two Normæ. Its spectrum was similar to that of the new star in Auriga, when it first appeared, and, like that object, the spectrum has now, according to Professor Campbell, "become distinctly nebular."

Another temporary star of about the eighth magnitude was also discovered by Mrs. Fleming in 1895, in that portion of the southern constellation Argo, known as Carina. It was in or close to the Milky Way—like so many of these new stars—between the variable star Eta Argûs and the star Lambda Centauri, near the Southern Cross, and close to a star of magnitude 5½. The photographic plates on which the discovery was made were taken at the Arequipa Station, in Peru. An examination of 62 photographs of the region showed no trace of the star on May 17, 1889, and March 5, 1895, although stars so faint as the fourteenth magnitude are visible on some of the plates. On nine plates, however, taken between April 8, 1895, and July 1, 1895, the star is visible, and during this interval the brightness diminished from the eighth to the eleventh magnitude. The spectrum showed the bright lines of hydrogen "accompanied by dark lines of slightly shorter wave length," and in all its "essential features" was "apparently identical" with the spectra of the temporary stars in Auriga and Norma.

With reference to this outburst, and the similarity of the stars spectrum to that of the new star in Auriga, Professor William H. Pickering points out "the improbability of two successive collisions between stars, occurring nearly in the line of sight, in both cases a bright and a dark-line star being involved, and in each case the bright-line star being the one to recede from us. The same remark applies to the theory of a collision of a star and a nebula. As a substitute I offered an explosion hypothesis, in which a dark sun suddenly gave out in all directions large quantities of hydrogen in an incandescent state. This would, of course, merely produce a spectrum with bright lines. But if the expulsion of hydrogen continued, the outer layers of gas would cool, producing absorption lines in the spectrum of the approaching hydrogen, but still leaving the spectrum lines of the receding hydrogen bright. Finally, when the expulsion ceased, we should find a heated spherical mass of gas, similar to a planetary nebula. It was shown that the velocities which were observed in the cases of these two *novæ* were less than fifty per cent. greater than had been observed in our own sun. The discovery of this third *nova*, with a spectrum identical with that of the two others, increases many times the improbability of the collision theories, and thereby strengthens the explosion hypothesis. If this latter is correct, we must look upon the phenomena presented by a *nova* not as indicating the birth of a new star, but rather as a cataclysm testifying to the death and final disrupture of an old one." [1]

Another apparently new star was detected by Mrs. Fleming in 1895, in the constellation Centaurus. It was situated about three degrees north-west of the double star 3 Centauri, and when at its brightest, seems to have been about the seventh magnitude. Mrs. Fleming's attention was first directed to it by its peculiar spectrum, as shown on a photographic plate taken at Arequipa on July, 1895. No trace of the star is visible on 55 plates taken from May 21, 1889, to June 14, 1895, but on plates taken on July 8 and 10, 1895, it appears of about the seventh magnitude. A photograph taken on December 16, 1895, shows it as a star of about the eleventh magnitude. On that date, and on December 19, it was seen about the same magnitude by Mr. O. C. Wendell, with a 15-inch telescope. The spectrum at first resembled that of the nebula 30 Doradus, and was unlike the spectra of the temporary stars in Auriga, Norma, and Carina. When it had faded to the eleventh magnitude, its spectrum seemed to be monochromatic, and very similar to that of a neighbouring nebula, N G C 5253, so that, like the new stars in Cygnus,

[1] *The Observatory*, December, 1895.

Auriga, and Norma, "it appears to have changed into a gaseous nebula."

It is a remarkable fact that the great majority of the temporary stars appeared in or near the Milky Way. The chief exceptions to this rule are:—the star of 76 B.C., in the Plough, the star recorded by Hepidannus in Aries, 1012, A.D., and the "Blaze Star" of 1866 in Corona Borealis.

CHAPTER VI.

CLUSTERS AND NEBULÆ.

CLUSTERS of stars and nebulæ are frequently classed together in one group. But this is incorrect. The term nebulæ should be restricted to those objects which the spectroscope shows to consist of gaseous matter, while the term cluster should be applied to those groups of stars in which the components are individually visible as distinct star-like points. There may be, of course, intermediate forms, like the Great Nebula in Andromeda, which, although not resolvable into stars with powerful telescopes, the spectroscope shows to be not gaseous. We will begin with clusters of stars, many of which can be seen with telescopes of moderate power, and some, like the Pleiades, even with the naked eye.

The Pleiades form perhaps the most remarkable group of stars in the heavens, and are probably familiar to most people, even to those whose knowledge of the constellations is limited to a few of the brighter stars. The cluster is a very remarkable and brilliant one, and forms a striking object in a clear sky. There is no other group visible to the naked eye in either hemisphere similar to it in the brightness and closeness of the component stars. It seems to have attracted the attention of observers since the earliest ages. Job says:

"Can'st thou bind the sweet influences of Pleiades, or loose the bands of Orion?"

Hesiod, writing nearly 1,000 years B.C., speaks of the Pleiades in words thus translated by Cooke:—

> "There is a time when forty days they lie,
> And forty nights conceal'd from human eye;
> But in the course of the revolving year,
> When the swain sharps the scythe, again appear."

This passage refers to the disappearance of the group in the sun's rays in summer, and their reappearance in the evening sky in the east at harvest time. Hesiod also speaks of them as the seven sisters, and in Cicero's "Aratus," they are represented as female heads, bearing the names Merope, Alcyone, Celæno, Electra, Taygeta, Asterope, and Maia, names by which they are still known to astronomers. The origin of the name Pleiades is somewhat doubtful. Some think that it is derived from the Greek word *pleia*, to sail. Others from the words *pleios*, full, a name perhaps suggested by the appearance of the cluster. Although seven stars are mentioned by Hipparchus and Aratus, Homer only speaks of six, and this is the number now visible to average eyesight. A larger number has, however, been seen with the naked eye by those gifted with exceptionally keen eyesight. Möstlin, Kepler's tutor, is said to have seen fourteen, and he actually measured and recorded the position of eleven, with wonderful accuracy, without the aid of a telescope! In recent years, Miss Airy, daughter of the late astronomer-royal, has seen twelve, and Carrington and Denning fourteen. But to most eyes probably six only are visible with any certainty. There is a tradition that, although seven stars were originally visible, one disappeared at the taking of Troy. Professor Pickering has recently discovered that the spectrum of Pleione, which forms a wide pair with Atlas, bears a striking resemblance to that of P Cygni, the so-called "temporary star"' of 1600. This similarity of spectra suggests the idea that Pleione may possibly —like the star in Cygnus—be subject to occasional fluctuations of light, which might perhaps account for its visibility to the naked eye in ancient times.

The grouping of even six stars visible to the naked eye in so small a space is very remarkable. Considering the total number of stars visible without optical aid, Mitchell—writing in 1767—calculated by the mathematical theory of probability that the chances are 500,000 to one against the close arrangement of six stars in the Pleiades being merely the result of accident. He therefore concludes "that this distribution was the result of design, or that there is reason or cause for such an assemblage."

Although to a casual observer the component stars may appear of merely equal magnitude, there is considerable difference in their relative brilliancy. Measures with a photometer show that Alcyone—the brightest of the group—is of the third magnitude, Maia, Electra, and Atlas of the fourth, Merope about $4\frac{1}{3}$, Taygeta $4\frac{1}{2}$, Celæno about $5\frac{1}{3}$, and Asterope about the sixth. Pleione is about $5\frac{1}{2}$, according to the photometric measures made at Oxford, but it lies so close to Atlas that to most eyes the two will probably appear as one star. About thirty more range from the sixth to the ninth magnitude, and this is about the number visible with an opera-glass. Galileo counted thirty-six stars with his small telescopes, but with modern instruments the number is largely increased. Some years since, M. Wolf, the distinguished French astronomer, published a chart of the Pleiades, showing about 500 stars made from his own observations. Photography has further added to the number of stars visible in this interesting group. On a photograph taken at the Paris Observatory in 1887, with an exposure of three hours, no less than 2,326 stars can be distinctly counted on a space of about three square degrees. The fainter stars on this photograph are supposed to be of the seventeenth magnitude. Now, as Alcyone, the brightest star of the group, is of the third magnitude, we have a difference of fourteen magnitudes between the brightest and the faintest. This implies that Alcyone is 398,100 times brighter than the faintest stars visible on the photographic plate. If we could conclude that the fainter stars really belonged to the cluster, they would be at practically the same distance from the earth, and the great difference of brightness would be very remarkable, and would suggest that Alcyone is a vastly larger body than the smallest stars of the group. The difference of brilliancy given above would indicate that the diameter of Alcyone is 631 times greater than that of the faintest stars revealed by photography. This is of course on the assumption that all the stars of the cluster are, surface for surface, of the same intrinsic brilliancy, and that this apparent brightness to the eye depends simply on their diameter. As spheres vary in volume as the cubes of their diameters, we have the volume of Alcyone equal to the cube of 631, or over 250 million times the volume of the faintest stars of the group. This startling result was very difficult to explain, for either we must assume that Alcyone is an enormously vast body, or else that the faint stars of the group are exceedingly small. If we take the diameter of Alcyone as 1,400,000 miles, then the diameter of the faintest stars in the group would be only 2,200 miles, or about the size of our moon, and it seems highly improbable, if not impossible, that such small bodies should shine with inherent light of their own. They would indeed be "miniature suns." On the other hand, if we assume that the faintest stars are of about the same size as the planet Jupiter, or about 87,000 miles, the diameter of Alcyone would be nearly 55 millions of miles, a result which is also highly improbable. The difficulty has, I think, been satisfactorily cleared up by some photographs recently taken by Professor Barnard at the Lick Observatory. A photograph taken with a lens of six inches aperture, and 31 inches focal length, and an exposure of 10 hours 15 minutes, shows that the sky surrounding the Pleiades is, on all sides, as thickly studded with small stars as the cluster itself. It seems clear, therefore, that the faint stars in the Pleiades are merely some of the "hosts of heaven" which happen to lie in that direction, and have probably no connection with the cluster, which is merely projected on a starry background of faint and distant stars.

The brilliancy of the Pleiades cluster would naturally suggest a comparative proximity to the earth. Attempts to determine their distance have, however, hitherto proved unsuccessful. This would indicate that the distance is very great, and would, of course, lead to the conclusion that the group is of vast dimensions. An effort has been made to determine the distance indirectly by a consideration of the "proper motion" of the principal stars. Professor Newcomb finds a proper motion for Alcyone of about 5·8 seconds of arc per century. This motion is in a direction nearly opposite to that of the sun's motion in space, and may possibly be due to that cause. If we assume that this apparent motion of Alcyone is wholly due to the effect of the sun's real motion at the rate of, say, fourteen miles a second, the distance of Alcyone would correspond to a "light journey" of about 267 years! Our sun, placed at this vast distance, would, I find, be reduced in brilliancy to a star of about the ninth magnitude, or six magnitudes fainter than Alcyone. This would imply that Alcyone is about 250 times brighter than the sun! As, however, the spectrum of Alcyone is of the first or Sirian type, it cannot properly be compared with the sun.

There are six other small stars in the Pleiades having proper motions similar in amount and direction to that of Alcyone. As the other bright stars of the group have much smaller motions, it has been suggested that the seven stars with comparatively large, proper motions do not really belong to the group, but are only optically associated with it. This would imply that the real cluster lies much farther from us than Alcyone, and the comparative brilliancy of some of its component stars would still denote enormous size.

In the year 1859, the well-known astronomer, Tempel, announced his discovery of a faint nebulosity extending in a southerly direction from Merope, the nearest bright star to Alcyone. This interesting discovery was practically confirmed by other astronomers; but from its visibility to some observers with small telescopes, and the failure of others to detect it with much larger instruments, the variability of its light was strongly suspected. The question remained in doubt for many years, but has now been finally set at rest by photography, which shows not only a mass of nebulous light surrounding Merope, but other nebulous spots involving Alcyone, Maia, and Electra. Indeed, a photograph taken by Dr. Roberts in 1889 shows that all the brighter stars of the group are more or less surrounded by nebulosity. The nebula surrounding Maia is of a somewhat spiral form, and its existence was not even suspected until it was revealed by photography. It was afterwards seen with the great 30-inch refractor of the Pulkowa Observatory. Had, however, its existence been unknown, it would probably have escaped detection, even with this large telescope, as it is one thing to see a faint object known to exist and another to discover it independently. Maia is surrounded by several faint stars of the twelfth to the fourteenth magnitude; and the Russian observers believe that one of these is variable in light, as it was seen distinctly on February 5, 1886, when its magnitude was carefully determined with reference to the neighbouring stars; but on February 24 of the same year, it could not be seen with a telescope of 15 inches aperture. Some of the other stars in the group seem to be connected by nebulous rays with the principal nebulous centres, and in looking at this wonderful Paris chart it seems impossible to avoid the conclusion that the stars and nebulous masses are actually mixed up together, and not merely placed accidentally in the same direction. Indeed, Professor Barnard's photograph referred to above shows the whole group involved in dense nebulosity.

Other well-known clusters or groups of stars are the Hyades, marked by the bright, reddish star, Aldebaran, the Præsepe, or Beehive, in Cancer, and Comæ Berenices, but these are larger and more scattered.

Of other irregular clusters, somewhat similar to the Pleiades, but not so bright, may be mentioned the double cluster in Perseus, which is visible to the naked eye on a clear night as a hazy spot of light in the midst of the Milky Way. Admiral Smyth says they form "one of the most brilliant telescopic objects in the heavens." They may be seen with a binocular field-glass, but, of course, a good telescope is necessary to see them well. They have been beautifully photographed at the Paris Observatory, the photograph showing no trace of nebulosity. They have also been photographed by Dr.

FIG. 9.—*The Double Star Cluster in Perseus.*
(From "Scenery of the Heavens.")

Roberts, who says, "The photograph presents to the eye the stars in the two clusters, and in the surrounding parts of the sky, with a completeness and accuracy of detail never before seen. The stars are shown in their true relative positions and magnitudes to about the sixteenth, and among them are many apparent double, triple, and multiple stars. They also appear to be arranged in clusters, curves, festoons, and patterns that are suggestive of some physical connexion existing between the groups ; but it is premature to assert that these appearances are not due to perspective effect by the eye arranging numerous close points of light into various patterns. Similar photographs to this, taken at intervals of several years be-

Fig. 10.—*Star Cluster in Gemini.*
(From "Scenery of the Heavens.")

tween them, will determine the reality, or otherwise, of these remarkable groupings of the stars."

A little north of the star Eta Geminorum is a pretty cluster of small stars known as 35 Messier, which is just visible to the naked eye. The component stars may be well seen with a telescope of moderate power. This cluster has been also photographed at the Paris Observatory, and shows a well-marked clustering tendency in the component stars. Admiral Smyth says: "It presents a gorgeous field of stars from the ninth to the sixteenth magnitude, but with the centre of the mass less rich than the rest. From the small stars being inclined to form curves of three or four, and often with a large

FIG. 11.—37 *Messier.*
(From "Worlds of Space.")

one at the root of the curve, it somewhat reminds one of the bursting of a sky rocket."

About ten degrees to the north of the cluster just described is another fine cluster known as 37 Messier. The accompanying photograph will show its telescopic appearance.

In the Southern Hemisphere there is a magnificent cluster of small stars surrounding the star Kappa Crucis, a reddish star of the seventh magnitude. It was thus described by Sir John Herschel: "A most vivid and beautiful cluster of 50 to 100 stars. Among the larger there are one or two evidently greenish. South of the red star is one, 13 minutes, also red, and near it one, 12 minutes, bluish . . . though neither a large nor a rich one, is yet an extremely brilliant and beautiful object when viewed through an instrument of sufficient aperture to show distinctly the very different colours of its constituent stars, which gives it the effect of a superb piece of fancy jewellery." He gives the positions of 110 stars, from the seventh to the sixteenth magnitude. It lies near the northern edge of the well-known "coal sack," and Dr. Gould says of it: "The exquisitely beautiful cluster, κ *Crucis*, contains a large number of stars of various tints and hues, contrasting wonderfully with each other, when viewed with a telescope of large aperture." Mr. Russell's drawing of this cluster, made at Sydney (N.S.W.) in 1872, shows several changes in the relative positions of the stars as laid down by Sir John Herschel, probably the result of proper motion.

About 2½° north of the star M Velorum, Sir John Herschel describes "an enormous cluster, of a degree and a half in diameter, very rich in stars of all magnitudes, from 8 minutes downwards, a sort of telescopic Præsepe."

Another fine cluster is that known as 11 Messier. It lies a little to the west of the star Lambda Aquilæ, and is just visible to the naked eye on a clear night. It consists of stars of about the eleventh magnitude, and Admiral Smyth compared it to a "flight of wild ducks." It has been beautifully photographed by Dr. Roberts, who says: "The negative shows the stars individually, though the print, owing to their closeness, does not separate them. . . . It is entirely free from nebulosity."

There are many other similar objects in both hemispheres too numerous to mention here, but those described are interesting objects of their class.

We now come to the "globular clusters." This term has been applied to those clusters of stars which evidently occupy a space of more or less spherical form. Some of these "balls of stars," as they have been called, are truly wonderful, and are among the most interesting objects visible in the sidereal heavens. Good specimens of the class are, however, rather rare objects, and there are not many in the Northern Hemisphere. The most remarkable, perhaps, is that called "the

FIG. 12.—*Star Cluster in Hercules.*
(From "Scenery of the Heavens.")

Hercules cluster," but known to astronomers as 13 Messier, it being No. 13 in the first catalogue of remarkable "nebulæ" formed by Messier, the famous discoverer of comets. It was discovered by Halley in 1714. This wonderful object lies between the stars Zeta and Eta in Hercules, nearer to the latter star. It may be seen with a binocular or good opera-glass as a hazy star of the sixth magnitude. Messier was certain that it contained no stars ; but when examined with a good telescope it is at once resolved into a multitude of small stars, which can be individually seen, and even counted, with large telescopes. According to Admiral Smyth, "No plate can give a fitting representation of this magnificent cluster. It is indeed truly glorious, and enlarges on the eye by studious gazing." And Dr. Nichol says : "Perhaps no one ever saw it for the first time through a telescope without uttering a shout of wonder." The number of stars included in the cluster was estimated by Sir William Herschel at 14,000; but the real number is probably much smaller. Were the number so great as Herschel supposed, I find that the cluster would form a much brighter object than it does. Assuming the average magnitude of the component stars at 12½, I find that an aggregation of 14,000 stars would shine as a star of about the second magnitude. But the cluster is only as bright as a star of about the sixth magnitude, and, with this magnitude, I find that the total number would be about 400. Examining it with his giant telescope, Lord Rosse observed three dark rifts radiating from the centre. These were afterwards seen by Buffham with a 9-inch reflector, and also by Webb. They were also observed at Ann Arbor Observatory (U.S.A.), in April, 1887, by Professor Harrington and Mr. Schaeberle, using telescopes of six and twelve inches aperture. It has been well photographed at the Paris Observatory, and also by Dr. Roberts and Mr. Wilson. In some of these photographs the dark rifts are perceptible to some extent, but owing to the over exposure of the central portion of the cluster, they are not so distinct as in drawings made at the telescope. Dr. Huggins, examining it with the spectroscope, finds that the spectrum is not gaseous ; but spectroscopic evidence is not necessary to prove that the cluster consists of small stars, as these are distinctly seen as points of light with telescopes of moderate power, and with the great Lick telescope the component stars are visible even in the central portion of the cluster. Its globular shape is evident at a glance, and we cannot doubt that the stars composing it form a gigantic system, probably isolated in space. Many people might think that this cluster was a mass of double and multiple stars; but this is not so. The components, close as they are, are too far apart to constitute true double stars. Mr. Burnham, the famous double star observer, finds *one* close double star near the centre, and notes the remarkable absence of close double stars in bright and apparently compressed clusters.

In the same constellation, Hercules, between the stars Eta and Iota, but nearer the latter, will be found another object of the globular class, but not so bright or so easily resolvable into stars as the cluster described above. It is known as 92 Messier. Buffham, with a 9-inch mirror, thought the component stars brighter and more compressed than in 13 Messier. Sir William Herschel found it seven or eight minutes of arc in diameter. The brighter components are easily visible in telescopes of moderate power, but even Lord Rosse's giant telescope failed to resolve the central blaze. This object was photographed by Dr. Roberts in May, 1891, with a 20-inch reflecting telescope, and an exposure of one hour. He says: "The photograph shows the cluster to be involved in dense nebulosity, which, on the negative, almost prevents the stars being seen through it, and on the print quite obscures the stars. The stars in this, as in all other globular clusters, are arranged in various patterns, and many of them appear to be nebulous."

About three degrees north preceding the star 9 Boötis, is another fine globular cluster, known as 3 Messier. Smyth describes it as "a brilliant and beautiful globular congregation of not less than 1,000 stars, between the southern Hound and the knee of Boötis; it blazes splendidly towards the centre, and has outliers. . . . This mass is one of those balls of compact and wedged stars, whose laws of aggregation it is so impossible to assign." The idea of the component stars being "compact and wedged" is, however, a mistake, as I have shown elsewhere.[1] Sir John Herschel described it as a

[1] "Planetary and Stellar Studies," p. 188.

remarkable object, exceedingly bright and very large, with stars of the eleventh magnitude. Buffham found it resolved even in the centre with a 9-inch mirror. It was photographed by Dr. Roberts in May, 1891, with an exposure of two hours, and the photograph confirms the general descriptions given of the cluster, though "the print fails to show the stars that on the negative crowd the space covered by the dense nebulosity." Dr. Roberts remarks that "nebulosity seems invariably to be present in globular clusters." From photographs of this cluster, taken at Arequipa in Peru, Professor Bailey finds 87 stars of the cluster to be variable in light, the variability amounting in some cases to two magnitudes, with usually short periods.

Another fine globular cluster is that known as 5 Messier. It lies closely north of the fifth magnitude star, 5 Serpentis. It was discovered by Kirch in 1702, and was observed in 1764 by Messier, who found he could see it with a telescope of one foot in length, but could not resolve it into stars. Smyth says: "This superb object is a noble mass, refreshing to the senses after searching for faint objects, with outliers in all directions, and a bright central blaze, which even exceeds 3 Messier in concentration." Sir William Herschel, with his 40-foot telescope, could count about 200 stars, but could not distinguish the stars near the central blaze. Sir John Herschel describes it as an excessively compressed cluster of a globular form, with stars from the eleventh to the fifteenth magnitude, condensed into a blaze at the centre. Lord Rosse found it more than seven or eight minutes of arc in diameter, with a nebulous appearance in the centre. This cluster was photographed by Dr. Roberts in April, 1892. "The photograph shows the stars to about the fifteenth magnitude, and the cluster is involved in dense nebulosity about the centre. The nebulosity hides the stars even on the negative." With reference to this latter remark, however, Dr. Common says[1] that, in photographs of this cluster taken with a larger instrument, "the stars are quite distinct, though the exposure was

[1] *Nature*, September 6, 1894.

much longer, a result that might fairly be expected." From photographs of this cluster taken at Arequipa, Peru, by Professor Bailey, he finds that the cluster contains about 750 stars, of which 46 are variable in light, or about 6 per cent. of the whole. This is remarkable, for, of the stars visible to the naked eye, less than 1 per cent. are variable, so far as is at present known. A further examination of the photographs made by Miss Leland shows that the periods of these variables are in general very short, not exceeding a few hours.[1] One star, situated about eight minutes of arc from the centre of the cluster, has a probable period of 11 hours, 7 minutes, 52 seconds, and varies from about magnitudes 13·50 to 14·73. The star remains at the minimum light for about half the period, and the maximum brightness is of comparatively short duration. The rate of increase is more rapid than the decrease—as in most short period variables—but in other respects the character of the light fluctuations does not seem to be similar to that of any other known variable star.

Another fine object of this class is that known as 15 Messier in Pegasus, discovered by Maraldi in 1745. Sir John Herschel describes it as a remarkable globular cluster, very bright and large, and blazing in the centre. Webb found it a glorious object with a nine and one-third inch mirror. It was photographed by Dr. Roberts in November, 1890, with an exposure of two hours. He says: "The photograph confirms the general descriptions, and the negative shows, separately, the stars of which the cluster is composed distinctly through the nebulosity in the centre. Many of the stars have a nebulous appearance, and they are arranged in curves, lines, and patterns of various forms, with lanes or spaces between them."

We may also mention the globular cluster known as 2 Messier, which is situated about five degrees north of the star Beta Aquarii. It was discovered by Maraldi in 1746 while looking for Cheseaux's comet. Sir William Herschel, with his forty-foot telescope, could "actually see and dis-

[1] *Nature*, June 4, 1896.

tinguish the stars even in the central blaze." Sir John Herschel compared it to a mass of luminous sand, and estimated the stars to be of the fifteenth magnitude. It is about five or six minutes of arc in diameter, and Smyth says: "This magnificent ball of stars condenses to the centre, and presents so fine a spherical figure that imagination cannot but figure the inconceivable brilliancy of the visible heavens to its

FIG. 13.—*The Star Cluster, Omega Centauri.*
(From "Worlds of Space.")

animated myriads." Taking Sir John Herschel's estimate of the component stars at fifteenth magnitude, and the total light of the cluster at sixth magnitude, I find that the total number of stars it contains would be about 4,000.

In the Southern Hemisphere there are some magnificent examples of globular clusters, and indeed, this hemisphere seems to be richer in these objects than the northern sky. Among these southern clusters is the truly marvellous object known as Omega Centauri. Its apparent size is very large—about two-thirds of the moon's diameter—and it is distinctly visible to the naked eye as a hazy star of the fourth magnitude, and I have often so seen it in the Punjab sky. Sir John Herschel, observing it with a large telescope at the Cape of Good Hope, describes it as "beyond all comparison, the richest and largest object of its kind in the heavens. The stars are literally innumerable. . . . All clearly resolved into stars of two sizes, *viz.*, 13 and 15; the larger lying in lines and ridges over the smaller. . . . The larger form rings like lace-work on it. One of these rings, 1½″ diameter, is so marked as to give the appearance of comparative darkness, like a hole in the centre. . . . On further attention, the hole is double, or an oval space crossed by a bridge of stars. . . . Altogether, this object is truly astonishing." This wonderful object has recently been photographed by Dr. Gill, at the Royal Observatory, Cape of Good Hope, and also at Arequipa, Peru, with a telescope of thirteen inches aperture. On the latter photograph, the individual stars can be distinctly seen and counted. The enumeration has been made by Professor and Mrs. Bailey, and a mean of their counts gives 6,389 for the number of stars in the cluster, but they consider that the real number is considerably greater.

Another wonderful object is that known as 41 Toucani, which lies near the smaller "Magellanic Cloud" in the Southern Hemisphere. Humboldt found it very visible to the naked eye in Peru, and mistook it for a comet.[1] Sir John Herschel describes it as "a most magnificent globular cluster. It fills the field with its outskirts; but within its more compressed part I can insulate a tolerably defined circular space of 90″ diameter, wherein the compression is much more decided, and the stars seem to run together, and this part has, I think, a pale pinkish or rose colour, . . . which contrasts evidently with the white light of the rest. . . . The stars are

[1] "Cosmos," vol. iii., Bohn's edition, p. 192.

equal, fourteen magnitude, immensely numerous, and compressed. . . . It is *completely insulated.* After it has passed, the ground of the sky is perfectly black throughout the whole breadth of the sweep. There is a double star of eleventh magnitude preceding the centre, . . . condensation in three distinct stages. . . . A stupendous object." Dr. Gould calls it one of the most impressive, and perhaps the grandest, of its kind in either hemisphere, and he estimated its apparent magnitude at 4½, as seen with the naked eye.

Another remarkable globular cluster is that known as 22 Messier, which lies about midway between Mu and Sigma Sagittarii. Sir John Herschel says: "The stars are of two sizes, *viz.*, 15 . . . 16 and 12m; and, what is very remarkable, the largest of these latter are visibly reddish, one in particular, the largest of all (12-11m) south following the middle, is decidedly a ruddy star, and so, I think, are all the other larger ones . . . very rich, very much compressed, gradually much brighter in the middle, but not to a nucleus . . . consists of stars of two sizes . . . with none intermediate, as if consisting of two layers, or one shell over another. A noble object." I saw the larger stars well with a 3-inch refractor in the Punjab.

Sir John Herschel remarks "the frequent association of nebulæ in pairs forming double nebulæ," and in his "Cape Observations" he figures several examples of this class. One of these is evidently a globular cluster, with two centres of condensation, one nucleus being much brighter than the other. Two others, much smaller, show two distinct nuclei. Another drawing shows apparently two globular clusters in contact. There are other examples in the Northern Hemisphere. Dr. See considers that some of these double nebulæ represent an early stage in the evolution of binary or revolving double stars, and certainly some of the drawings of these nebulæ are very remarkable and suggestive.

The actual dimensions of the globular clusters is an interesting question. Are they composed of stars comparable in size and mass with our sun? or are the component stars really small and comparatively close together? This is a difficult question to answer satisfactorily, as the distance of these objects from the earth has not yet been determined. They may, on the one hand, be collections of suns similar to ours in size and brightness, and situated at vast distances from the earth; or, on the other hand, the stars composing them may be comparatively small objects, lying at a distance from the earth not exceeding that of some stars visible to the naked eye. Perhaps the latter hypothesis may be considered the more probable of the two. But there is really no reason to suppose that these collections of suns are comparatively near our system. The probability seems to be in favour of their great distance from the earth. The question of the absolute size of the component stars is one which, I think, has not been hitherto sufficiently considered. Let us examine both alternatives, and let us take the cluster Omega Centauri as one in which the number of the component stars has been *actually counted.* Assuming that the real number of stars in this cluster is 10,000, and that they are individually equal, on an average, to our sun in mass and volume, we may estimate the probable distance and dimensions of the cluster. Taking the stellar magnitude of Omega Centauri as four (as estimated at the Cordoba Observatory), I find that, with the number 10,000, the average magnitude of the component stars would be fourteen. This agrees with Sir John Herschel's estimate of thirteenth to fifteenth magnitude. Now, to reduce the sun to a star of the fourteenth magnitude, I find that, assuming the sun to be 28 magnitudes brighter than an average star of the first magnitude, it would be necessary to remove it to a distance of about 158,500,000 times the sun's distance from the earth—a distance so great that light would take no less than 2,500 years to reach us from the cluster! Taking the apparent diameter of the cluster at twenty minutes of arc, I find that its real diameter would be 922,000 times the sun's distance from the earth—a distance so great that light would take over 14 years to pass across the cluster. These results are certainly very startling, and might lead us to suspect that these globular clusters are external universes. Judging, however, from the average distance recently found for stars of the first and second magnitude (see p. 423), the distance of ordinary stars of the first magnitude—on the supposition that they are of the same size and brightness as the sun, and that their light is simply reduced by distance—would be about five times greater than that found above for Omega Centauri. If, then, we increase the distance of the cluster five times, it would be necessary to increase the diameters of the component stars to five times that of the sun. This would give them a volume 125 times that of our sun—a result which seems improbable. If, on the other hand, we do not like to admit that each of the faint points of light composing the cluster is equal in volume to our sun, let us diminish the distance ten times. If we do so, we must also diminish the diameter of the component stars ten times. This would make them about the size of the planet Jupiter, and it seems improbable that such comparatively small bodies could retain their solar heat for any great length of time. They would probably have cooled down, as Jupiter has done—at least to a great extent—ages ago, and would not now be visible as a cluster of stars. Even this reduction of the distance to one-tenth of the value first found would still leave the cluster at an immense distance from the earth, a distance represented by 250 years of light travel! A reduction of the distance to one-tenth of this again, or 25 years of light travel, would make the components about the size of the earth, and that bodies of this small size could shine with stellar light seems to be an untenable hypothesis. We seem, therefore, forced to conclude that these globular star clusters lie at an immense distance from the earth.

There is, however, another point to be considered with reference to the size of the bodies composing a globular cluster. This is the character of their light. I am not aware that the spectrum of a globular cluster has yet been thoroughly examined, but if that of Omega Centauri is of the first or Sirian type, it would modify the above conclusions to some extent. It now seems probable that stars having a spectrum of the Sirian type are intrinsically brighter than our sun, and I have shown already that Sirius is considerably brighter than the sun would be if placed at the same distance, although the mass of Sirius is but little more than twice the sun's mass. The components of a star cluster, therefore—if of the Sirian type of stars—might be as bright as the sun, and at the same time have a smaller mass and volume. This, however, would not make a very great difference in the computed vast distance of the cluster, and the calculations given above seem to point to the conclusion that these globular clusters are probably composed of stars of average size and mass, and that the faintness of the component stars is simply due to their immense distance from the earth.

We will now consider the nebulæ, properly so-called, that is to say, objects which the spectroscope shows to consist of glowing gas. These are sometimes large and irregular in form, like the great nebula in the "Sword" of Orion, sometimes with spiral convolutions, and sometimes of a definite shape, like the planetary and annular nebulæ.

Of the large and irregular nebulæ, one of the most remarkable is that known as "the great nebula in Orion." It surrounds the multiple star, Theta Orionis, which has been already referred to in a preceding chapter. It is a curious fact that it escaped the searching eye of Galileo, although he gave special attention to the constellation of Orion, for even with a good opera-glass a nebulous gleam is distinctly visible round the central star of the "Sword." The nebula seems to have been discovered by Cysat, a Swiss astronomer, in the year 1618, and it was sketched by Huygens in 1656. Huygens says: "While I was observing with a refractor of twenty-five feet focal length, the variable belts of Jupiter, a dark central belt in Mars, and some phases of this planet, my attention was attracted by an appearance among the fixed stars, which, as far as I know, has not been observed by anyone else, and which, indeed, could not be recognised, except by such powerful instruments as I employ. Astronomers enumerate three stars in the Sword of Orion, lying very near one another. On one occasion when, in 1656, I was accidentally observing the middle one of these stars through my telescope, I saw twelve stars instead of a single one, which, indeed, not unfrequently happens in using the telescope. Three of this number were almost in contact with one another, and *four* of them shone as if through a mist, so that the space around them, having the form drawn in the appended figure, appeared much brighter than the rest of the sky, which was perfectly clear, and looked almost black. This appearance looked, therefore, almost as if there were a *hiatus* or interruption. I have frequently observed this phenomenon, and up to the present time, as always unchanged in form; whence it would appear that this marvellous object, be its nature what it may be, is very probably permanently situated at this spot. I never observed anything similar to this appearance in the other fixed stars."[1] It has been called the "fish-mouth" nebula, from the fancied resemblance of the centre portion to the mouth of a fish. A number of small stars are visible over the surface of the nebula, and at one time, Lord Rosse thought it showed indications of resolution into stars when examined with his giant telescope; but this is now known to have been a mistake, for Dr. Huggins finds, with the spectroscope, that it consists of nothing but glowing gas, of which hydrogen is certainly one constituent, and he has succeeded in photographing the complete series of lines of this gas in the spectrum of the nebula.

Referring to his earlier observations, Dr. Huggins says:—"The light from the brightest parts of the nebula near the trapezium was resolved by the prisms into three bright lines, in all respects similar to those of the gaseous nebulæ. The whole of this great nebula, as far as lies within the power of my instrument, emits light which is identical in character. The light from one part differs from the light of another in intensity alone." The brightest line in the nebular spectrum—the "chief nebular line," as it is called—has not yet been

[1] Humboldt's "Cosmos," Bohn's edition, vol. iv., pp. 327, 328.

identified with that of any terrestrial substance. It was at first supposed to be identical with a line of nitrogen, but this was afterwards disproved. It was then incorrectly identified with a line of lead, and more recently by Lockyer with the edge of a "fluting" in the magnesium spectrum. Dr. Huggins and Professor Keeler, however, have shown conclusively that the nebular line does not coincide with the magnesium fluting, although very close to it. Observations by Dr. Copeland in 1886 showed the existence of the yellow line, know as D_3, which is visible in the solar spectrum during total eclipses of the sun, and indicates the existence of a gas in the sun's surroundings, to which the name "helium" has been given. Dr. Copeland says:—"The recurrence of this line in the spectrum of a nebula is of great interest, as affording another connecting link between gaseous nebula and the sun and stars with bright line spectra, especially with that remarkable class of stars of which the finest examples were detected by M. M. Wolf and Rayet in the constellation of Cygnus."[1] As has been already mentioned in the chapter on variable and new stars, the bright lines of hydrogen and helium have also been observed in the spectra of these remarkable objects. The gas, giving the line D_3 in its spectrum, has quite recently been discovered by Professor Ramsay in gases obtained by heating certain terrestrial minerals, so that the objective existence of the gaseous element "helium"—previously only suspected—is now definitely established. From recent spectroscopic observations of the Orion nebula, Dr. Huggins thinks that "the stars of the 'trapezium' are not merely optically connected with the nebula, but are physically bound up with it, and are very probably condensed out of the gaseous matter of the nebula." With reference to this point, Professor Keeler, who has carefully examined the spectra of the nebula and the associated stars, says:—"The trapezium stars have spectra marked by strong absorption bands; they have not the direct connexion with the nebula that would be indicated by a bright line spectrum, but are, in fact, on

[1] *Monthly Notices*, Royal Astronomical Society, June, 1888.

precisely the same footing (spectroscopically) as other stars in the constellation of Orion. While their relation to the nebula is more certain than ever, they can no longer be regarded as necessarily situated *in* the nebula, but within indefinite limits they may be placed anywhere in the line of sight." These results were confirmed by Professor Campbell. He finds, "that of the twenty-five bright lines known to exist in the spectrum of the Orion nebula, at least nineteen are definitely matched by dark lines in the Orion stars, and at least fifteen by dark lines in the six faint stars situated in the dense parts of the nebula."

Numerous drawings of this wonderful nebula have been made. Of these, the best are those by Sir John Herschel, made at the Cape of Good Hope in the years 1834-38, by Bond in America, and by Lassell at Malta. The difficulty of accurately delineating so difficult and delicate an object has given rise to discrepancies in the drawings, which have led to the idea that changes of form have occurred, but this seems improbable. The nebula has been very successfully photographed by Dr. Common and Dr. Roberts, and these photographs confirm the general accuracy of the later drawings.

From a consideration of the apparent size of the Orion nebula and its probable mass and distance from the earth, the late Mr. Ranyard came to the conclusion that its average density "cannot exceed one ten thousand millionth of the density of atmospheric air at the sea level."[1]

Mr. W. H. Pickering and Dr. Max Wolf have photographed another nebula surrounding the star Zeta Orionis—the southern star of the "Belt," which seems to be connected with the nebula in the "Sword"; and, Prof. Barnard, using the "lens of a cheap oil lantern" of 1½ inch aperture, and 3½ inches focal length, has photographed "an enormous curved nebulosity" stretching over nearly the whole of the constellation of Orion, and involving the "great nebula."

Prof. Keeler has recently found, with the spectroscope, that the Orion nebula is apparently receding from the earth at the

[1] "Old and New Astronomy," p. 794.

rate of nearly eleven miles a second, but this motion may be, in part at least, due to the sun's motion in space in the

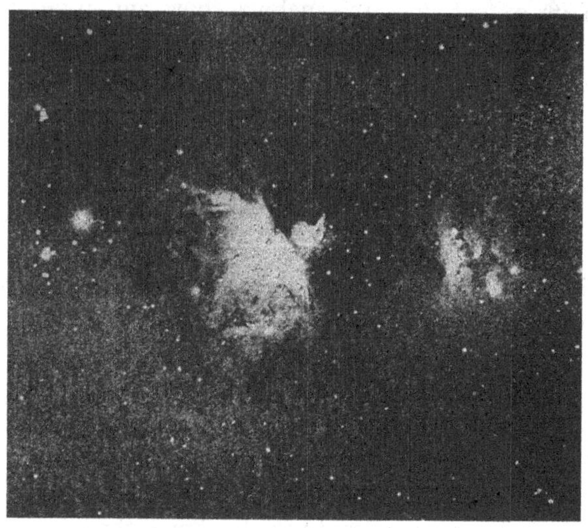

FIG. 14.—*The Orion Nebulæ.*
(From "Worlds of Space.")

opposite direction. Prof. Pickering considers that the parallax of the nebula is probably not more than 0″·003, which corresponds to a thousand years' journey for light!

In the southern constellation, Argo is a magnificent nebula, somewhat similar in appearance to the great nebula in Orion. It surrounds the famous variable star Eta Argûs, whose remarkable fluctuations in light have been already described in the chapter on variable stars. It is sometimes spoken of as the "key-hole" nebula, owing to a curious opening of that shape near its centre. It was carefully drawn by Sir John Herschel at the Cape of Good Hope in the years 1834-38. It lies in a very brilliant portion of the Milky Way, and Sir John Herschel thus describes it: "It is not easy for language to convey a full impression of the beauty and sublimity of the spectacle which the nebula offers as it enters the field of view of a telescope, fixed in right ascension, by the diurnal motion, ushered in as it is by so glorious and innumerable a procession of stars, to which it forms a sort of climax, and in a part of the heavens otherwise full of interest," and he adds: "In no part of its extent does this nebula show any appearance of resolvability into stars, being, in this respect, analogous to the nebula of Orion. It has, therefore, nothing in common with the Milky Way, on the ground of which we see it projected, and may therefore be, and not improbably is, placed at an immeasurable distance behind that stratum." Sir John Herschel's conclusion as to its physical constitution has been fully confirmed by the spectroscope, which shows it to consist of luminous gas. As in the Orion nebula, there are numerous stars scattered over it. Some of these may possibly have a physical connexion with the nebula, while others may belong to the Milky Way. The nebula is of great extent, covering an apparent space about five times the area of the full moon, and its real dimensions must be enormous. It was photographed by Mr. Russell, director of the Sydney Observatory, in July, 1890, and the photograph shows that "one of the brightest and most conspicuous parts of the nebula"—the swan-shaped form near the centre of Herschel's drawing—has "wholly disappeared," and its place is now occupied by "a great, dark oval." Mr. Russell first missed the vanished portion of the nebula in the year 1871, while examining it with

FIG. 15.—*Sir John Herschel's drawing of the Nebula round Eta Argus.*
[(From Flammarion's "Popular Astronomy.")

a telescope of 11½ inches aperture, and the photograph now confirms the disappearance, which is very remarkable, and shows that changes are actually in progress in these wonderful nebulæ, changes which may be detected after a comparatively short interval of time.

Smaller than the nebula in Argo, but somewhat similar in general appearance, is that known as 30 Doradus, which forms one of the numerous and diverse objects which together constitute the greater Magellanic Cloud. Sir John Herschel drew it carefully at the Cape of Good Hope, and describes it as "one of the most singular and extraordinary objects which the heavens present," and he says "it is unique even in the system to which it belongs, there being no other object in either nubecula to which it bears the least resemblance." It is sometimes called the "looped nebula," from the curious openings it contains. One of these is somewhat similar to the "key-hole" opening in the Argo nebula. Near its centre is a small cluster of stars, and scattered over the nebula are many faint stars, of which Sir John Herschel gives a catalogue of 105 ranging from the ninth to the seventeenth magnitude. I do not know whether this nebula has been examined with the spectroscope, but its appearance would suggest that it is gaseous. It is remarkable as being the only object of its class which is found outside the zone of the Milky Way.

Among the nebula of irregular shape, although its spectrum is said to be not gaseous, may be mentioned that known as the "trifid nebula," or 20 Messier. It lies closely north of the star 4 Sagittarii in a magnificient region of the heavens. As will be seen in the drawing made by Sir John Herschel at the Cape of Good Hope, the principal portion consists of three masses of nebulous matter separated by dark "lanes" or "rifts." Near the junction of the three "rifts" is a triple star. A beautiful drawing of this nebula has also been made by Trouvelot. It agrees fairly well with that of Sir John Herschel, but shows more detail.

Among other gaseous nebula may be mentioned that called by Sir John Herschel the "dumb-bell" nebula. It lies a little south of the sixth magnitude star 14 Vulpeculæ, and was discovered by Messier in 1779, while observing Bode's comet of that year. In small telescopes it has the appearance of a dumb-bell, or hour-glass, but in larger telescopes the outline is filled in with fainter nebulous light, giving to the whole an

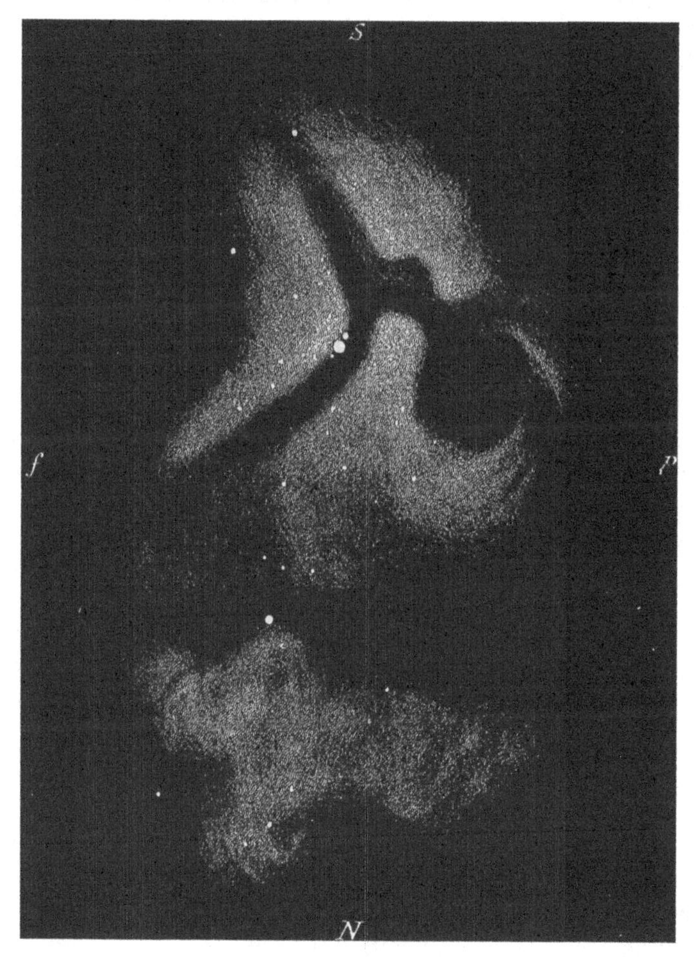

FIG. 16.—*The Trifid Nebula, Sagittarius.*
(From "Scenery of the Heavens.")

elliptical form. Several faint stars have been seen in it, but these probably belong to the Milky Way, as Dr. Huggins finds the spectrum gaseous. Dr. Roberts has photographed it, and he thinks that "the nebula is probably a globular mass of nebular matter, which is undergoing the process of condensation into stars, and the faint protusions of nebulosity in the *south following* and *north preceding* ends are the projections of a broad ring of nebulosity which surrounds the globular mass. This ring, not being sufficiently dense to obscure the light of the central region of the globular mass, is dense enough to obscure those parts of it that are hidden by the increased thickness of the nebulosity, thus producing the 'dumb-bell' appearance. If these inferences are true, we may proceed yet a step, or a series of steps, farther, and predict that the consummation of the life history of this nebula will be its reduction to a globular cluster of stars."

Among the gaseous nebula may also be included those known as "annular nebulæ." These are very rare objects, only a few being known in the whole heavens. The most remarkable is that known as 57 Messier, which lies between the stars Beta and Gamma Lyræ, south of the bright star Vega. It was discovered by Darquier, at Toulouse, in 1779, while following Bode's comet of that year. Lord Rosse thought it resolvable into stars, and so did Chacornac and Secchi, but no stars are perceptible with the great American telescopes, and Dr. Huggins finds it to be gaseous. The central portion is not absolutely dark, but contains some faint nebulous light. Examined with the great telescope of the Lick Observatory, Professor Barnard finds that the opening of the ring is filled in with fainter light "about midway in brightness between the brightness of the ring and the darkness of the adjacent sky." [1] "The aperture was more nearly circular than the outer boundary of the nebula, so that the ends of the ring were thicker than the sides." The entire nebula was of a milky colour. A central star, noticed by some observers, was usually seen by Professor Barnard, but was never a conspicuous object. He found the extreme dimensions of the nebula about 81″ in length by about 59″ in width, or more than double the apparent area of Jupiter's disc. It has been

[1] *Nature*, June 4, 1896.

beautifully photographed by Dr. Roberts, and he says "the photograph shows the nebula and the interior of the ring more elliptical than the drawings and descriptions indicate ; and the star of the *following* side is nearer to the ring than the distance given. The nebulosity on the *preceding* and *following* ends of the ring protrudes a little, and is less dense than on the *north* and *south* sides. This probably suggested the filamentous appearance which Lord Rosse shows. Some photographs of the nebula have been taken between 1887 and 1891, and the central star is strongly shown on some of them, but on others it is scarcely visible, which points to the star being variable." On a photograph taken by MM. Androyer and Montaugerand of the Toulouse Observatory, with an exposure of nine hours (in multiple exposures), about 4,800 stars are visible on and near the nebula in an area of three square degrees.

Another object of the annular class will be found a little to the south-west of the star Lambda Scorpii. It is thus described by Sir John Herschel : "A delicate, extremely faint, but perfectly well defined, annulus. The field crowded with stars, two of which are on the nebula. A beautiful, delicate ring, of a faint, ghost-like appearance, about 40″ in diameter in a field of about 150 stars, eleven and twelve magnitude and under."

Near the stars 44 and 51 Ophiuchi is another object of the annular class, which Sir John Herschel describes as "exactly round, pretty faint, 12″ diameter, well terminated, but a little cottony at the edge, and with a decided darkness in the middle, equal to a tenth magnitude star at the most. Few stars in the field, a beautiful specimen of the planetary annular class of nebula."

The Planetary Nebulæ form an interesting class. They were so named by Sir William Herschel from their resemblance to the discs of the planets, but, of course, much fainter. They are generally of uniform brightness, without any nucleus or brighter part in the centre. There are numerous examples of this class, one of the most remarkable being that known as 97 Messier, which is situated about two degrees south-east of

Beta Ursæ Majoris—the southern of the two "pointers" in the Plough. It is of considerable apparent size, and even supposing its distance to be not greater than that of 61 Cygni, its real dimensions must be enormous. Lord Rosse observed two openings in the centre with a star in each opening, and from this appearance he called it the "owl nebula." One of the stars seems to have disappeared since 1850, and a photograph recently taken by Dr. Roberts confirms the disappearance.

Another fine object of the planetary class is one which lies close to the pole of the ecliptic. Webb saw it "like a considerable star out of focus." Smyth found it pale blue in colour. Dr. Huggins finds a gaseous spectrum, the first discovery of the kind made. Professor Holden, observing it with the great Lick telescope, finds its structure extraordinary. He says it "is apparently composed of rings overlying each other, and it is difficult to resist the conviction that these are arranged in space in the form of a true helix," and he ranks it in a new class which he calls "helical nebulæ."

A somewhat similar nebula lies a little to the west of the star Nu Aquarii. Secchi believed it to be in reality a cluster of small stars, but Dr. Huggins finds its spectrum gaseous. A small nebula on each side gives it an appearance somewhat similar to the planet Saturn, with the rings seen edgeways. The great Lick telescope shows it as a wonderful object—"a central ring lies upon an oval of much fainter nebulosity." Professor Holden says "the colour is a pale blue," and he compares the appearance of the central ring "to that of a footprint left in the wet sand on a sea beach."

About two degrees south of the star Mu Hydræ is another planetary nebula, which Smyth describes as resembling the planet Jupiter in "size, equable light and colour." Webb saw it of "a steady, pale blue light," and Sir John Herschel, at the Cape of Good Hope, speaks of its colour as "a decided blue—at all events, a good sky-blue," a colour which seems characteristic of these curious objects. Although Sir William Herschel, with his large telescopes, failed to resolve it into stars, Secchi thought he saw it breaking up into stars with a "sparkling ring." Dr. Huggins, however, finds the spectrum to be gaseous, so that the luminous points seen by Secchi could not have been stellar.

Sir John Herschel, in his "Cape Observations," describes a planetary nebula which lies between the stars Pi Centauri and Delta Crucis. He says it is "perfectly round, very planetary, colour fine blue . . . very like Uranus, only about half as large again, and blue. . . . It is of the most decided independent blue colour when in the field by itself, and with no lamplight and no bright star. About 10′ north of it is an orange-coloured star, eighth magnitude. When this is brought into view, the blue colour of the nebula becomes intense . . . colour, a beautiful rich blue, between Prussian blue and verditer green."

There are some rare objects called "nebulous stars." The star Epsilon Orionis—the centre star of Orion's Belt—is involved in a great nebulous atmosphere. The triple star Iota Orionis is surrounded by a nebulous haze. The star Beta in Canes Venatici is a 4½ magnitude star surrounded by a nebulous atmosphere.

The term elliptical nebulæ has been applied to those of an elliptical or elongated shape. This form is probably due in many cases to the effect of perspective, their real shape being circular, or nearly so. Perhaps the most remarkable object of this class is the well-known "nebula in Andromeda," known to astonomers as 31 Messier. It can be just seen with the naked eye, on a clear moonless night, as a hazy spot of light near the star Nu Andromedæ, and it is curious that it is not mentioned by the ancients, although it must have been very visible to their keen eyesight in the clear Eastern skies. It was, however, certainly seen so far back as 905 A.D., and it is referred to as a familiar object by the Persian astronomer, Al-Sûfi, who wrote a description of the heavens about the middle of the tenth century. Tycho Brahé and Bayer failed to notice it, but Simon Marius saw it in December, 1612, and described it "as a light seen from a great distance through half-transparent horn plates." It was also observed by Bullialdus, in 1664, while following the comet of that year. It has frequently been mistaken for a comet by amateur observers in recent years. Closely north-west of the great nebula is a smaller one discovered by Le Gentil in 1749, and another to the south, detected by Miss Caroline Herschel in 1783. The great nebula is of an elliptical shape and considerable apparent size. The American astronomer, Bond, using a telescope of 15 inches aperture, traced it to a length of about four degrees, and a width of two and a half degrees. A beautiful photograph taken by Dr. Roberts in December, 1888 (see p. 398), shows an extension of nearly two degrees in length, and about half a degree in width, or considerably larger than the apparent size of the full moon. Bond could not see any symptom of resolution into stars, but noticed two dark rifts or channels running nearly parallel to the length of the nebula. In Dr. Roberts' photograph these rifts are seen to be really dark intervals between consecutive nebulous rings into which the nebula is divided. Dr. Roberts says: "A photograph which I took with the 20-inch reflector on October 10, 1887, revealed for the first time the true character of the great nebula, and one of the features exhibited was that the dark bands, referred to by Bond, formed parts of divisions between symmetrical rings of nebulous matter surrounding the large diffuse centre of the nebula. Other photographs were taken in 1887, November 15; 1888, October 1; 1888, October 2; 1888, December 29; besides several others taken since, upon all of which the rings of nebulosity are identically shown, and thus the photographs confirm the accuracy of each other, and the objective reality of the details shown of the structure of the nebula." Dr. Roberts adds: "These photographs throw a strong light on the probable truth of the *Nebular Hypothesis*, for they show what appears to be the progressive evolution of a gigantic stellar system."

The largest telescopes have hitherto completely failed to resolve this wonderful object into stars. Dr. Huggins, however, finds that the spectrum is *not* gaseous, so that if the nebula really consists of stellar points, they must be of very small dimensions. Assuming a parallax of one-fiftieth of a second of arc—corresponding to 163 years of "light travel"—I find that our sun, placed at this distance, would be reduced in brightness to a star of about the eighth magnitude. If we assume the components to have only one-hundredth of the sun's diameter, they would shine as stars of only the eighteenth magnitude, which no telescope yet constructed would show as separate points of light. A more probable explanation, however, seems to be that the nebula may consist of masses of nebulous matter partially condensed into the solid form, but not yet arrived at the stage in which our sun is at present. In other words, the whole nebulous mass may be in a fluid or viscous state, which might perhaps account for the continuous spectrum found by Dr. Huggins.

The question may be asked, What is the probable size and distance of this wonderful nebula? and could it be an external universe? Possibly its distance from the earth may be even greater than that indicated by the small parallax I have assumed above, but taking this parallax and the apparent dimensions of the nebula as shown by Dr. Huggins' photograph, I find that its real distance would be no less than 330,000 times the sun's diameter from the earth, a diameter so great that light would take over five years to pass from one side of the nebula to the other! This result might lead us to imagine that the nebula may be really an external universe. But let us consider the matter a little further. The diameter found above is not very much greater than the distance of the *nearest* fixed star, Alpha Centauri, from the earth, and the limits of *our* universe are certainly far beyond Alpha Centauri. If we diminish the parallax to, say, $\frac{1}{200}$th of a second, or a "light journey" of 652 years, the diameter of the nebula would be increased to 1,320,000 times the sun's distance from the earth, or about five times the distance of Alpha Centauri, and there are probably many faint stars belonging to our system much farther from the earth than this.

The temporary star which appeared near the nucleus of the nebula in August, 1885—already referred to in the chapter on variable stars—was of the seventh magnitude. I find that our sun, if placed at the distance indicated by a parallax of $\frac{1}{200}$th of a second, would be reduced to a star of about the eleventh magnitude, or four magnitudes fainter than the temporary star appeared to us. That is to say, the star would have been—with the assumed distance—about forty times brighter than the sun. With any greater distance, the star would have been proportionately brighter, compared with the sun. This seems improbable, and tends to the conclusion that the nebula is *not* an external galaxy, but a member of our own sidereal system, a system which probably includes all the stars and nebulæ visible in our largest telescopes. Dr. Common, indeed, suggests that it may be comparatively near our system. He says: "It is difficult to imagine that such an enormous object, as the Andromeda nebula must be, is not very near to us; perhaps it may be found to be the nearest celestial object of all beyond the solar system. It is one that offers the best chance of the detection of parallax, as it seems to be projected on a crowd of stars, and there are well defined points that might be taken as fiducial points for measurement," and he adds: "Apart from the great promise this nebula seems to give of determining parallax, there is a fair presumption that in the course of time, the rotation of the outer portion may perhaps be detected by observation of the positions of the two outer detached portions in relation to the neighbouring stars."[1] Prof. Hall's failure to detect any parallax in the temporary star, as mentioned in the last chapter, is, of course, against Dr. Common's idea of its proximity to the earth. Referring to the latter portion of Dr. Common's remarks, Mr. C. Easton points out[2] that a comparison of a drawing by Trouvelot, in 1874, with Dr. Roberts' photograph, suggests that the small elongated nebula —*h* 44—which lies to the north of the great nebula, "has

[1] *Nature*, September, 1894.
[2] *Ibid.*, October 4, 1894.

turned about 15° from left to right. The globular nebula (M 32), to the other side of M 31, seems to have slightly shifted its position."

The spiral nebulæ are wonderful objects, and were discovered by the late Lord Rosse, with his great six-foot telescope. Their character has been fully confirmed by photo-

FIG. 17.—*Spiral Nebula*, 51 *Messier*.
(From "The Visible Universe.")

graphs taken by Dr. Roberts. One of the most remarkable of these extraordinary objects is that known as 51 Messier. It lies about three degrees south-west of the bright star Eta Ursæ Majoris—the star at the end of the Great Bear's tail. It was discovered by Messier while comet-hunting on October 13, 1773. Telescopes of moderate power merely show two nebulæ nearly in contact, but Lord Rosse saw it as a wonderful spiral, and his drawing agrees fairly well with a photograph taken by Dr. Roberts in April, 1889. The nebula has also been photographed by Dr. Common. Dr. Roberts says: "The photograph shows both nuclei of the nebula to be stellar, surrounded by dense nebulosity, and the convolutions of the spiral in this as in other spiral nebulæ are broken up into starlike condensations with nebulosity around them. Those stars that do not conform to the trends of the spiral have nebulous trails attached to them, and seem as if they had broken away from the spirals." A tendency to a spiral structure in the smaller nebula is also visible on the original negative. Dr. Huggins finds that the spectrum is *not* gaseous.

The nebulæ known as 99 Messier is of the spiral form. It lies on the borders of Virgo and Coma Berenices, near the star 6 Comæ. In large telescopes it somewhat resembles a "Catherine wheel." D'Arrest and Key thought it resolvable into stars. It has been photographed by M. Von Gothard.

Among the clusters and nebulæ, we may class the Magellanic Clouds, or Nubeculæ in the Southern Hemisphere, as they consist of stars, clusters, and nebulæ. These very remarkable objects form two bright spots of milky light, which, at first sight, look like luminous patches of the Milky Way, but are in no way connected with the Galaxy. Sir John Herschel, speaking of the larger cloud, says: "The immediate neighbourhood of the Nubecula Major is somewhat less barren of stars than that of the Minor, but it is by no means rich, nor does any branch of the Milky Way whatever form any certain or conspicuous junction with, or include, it," and again he says, with reference to the smaller cloud: "Neither with the naked eye, nor with a telescope, is any connexion to be traced either with the greater Nubecula, or with the Milky Way." The Nubeculæ are roughly circular in form, and, viewed with the naked eye, they very much resemble irresolvable nebulæ as seen in a telescope. The larger cloud, or Nubecula Major, as it is called, is of considerable extent, and covers about 42 square degrees, or over two hundred times the apparent size of the full moon. It was called by the Arabs *el-baker*, or "the White Ox," and is referred to by Al-Sûfi in his "Description of the Heavens," written in the tenth century. When examined with a good telescope, it is found to consist of about six hundred stars of the sixth to the tenth magnitude, with many fainter ones, and about three hundred clusters and nebulæ. Sir John Herschel, in his "Cape Observations," says: "The Nubeculæ Major, like the Minor, consists partly of large tracts and ill-defined patches of irresolvable nebula, and of nebulosity in every stage of resolution, up to perfectly resolved stars like the Milky Way, as also of regular and irregular nebulæ properly so-called, of globular clusters in every stage of resolvability, and of clustering groups sufficiently insulated and condensed to come under the designation of 'clusters of stars.' . . . It is evident, from the intermixture of stars and unresolved nebulosity, which probably might be resolved with a higher optical power, that the nubeculæ are to be regarded as systems *sui-generis*, and which have no analogues in our hemisphere."

The smaller Magellanic Cloud, or Nubecula Minor, is fainter to the eye, and not so rich in the telescope. It covers about 10 square degrees, or about fifty times the area of the full moon. Sir John Herschel, in his "Cape Observations," describes it as "a fine large cluster of very small stars, 12 . . . 18 magnitude, which fills more than many fields, and is broken into many knots, groups, and straggling branches, but *the whole* (*i.e.*, the whole of the clustering part) is clearly resolved." It is surrounded by a barren region remarkably devoid of stars. Sir John Herschel says: "The access to the Nubecula Minor is on all sides through a desert." . . . "It is preceded at a few minutes in R. A. by the magnificent globular cluster, 47 Toucani (Bode), but is completely cut off from all connexion with it; and with this exception, its situation is in one of the most barren regions in the heavens." Herschel found the middle of the cloud clearly resolved into stars, while its edges remained irresolvable with his large reflector. He says: "The edge of the smaller *cloud* comes on as a mere nebula.

. . . We are now *in the cloud.* The field begins to be full of a faint light perfectly irresolvable. . . . I should consider about this place to be the body of the cloud which is here fairly resolved into excessively minute stars. . . . It is not like the stippled ground of the sky. The borders fade away, quite insensibly, and are less or not at all resolved." Herschel gives a catalogue of 244 objects in the Nubecula Minor. Of these about 200 are stars, and the remainder nebula and clusters. From this it appears that the smaller nubecula contains a much larger proportion of stars than the larger cloud.

Judging from their roughly globular form, the dimensions of the Magellanic Clouds are probably small compared with their distance from the earth, so that in these remarkable objects—particularly in the larger cloud—we see stars of the seventh, eighth, ninth, and tenth magnitude, apparently mixed up with fainter stars, and "clusters of all degrees of resolvability," and Sir John Herschel says: "It must therefore be taken as a demonstrated fact, that stars of the seventh or eighth magnitude, and irresolvable nebulæ, may co-exist within limits of distance not differing in proportion more than as 9 to 10."[1] It should be remembered, however, that possibly some of the fainter stars may—as in the Pleiades—lie far out in space beyond the greater Magellanic Cloud.

The Magellanic Clouds have recently been photographed by Mr. Russell at the Sydney Observatory. He finds the larger cloud—the Nubecula Major—to be of a most complex form, with evidence of a spiral structure, a feature also traceable, but not so clearly, in a photograph of the Nubecula Minor, or smaller cloud.

Dr. Dreyer's new index catalogue of recent discoveries of nebulæ, together with the general catalogue previously published, gives the position of 9,369 nebulæ.[2] A very small proportion of the new discoveries have been made by photography, and more than half of them were found by M. Javelle with the great refractor of the Nice Observatory. Most of the

[1] "Outlines of Astronomy," tenth edition, p. 657.

[2] *Nature*, November, 21, 1895.

new objects are very small and faint, and form probably "only a small portion of the number visible in large telescopes."

Several nebulæ have been suspected of variation in light.

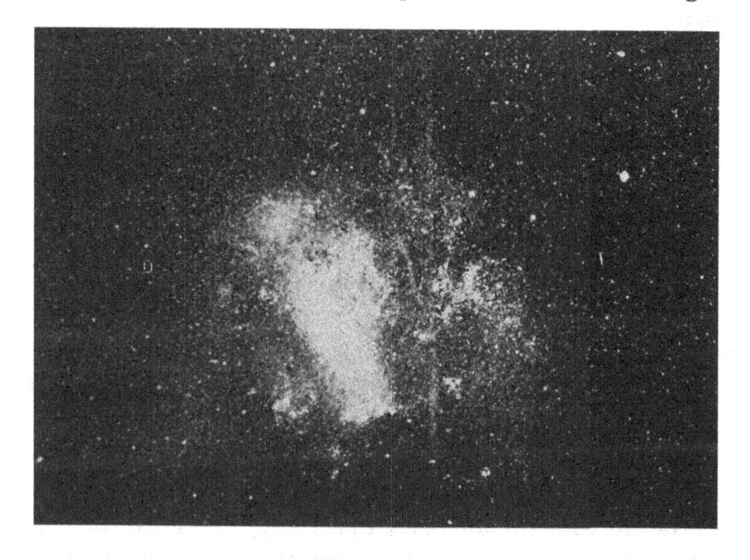

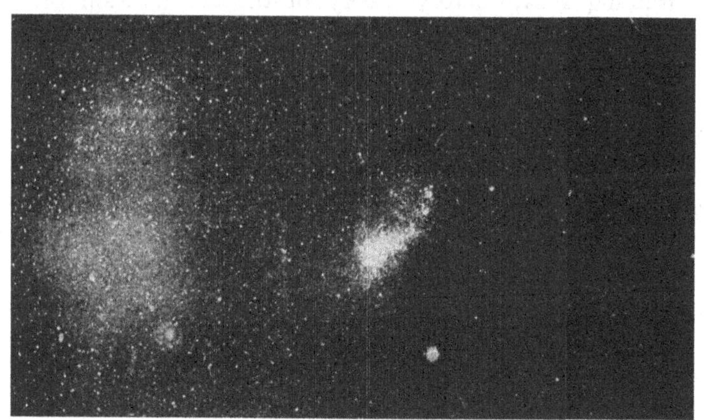

FIG. 18.—*Magellanic Clouds.*
(From "Worlds of Space.")

One discovered by Dr. Hind in 1852 near the variable star T Tauri was found to be an easy object with the great Lick telescope in February, 1895, but in September of the same year it had "entirely vanished." In the same instrument, "T Tauri was involved in a small hazy nebulosity, but the definite nebula in which it shone in 1890 did not exist in September, 1895."[1]

CHAPTER VII.

THE CONSTRUCTION OF THE HEAVENS.

THE construction of the visible universe is one of great interest, but of considerable difficulty. If we reflect that in viewing the starry heavens we are placed at the centre of a hollow sphere of indefinite extent, and that the distance of only a few of the stars from the earth has hitherto been ascertained with any approach to accuracy, the great difficulty of framing a satisfactory theory of the construction of the heavens will be easily understood.

In considering the subject, let us first inquire as to the probable number of stars visible in our largest telescopes. Are the visible stars infinite or limited in number? The reply to this question is easy. As the number of stars visible to the naked eye is limited, so the number of stars visible in the largest telescopes is limited also. Those who do not give the subject sufficient consideration seem to think that the number of the stars is practically infinite, or at least that the number is so great that it cannot be estimated. But this idea is totally incorrect, and due to complete ignorance of telescopic revelations. It is certainly true that, to a certain extent, the larger the telescope used in the examination of the heavens, the more the number of the stars seems to increase; but we now know that there is a limit to this increase of telescopic vision. And the evidence clearly shows that we

[1] *Nature*, January 16, 1896.

are rapidly approaching this limit. Although the number of stars visible in the Pleiades rapidly increases at first with increase in the size of the telescope used, and although photography has still further increased the number of stars in this remarkable cluster, it has recently been found that an increased length of exposure—beyond three hours—adds very few stars to the number visible on the photograph taken at the Paris Observatory in 1885, on which over 2,000 stars can be counted. Even with this great number on so small an area of the heavens, comparatively large vacant spaces are visible between the stars, and a glance at the original photograph is sufficient to show that there would be ample room for many times the number actually visible. I find that, if the whole heavens were as rich in stars as the Pleiades, there would be only 33 millions in both hemispheres.

On a photograph of the region surrounding Gamma Cassiopeiæ, taken by Dr. Roberts in December, 1895, with a reflecting telescope of 20 inches aperture, and an exposure of two hours and twelve minutes, he finds 17,100 stars on an area of four square degrees. This would give for the whole area of the heavens—if equally rich in stars—a total of about 176 millions; but Gamma Cassiopeiæ lies in a rich region of the Milky Way, and probably the great majority of the stars shown on Dr. Roberts' photograph belong to the Galaxy, which we know to be especially rich in stars. One thing is certain, that the heavens as a whole are not nearly so rich as this particular spot. There may, perhaps, be richer spots elsewhere in the Milky Way, but in other parts of the sky there are many regions considerably poorer.

Let us consider a still more extreme case of stellar richness. On a photograph of the great globular cluster, Omega Centauri, recently taken in Peru, a count of the stars has been carefully made by Professor and Mrs. Bailey, and, as stated in the last chapter, the number of stars contained in the cluster may be taken as 10,000. Now, if the whole sky were as thickly studded with stars as in this cluster, the total number visible in the whole heavens would be 1,650 millions, a very large number, of course, but not much in excess of the present population of the earth, and I am not aware that the number of the earth's inhabitants has ever been described as "infinite."

Clusters, such as the Pleiades and Omega Centauri, are, of course, remarkable, and rare exceptions to the general rule of stellar distribution, and the heavens in general are not—even in the richest portions of the Milky Way—nearly so rich in stars as the globular clusters. The fact of these clusters being remarkable objects, proves that they are unusually rich in stars, and there is strong evidence—evidence amounting to absolute proof in the case of the globular clusters—that these collections of stars are really, and not apparently, close, and that they are actually systems of suns, and occupy a comparatively limited volume in space. We cannot, then, estimate the probable number of the visible stars by counting those visible in one of the globular clusters.

That the number of the visible stars will not probably be largely increased by any increase in telescopic power, is indicated by the fact that Celoria, using a small telescope, of power barely sufficient to show stars to the eleventh magnitude, found that he could see almost exactly the same number of stars near the north pole of the Milky Way as were visible in Sir William Herschel's great telescope! thus indicating that, here at least, no increase of optical power will materially increase the number of stars visible in that direction; for Herschel's large telescope certainly showed far fainter stars than those of the eleventh magnitude in other portions of the heavens. It should therefore have shown fainter stars at the pole of the Milky Way also, if such stars existed in that region of space. Their absence, therefore, seems certain proof that very faint stars do *not* exist in that direction, and that, here at least, our sidereal universe is limited in extent. A photograph, taken by Dr. Roberts not very far from the spot in question, shows only 178 stars to the square degree. This rate of distribution would give a total of only 7,343,000 stars for both hemispheres!

An examination by Miss Clerke of Professor Pickering's catalogue of stars surrounding the north pole of the heavens shows that "the small stars are overwhelmingly too few for the space they must occupy, if of average brightness; and they are too few in a constantly increasing ratio."[1] Here again, a "thinning out" of the stellar hosts seems clearly indicated, and suggests that a limit will soon be reached, beyond which our most powerful telescopes and photographic plates will fail to reveal any further stars.

Let us now consider the number of stars actually visible. Maps of the northern portion of the heavens have been published by Argelander and Heis, and charts of the southern sky by Behrmann and Gould. Heis shows stars to about magnitude 6⅓, and Behrmann to about the same brightness. I find that the total number shown by both observers, as visible to the naked eye, is 7,249. The total number, to the sixth magnitude inclusive, shown by both observers, is 4,181. Argelander gives 5,000 stars to the sixth magnitude inclusive, and for stars to the ninth magnitude, the following numbers in each magnitude:—First magnitude, 20; second magnitude, 65; third magnitude, 190; fourth magnitude, 425; fifth magnitude, 1,100; sixth magnitude, 3,200; seventh magnitude, 13,000; eighth magnitude, 40,000; and ninth magnitude, 142,000, or a total of "200,000 for the entire number of stars from the first to the ninth magnitude inclusive."[2] This result agrees closely with an estimate previously made by Struve. From a formula given by Dr. Gould, deduced from observations in the Southern Hemisphere, I find the number of stars to the ninth magnitude inclusive would be 215,674, so that Argelander's estimate of 200,000 stars to the ninth magnitude inclusive cannot be far from the truth. It will be seen from Argelander's figures that the number of stars in each class of magnitude is roughly three times that in the class one magnitude brighter. Supposing this progressive increase continued to the seventeenth magnitude—the faintest

[1] *Nature*, August 9, 1888.

[2] Humboldt's "Cosmos," Bohn's edition, vol, iii., p. 143.

visible in the great Lick telescope—I find that the total number of stars would be nearly 1,400 millions, or less than the number found from a consideration of the cluster Omega Centauri. But it is evident from Celoria's observation, referred to above, and from Professor Pickering's photographs of stars near the North Pole, that the fainter stars do *not* increase in the ratio assumed above. We must therefore conclude that there is a "thinning out" of the fainter stars at some point below the ninth magnitude. Taking into consideration the rich regions of the Milky Way, and the comparatively poor portions of the sky, it is now generally admitted by astronomers, who have studied this particular question, that the probable number of stars visible in our largest telescopes does not exceed 100 millions, a number which, large as it absolutely is, may be considered as relatively very small, and even utterly insignificant, when compared with an "infinite number."

Let us see what richness of stellar distribution is implied by this number of 100 millions of visible stars. It may be easily shown that the area of the whole sky, in both hemispheres, is 41,253 square degrees, or about 200,000 times the area of the full moon. This gives 2,424 stars to the square degree. The moon's apparent diameter being slightly over half a degree (31 5″), the area of its disc is about one-fifth of a square degree. Hence, for 100 millions of stars in the whole star sphere, we have 485 stars to each space of sky, equal in area to the full moon. This seems a large number, but stars scattered even as thickly as this would appear at a considerable distance apart when viewed with a large telescope and a high power. As the area of the moon's disc contains about 760 square minutes of arc, there would not be an average of even one star to each square minute. A pair of stars half a minute, or 30 seconds, apart, would form a very wide double star, and with stars placed at even this distance, the moon's disc would cover about 3,000, or over six times the actual number visible in the largest telescopes. In Dr. Roberts' photograph of the region surrounding Gamma

Cassiopeiæ, which shows over 17,000 stars, on four square degrees, or over 4,000 stars to the square degree, the stars do not seem very crowded, and there is a good deal of black sky visible between them.

But, in addition to the conclusive evidence as to the limited number of the visible stars derived from actual observation and the results of photography, we have indisputable evidence from mathematical considerations that the number of the visible stars *must necessarily* be limited. For were the stars infinite in number, and scattered through infinite space with any approach to uniformity, it may be proved that the whole heavens would shine with the brightness of the sun. As the surface of a sphere varies as the square of its radius, and light inversely as the square of the distance (or radius of the star sphere at any point), we have the diminished light of the stars exactly counterbalanced by the increased number at any given distance. For a distance of say ten times the distance of the nearest fixed star, the light of each star would be diminished by the square of 10 or 100 times, but the total number of stars would be 100 times greater, so that the total star light would be the same. This would be true for *all* distances. The total light would therefore—by addition—be proportional to the distance, and hence, for an infinite distance we should have an infinite amount of light. For an infinite number of stars, therefore, we should have a continuous blaze of light over the whole surface of the visible heavens. Far from this being the case, the amount of light afforded by the stars on the clearest nights is, on the contrary, comparatively small, and the blackness of the background, "the darkness behind the stars," is very obvious. According to Miss Clerke ("System of the Stars," p. 7), the total light of all the stars, to magnitude 9½, is about one-eightieth of full moonlight. M. G. l'Hermite found for the total amount of starlight one-tenth of moonlight; but this estimate is evidently too high. Assuming the sun's brightness as 28 magnitudes brighter than a star of the first magnitude,[1] and Zöllner's estimate that sunlight is

[1] See *Knowledge*, June, 1895.

618,000 times that of moonlight, I find that the total light of the stars to magnitude 9½, as stated by Miss Clerke, would be equivalent to the combined light of about 320,000 stars of the sixth magnitude, or 3,200 stars of the first magnitude. Even taking M. l'Hermite's high estimate of one-tenth of moonlight, the total starlight would be represented by 25,600 stars of the first magnitude.

To explain the limited number of the visible stars, several hypothesis have been advanced. If space be really infinite, as we seem compelled to suppose, it would be reasonable to expect that the number of the stars would be practically infinite also. But, as I have shown above, the number of the *visible* stars is certainly finite, and the number visible and invisible must be finite also, for otherwise the amount of starlight would be much greater than it is. To account for the limited number of visible stars, it has been suggested that beyond a certain distance in space, there may be an "extinction of light," caused by absorption in the luminiferous ether. In a recent paper on this subject, Schiaparelli, the famous Italian astronomer, suggests that if any extinction of light really takes place, it may probably be due, not to absorption in the ether, but to fine particles of matter scattered through interstellar space. In support of this hypothesis, he refers to the supposed constitution of comets' tails, of falling stars, and meteorites, and he shows that the quantity of matter necessary to produce the required extinction would be very small—so small, indeed, that a quantity of this matter scattered through a volume equal to that of the earth, if collected into one mass, would only form a ball of less than one inch in diameter. We can readily admit the existence of such a minute quantity of matter in a fine state of sub-division scattered through space, but it seems to me much more probable that the limited number of the visible stars is due, not to any extinction of their light by absorption in the ether, or by fine particles scattered through space, but to a real thinning out of the stars as we approach the limits of our sidereal universe. Celoria's observation, mentioned above, seems to prove that near the pole of the Milky Way very few stars fainter than the eleventh magnitude are visible, even in a large telescope, and Dr. Roberts' photographs, taken in the vicinity of the celestial pole, confirm this conclusion. Now, this paucity of stars of the fainter magnitudes cannot be due to any absorption of light in the ether, for numerous stars of the sixteenth magnitude, or perhaps fainter, are visible in other parts of the heavens, and if in one place, why not in another? Sir John Herschel's observations of the Milky Way in the Southern Hemisphere appear to render the hypothesis of any extinction of light very improbable. He says that the hypothesis, "if applicable to any, is equally so to every part of the Galaxy. We are not at liberty to argue that at one part of its circumference our view is limited by this sort of cosmical veil, which extinguishes the smaller magnitudes, cuts off the nebulous light of distant masses, and closes our view in impenetrable darkness; while at another we are compelled, by the clearest evidence telescopes can afford, to believe that star-strewn vistas *lie open*, exhausting their powers, and stretching out beyond their utmost reach, as is proved by that very phænomenon which the existence of such a veil would render impossible, *viz.*, infinite increase of number and diminution of magnitude, terminating in complete irresolvable nebulosity."

How then are we to explain the limited number of the visible stars? If space be infinite, as we seem compelled to suppose, the number of the stars would probably be infinite also, or at least vastly greater than the number actually visible. It has been suggested that, owing to the progressive motion of light, the light of very distant stars may probably not yet have reached the earth, although travelling through space for thousands of years. But considering the vast periods of time during which the stellar universe has probably been in existence, this hypothesis seems very unsatisfactory. The most probable hypothesis seems to be that all the stars, clusters and nebulæ, visible in our largest telescopes, form together one vast system, which constitutes our visible universe, and that this system is isolated by a starless void from other similar systems which probably exist in infinite space. The distance between these separate systems—or "island universes," as they have been called—may be very great, compared with the diameter of each system, in the same way that the diameter of our visible universe is very great compared with the diameter of the solar system. As the sun is a star, and the stars are suns, and as our sun is separated from his neighbour suns in space by a sunless void, so may our universe be separated from other universes by a vast and starless abyss. On this hypothesis, the supposed extinction of light—which may have little or no perceptible effect within the limits of our visible universe—may possibly come into play across the vast and immeasurable distances which probably separate the different universes from each other, and may perhaps extinguish their light altogether.

Another hypothesis which also seems possible is that the luminiferous ether which extends throughout our visible universe may perhaps be confined to this universe itself, and that beyond its confines, the ether may thin out, as our atmosphere does at a certain distance from the earth, and finally cease to exist altogether, ending in an *absolute* vacuum, which would, of course, arrest the passage of all light from outer space, and thus produce " the darkness behind the stars."

Let us now consider the apparent distribution of the stars and nebulæ on the celestial vault, and their probable relation to each other in space. As already stated, Argelander considered the number of stars of the first magnitude to be about twenty, but modern photometric measures have reduced this number to thirteen or fourteen. According to the Harvard measures, the fourteen brightest stars in the heavens, in order of magnitude, are: Sirius, Canopus, Arcturus, Capella, Vega, Alpha Centauri, Rigel, Procyon, Achernar, Beta Centauri, Betelgeuse, Altair, Aldebaran and Alpha Crucis. Seven of these are in the Northern Hemisphere, namely: Arcturus, Capella, Vega, Procyon, Betelgeuse, Altair, and Aldebaran; and seven in the Southern Hemisphere: Sirius, Canopus, Alpha Centauri, Rigel, Achernar, Beta Centauri, and Alpha Crucis, so that the brightest stars are pretty evenly distributed between the two hemispheres. Of these bright stars, no less than twelve lie in or near the Milky Way, Arcturus and Achernar being the only two at any considerable distance from the Galaxy. This is very remarkable and suggestive, as the area covered by the Milky Way is probably not more than one-fourth of the whole star sphere.

Of the stars fainter than the first magnitude, but brighter than magnitude 2·0, there are about 10 in the Northern Hemisphere, of which 4 lie in or near the Milky Way, and about 19 in the Southern Hemisphere, of which no less than 14 are situated in or near the Galaxy.

Of those brighter than magnitude 3·0, I find 33 stars in or near the Milky Way out of a total of about 95 in both hemispheres. To extend this investigation to all stars visible to the naked eye, I made, some years since, an examination of all the stars in Heis' atlas that lie in the Milky Way, and found that number to be 1,186 out of a total of 5,356, or a percentage of about 22. At my request, Col. Markwick, F.R.A.S., made a similar count for the stars in Dr. Gould's charts of the Southern Hemisphere (*Uranometria Argentina*), and found that, down to the fourth magnitude, there are 121 stars on the Milky Way out of 228, or a percentage of 53, and for all stars to the seventh magnitude inclusive, there are 3,072 on the Milky Way out of a total of 6,694, or a percentage of nearly 46. Col. Markwick finds that the Milky Way in the Southern Hemisphere, as shown on Gould's charts, covers about one-third of the whole hemisphere. As will be seen by the above figures, the percentage of stars, even to the fourth magnitude, lying on the Milky Way is considerably greater than this proportion.

The above results show that the brighter stars which are apparently projected on the Milky Way probably belong to that zone, and are not merely fortuitously scattered over the surface of the heavens.

To extend the investigation still further, and include stars to the eighth magnitude, I made an examination of the stars shown on Harding's charts to that magnitude, in a zone of 30° in width—15° degrees on each side of the Equator—and found a marked increase in the number of stars where the zone crossed the Milky Way. The numbers per hour of Right Ascension varied from a minimum of 275 (hours I. and II.) to maxima of 601 in the Milky Way in Monoceros, and 611 in the Galaxy in Serpens and Aquila. A valuable investigation by the late Mr. Proctor went further still. He plotted all the stars shown in the charts of Argelander's *Durchmasterung*, which contains stars to 9½ or 10th magnitude. In this remarkable chart the course of the Milky Way is clearly defined by a marked increase of stellar density. Proctor says: "In the very regions where the Herschelian gauges showed the minutest telescopic stars to be most crowded, my chart of 324,198 stars shows the stars of the higher orders (down to the eleventh magnitude) to be so crowded that, by their mere aggregation within the mass, they show the Milky Way with all its streams and clusterings. This evidence, I venture to affirm, is altogether decisive as to the main question, whether large and small stars are really intermixed in many regions of space, or whether the small stars are excessively remote. It is utterly impossible that excessively remote stars could seem to be clustered exactly where relatively near stars are richly spread. This might happen, no doubt, in a single instance; but that it could be repeated over and over again, so as to account for all the complicated features seen in my chart of 324,198 stars, I maintain to be utterly incredible."[1]

From a careful examination of the Milky Way in Aquila and Cygnus, Mr. Easton finds that "(1) In the zones considered, the distribution of stars down to 9·5 magnitude corresponds to the greater or less intensity of galactic light. (2) There is a real correspondence of the general outlines of the galactic forms with the distribution of 11 magnitude stars, and with those of stars between 10 and 15 magnitude. (3) Thus, in general, for the zones considered, the faint stars which form the Milky Way are thickly or sparsely scattered

[1] "The Universe and the Coming Transits," p. 200.

in respectively the same regions as the stars in Argelander's last class; it follows, therefore, with a great degree of probability, that there is a real connexion between the distribution of 9 and 10 magnitude stars and that of the very faint stars of the Milky Way. Consequently, the very faint stars are at a distance which does not greatly exceed that of 9-10 magnitude stars. If stars of 13-15 magnitude were at their theoretical distance, there would be no reason why they should have the same apparent distribution in galactic latitude and longitude as 9-10 magnitude stars, separated from them by enormous intervals."[1]

There are some regions in both hemispheres especially rich in naked eye stars. Of these the following may be mentioned in the Northern Hemisphere:—the region including the Pleiades, and Hyades in Taurus, the Northern portion of Orion, and the adjoining part of Gemini, the constellation Lyra, the northern portion of Cygnus, Cassiopeia's Chair, and Coma Berenices. In the Southern Hemisphere there are several rich spots. A rich region extends from Canis Major to the Southern Cross, and nearly coincides with the course of the Milky Way. The richest spot of all, and perhaps the richest in the whole heavens in naked eye stars—with exception of the Pleiades—is that including the Southern Cross. This spot has an average of three stars to five square degrees, and if the whole heavens were as richly studded with stars there would be about 24,000 visible to the naked eye! The poverty of the adjoining "coal sack" is very remarkable. Another rich spot surrounds the variable star Eta Argûs, and the great nebula in Argo. There is another rich spot in the constellation Hydrus, not far from the greater Magellanic Cloud, and another will be found in Centaurus and Lupus, with its centre about Alpha of the latter constellation. According to Gould's maps of the Southern Hemisphere, the richest region in stars down to the seventh magnitude is the southern portion of that part of the constellation Argo, known as Puppis.

[1] *Journal of the British Astronomical Association*, May, 1895, p. 383.

In contrast to these rich regions, and in many cases closely adjoining them, are some barren regions, very poor in naked eye stars. For example, closely following the rich spot in Cassiopeia and between Iota Cassiopeiæ and Eta Persei is a remarkably poor spot, where a space of some sixty square degrees does not contain a single star brighter than the sixth magnitude! There is another poor region south of Alpha Hydræ, and another in the southern portion of the constellation Cetus.

A region of considerable extent, remarkably deficient in bright stars, will be noticed in the Northern Hemisphere. This comparatively barren region, which contains no star brighter than the fourth magnitude, is bounded by Cepheus, Cassiopeia, Perseus, Auriga, Gemini, Ursa Major, Draco, and Ursa Minor, and forms a conspicuous feature in the north-eastern portion of the sky in the early winter evenings. It will be noticed that the surrounding constellations all contain bright stars.

Whether the apparent crowding of stars in certain regions of the heavens is caused by a real proximity in space, or whether it is merely due to their being placed accidentally in the line of sight, is a question difficult to determine. In the case of star clusters, and especially the globular clusters, there is a high mathematical probability, amounting almost to absolute certainty, that they are comparatively close together, but in groups scattered over a considerable area, like those referred to above, the probability in favour of proximity is not so great. As we know the distance of so few stars from the earth, it is impossible to say whether the crowding is real or only apparent, but the probability seems to be that it is to some extent real.

A tendency to an arrangement of stars in streams was pointed out by Proctor in his "Universe and the Coming Transits." This tendency to stream formation may be noticed on a large scale among the naked eye stars, for example, in Pisces, Scorpio, the River Eridanus, Aquarius, and the festoon of stars in Perseus. In some of these cases, of course, the stars are so far apart that the formation may be more apparent than real, but the tendency can also be clearly recognised among the fainter stars, and even among those only visible in telescopes and stellar photographs. This tendency to run in streams is well marked on the photographs taken at the Paris Observatory, and on those taken by Professor Barnard, Dr. Max Wolf, and others. It is a suggestive fact that these star streams are also very noticeable in star clusters, where there can be little or no doubt of a physical connexion between the component stars. With reference to a photograph of the southern portion of Aquila taken by Dr. Max Wolf in July, 1892, the late Mr. Ranyard, remarked: "Some of the streams of fainter stars in this region are very striking, and must convince the most sceptical of their reality. It is possible to draw an arc of a circle through any three stars, and a conic section through any five; but where we find ten or twenty stars falling into line, not once, but in many cases, and that there is a curious similarity between the strange curves and branching streams which these phalanges of stars mark out on the heavens, there is no room left for doubt that the mind is not being led away by a tendency of the imagination similar to that which finds faces in the fire, or sees a man carrying sticks on the face of the moon. If it is proved that a group of stars is arranged in line or marshalled in any order, it would follow that the individuals of the group must be actually as well as apparently close to one another, and that they form some kind of system, having all of them had a common origin, or been subject to some common influence." [1]

The great majority of the star clusters are found along the course of the Milky Way, while the irresolvable nebulæ seem to congregate towards the poles of the galactic zone.

Dr. Gould is of opinion that "a belt or stream of bright stars appears to girdle the heavens very nearly in a great circle, which intersects the Milky Way at about the points of its highest declination, and forms with it an angle not far from 20°; the southern node being near the margin of the

[1] *Knowledge*, May, 1896.

Cross, and the northern in Cassiopeia." According to Gould, this belt covers Orion, Canis Major, Columba, Puppis, Carina, the Southern Cross, Centaurus, Lupus, and the head of Scorpion in the Southern Hemisphere, its northern course being indicated by the brightest stars in Taurus, Perseus, Cassiopeia, Cepheus, Cygnus, and Lyra. Dr. Gould considers that our sun may possibly be a member of this belt of stars, which perhaps numbers less than 500, and which constitute "a small cluster, distinct from the vast organisation of that which forms the Milky Way, and of a flattened and somewhat bifid form. The southern portion of this supposed stream of bright stars had been previously recognised by Sir John Herschel, who says in his 'Cape Observations,' (p. 385), 'It is about this region, or, perhaps, somewhat earlier, in the interval between η Argus and α Crucis, that the galactic circle, or medial line of the Milky Way may be considered as crossed by that zone of large stars, which is marked out by the brilliant constellation of *Orion*, the bright stars of Canis Major, and almost all the more conspicuous stars of *Argo*, the Cross, the Centaur, Lupus, and *Scorpion*. A great circle passing through ϵ Orionis and α Crucis will mark out the axis of the zone in question, whose inclination to the galactic circle is, therefore, about 20°, and whose appearance would lead us to suspect that our nearest neighbours in the sidereal system (if really such) form part of a subordinate sheet or stratum deviating to that extent from parallelism to the general mass which, seen projected on the heavens, forms the Milky Way.'"

These conclusions might seem probable enough when we compare the supposed zone of bright stars with the very diagrammatic drawings of the Milky Way as shown in many star maps; but when we consider the stars referred to with reference to the more artistic and accurate delineations of the Milky Way as drawn by Boeddicker, and even by Gould himself, we see that most of them are involved in the milky light of the Galaxy, and their connection with the Milky Way itself seems quite as probable as that they form a belt distinct from the galactic zone. The apparent connexion of the stars in question with the Milky Way does not, however, disprove the existence of Dr. Gould's belt or zone of bright stars. If the plane of the supposed belt nearly coincided with that of the Milky Way, the apparent connexion might not be real.

Mr. J. R. Sutton advances the theory[1] that the Milky Way consists of "a great ring of large stars"—Dr. Gould's solar cluster above referred to—"intersecting an equal ring of small ones (the Milky Way) at the extremities of a common diameter." He considers that "the great star belt is a genuine girdle of stars in space, in which also the foundations of the sidereal system are laid, the Milky Way being an appendant to it of lesser rank."

That the Milky Way really forms a ring of stars in space there is strong evidence to show. Sir William Herschel's original theory that the galactic gleam is due to our sun being situated near the centre of an indefinite stratum of stars—the "disc theory," as it is termed—was abandoned by its illustrious author in his later writings, and is now considered to be wholly untenable by nearly all astronomers who have studied the subject. Sir John Herschel remarks that the general aspect of the galaxy near the Southern Cross indicates "that the Milky Way, in this neighbourhood, at any rate, is really what it appears to be, a belt or zone of stars separated from us by a starless interval." It certainly seems utterly improbable that the nearly circular blank space near the Southern Cross, known as "the coal sack," should represent a tunnel through a disc, of which the thickness is comparatively small, while its diameter, on the "disc theory," stretches out almost to infinity. A straight, tunnel-shaped opening of great length, pointing directly towards the earth, would form an extraordinary phenomenon even in a solitary instance ; yet there are several somewhat similar openings to be found in the Milky Way, as viewed both with the naked eye and with a telescope. That *all* these openings should represent tunnels radiating from a common centre is quite beyond the bounds of probability, and, indeed, such an hypothesis does not deserve serious considera-

[1] *Knowledge*, July, 1891.

tion. With reference to a photograph of the Milky Way in the constellation Cepheus, Professor Barnard says, "the sky (or Milky Way) is broken up into numerous black cracks or crevices. Looking at these peculiar features, I cannot well see how one can avoid the conclusion that they are necessarily real vacancies in the Milky Way, through which we look out into the blackness of space."[1] Using a telescope with a low power, Mr. S. M. Baird Gemmill says, "December 1, 1886. In sweeping over the constellation of Monoceros, I was much struck with the reticulated character of the arrangement of the brighter stars upon the glimmering background, and the way in which this background seemed to follow the reticulation. By 'brighter stars' are meant stars of from 8 to 10 magnitude, for it was among these that I noticed this peculiarity of arrangement. It put me in mind of M. M. Henry's photographs of Cygnus. The region seemed, in fact, a vast network of stars, the reticulations of which were separated by desert, or comparatively desert spaces."[2] I have noticed the same thing myself while examining the Milky Way with a binocular field-glass. On October 26, 1889, I noted as follows: "North of Alpha Cygni, and near Xi and Nu Cygni, the nebulous light of the Milky Way seems to cling round and follow streams of small stars in a very remarkable way; numerous small 'coal sacks' and rifts are visible, in which comparatively few stars are to be seen with the binocular." This observation has been fully confirmed by photographs of this region, taken by Dr. Max Wolf in 1891.

That the Milky Way is not indefinitely extended in the line of sight seems clearly shown by Sir John Herschel's observations in the Southern Hemisphere. In his "Outlines of Astronomy" (p. 578), he says: "When examined with powerful telescopes, the constitution of this wonderful zone is found to be no less various than its aspect to the eye is irregular. In some regions, the stars of which it is wholly composed are scattered with remarkable uniformity over

[1] *Knowledge*, January, 1894, p. 17.

[2] *Journal of the British Astronomical Association*, April, 1895, p. 304.

immense tracts, while in others the irregularity of their distribution is quite as striking, exhibiting a rapid succession of closely clustering rich patches, separated by comparatively poor intervals, and indeed, in some instances, by spaces absolutely dark *and completely void of any star*,[1] even of the smallest telescopic magnitude. . . . In some, for instance,

FIG. 19.—*Photograph of Milky Way, Sagittarius.*
(From "Visible Universe.")

extremely minute stars, though never altogether wanting, occur in numbers so moderate, as to lead us irresistibly to the conclusion that, in those regions, we see *fairly through* the starry stratum, since it is impossible otherwise (supposing their light not intercepted), that the members of the smaller magnitude

[1] The Italics are Herschel's.

should not go on increasing *ad infinitum*. In such cases, moreover, the ground of the heavens, as seen between the stars, is for the most part perfectly dark, which again would not be the case if innumerable multitudes of stars, too minute to be individually discernible, existed beyond. In other regions we are presented with the phænomenon of an almost uniform degree of brightness of the individual stars, accompanied with a very even distribution of them over the ground of the heavens, both the larger and smaller magnitudes being strikingly deficient. In such cases it is equally impossible not to perceive that we are looking *through* a sheet of stars nearly of a size and of no great thickness compared with the distance which separates them from us. Were it otherwise, we should be driven to suppose the more distant stars uniformly the larger, so as to compensate by their greater intrinsic brightness for their greater distance, a supposition contrary to all probability. In others again, and that not unfrequently, we are presented with a double phænomenon of the same kind, *viz.*, a tissue, as it were, of large stars spread over another of very small ones, the intermediate magnitude being wanting. The conclusion here seems equally evident that in such cases we look through two sidereal sheets separated by a starless interval."

An examination of the evidence at present available, with reference to the distribution of the visible stars in space, has recently been undertaken by Professor Kapteyn of Groningen, and an account of the conclusions he has arrived at may prove of interest to the reader.

We must first explain that in order to obtain a clear view of the construction of the visible universe, it would be necessary to know the relative distances of a large number of stars; but as the distances of only a few stars from the earth have yet been determined by actual measurement, and the results hitherto obtained are open to much uncertainty, we must have recourse to some other method of estimating the distances. While travelling in a railway carriage, if we fix our attention on trees, buildings, and other objects we pass on our journey,

Fig. 20.—*The Milky Way.*
(From *Knowledge*, Nov., 1894.")

it will be noticed that all objects apparently move past us in the opposite direction to that in which we are travelling, and that the nearer the object is the faster it seems to move with reference to distant objects near the horizon. So it is with the stars. As we showed in Chapter III., the sun is moving through space, carrying along with the earth all the planets, satellites, and comets, forming the solar system. The effect of this motion is to cause an apparent small motion of the stars in the opposite direction, and the nearer the star is to the earth, the greater will this apparent motion seem to be as in the case of the railway train. In addition to this apparent motion, the stars are themselves—like the sun—moving through space, and this *real* motion is also visible. If this real motion takes place in the *opposite* direction to that in which the sun and earth are moving, it will add to the apparent motion, and will increase the star's "proper motion," as it is termed. If, on the other hand, the real motion is in the *same* direction as the earth's motion, the proper motion will be diminished. In either case, the nearer the star is to the earth, the greater will be its apparent annual displacement on the background of the heavens. The amount of the "proper motion" is, therefore, considered by astronomers to form a reliable criterion of the star's distance from the earth, and the actual measures of distance which have been made show that this assumption is approximately true. Of fourteen stars which have proper motion of over three seconds of arc per annum, eleven have yielded a measurable parallax, or displacement, due to the earth's annual motion round the sun; that is to say, eleven out of fourteen fast-moving stars are within a measurable distance of the earth, and are, therefore, near us, when compared with the great majority of stars which are not within measurable distance, or, at least, are beyond the reach of our present methods of measurement.

In the case of small groups of stars, we may assume that the real motions of the individual stars take place indifferently in all directions, and that consequently, taking an average of all the motions of the stars composing the group, the effects due to the real motions will destroy each other, and there will remain, as the most reliable criterion, the effect due to the sun's motion in space. If, however, we compare the proper motions of groups situated in *different parts* of the sky, there is a consideration which, to a great extent, vitiates this conclusion. For, near the point of the heavens, towards which the sun and earth are moving, known as the "apex of the solar way," and probably situated not far from the bright star Vega, as indicated by recent researches, and near the point away *from* which the sun is moving known as the *ant-apex*, about 15° south of Sirius, there will be no apparent displacement due to the solar motion through space, as this motion takes place in the line of sight with reference to these points of the sky. The observed proper motion at these points will, therefore, be solely due to the real motions of the stars themselves in those regions. In other parts of the heavens, however, the total proper motion will be a combination of the apparent and real motions of the stars, and for stars in different parts of the sky, it will not follow that stars having equal proper motions are necessarily at the same distance from the earth. To make this point clearer, let us suppose that there are two stars at absolutely the same distance from the earth, one situated at or near the solar "apex," and the other at a point 90° from the apex, and let us suppose that both stars are moving through space with exactly the same velocity and in the same direction, say at right angles to the direction of the solar motion. Then in the case of the star near the apex, the observed "proper motion" will be solely due to the star's *real motion*, and in the star 90° distant from the apex, the proper motion will be solely due to the solar motion, as the star's real motion, being in the line of sight, will not be visible. Now, unless the stellar motion and the solar motion happen to be equal, the observed "proper motions" will not be equal, although both stars are at the same distance from the earth. If both the stars are really at rest, the star at the apex will have no proper motion, while the star 90° distant will have an apparent proper motion due to the sun's motion. To overcome this source of error in estimating the distance of a star from its proper motion, Professor Kapteyn made use of another measure, which is independent of the solar motion. This is the component of the proper motion measured at right angles to a great circle of the sphere passing through a star and the solar apex. The amount of motion in this direction will evidently not be affected by the sun's motion, and from a discussion of the stars, contained in the Draper "Catalogue of Stellar Spectra," which were observed by Bradley (and of which the proper motions are now known with accuracy), Professor Kapteyn finds that this motion is "nearly inversely proportional to the distance," that is, the greater the motion, the less the distance of the stars, and the smaller the motion, the greater the distance. Excluding stars with proper motions greater than half a second of arc per annum, Professor Kapteyn found that for stars at various distances from the Milky Way this component of the "proper motion" forms a good measure of distance.

As the result of his investigations on the subject, Professor Kapteyn arrives at the following conclusions. Neglecting stars with small or imperceptible proper motions, we have a group of stars which no longer show any condensation in a plane. Stars with very small or no proper motions show a condensation towards the plane of the Milky Way. This applies to stars of the second or solar type, as well as to those of the first or Sirian type of spectrum, and evidently indicates that the stars composing the Milky Way lie at a great distance from the earth. The extreme faintness of the majority of the stars composing the Galaxy seems in favour of this conclusion. The condensation of stars of the first type is more marked than those of the second, and this agrees with the fact which has been noticed by Professor Pickering, that the majority of the brighter stars of the Milky Way have spectra of the Sirian type.

Professor Kapteyn finds that this condensation of stars with small proper motions is very perceptible even for stars visible to the naked eye, and is as well marked in those stars which have spectra of the second type as for all the stars of the ninth magnitude; but for stars of the first type the condensation is still more marked. He considers that this condensation is either partly real, or that there is a real thinning out of stars near the pole of the Milky Way. As already mentioned (in the beginning of this chapter), Celoria's observations with a small telescope, compared with Sir William Herschel's observations with a large telescope, indicate clearly that there *is a real thinning out* of stars near the poles of the Galaxy.

Professor Kapteyn concludes that the arrangement of the stars suggested by Struve—a modification of the "disc theory"—has no real existence.[1] He attributes the fallacy in Struve's hypothesis to the fact that the mean distance of stars of a given magnitude in the Milky way, and outside it, is not the same.

Professor Kapteyn finds that the vicinity of the sun is almost exclusively occupied by stars of the second or solar type, a conclusion which evidently tends to strengthen Dr. Gould's theory of a "solar cluster." He finds that the number of Sirian type stars increases gradually with the distance, and that beyond a distance corresponding to a proper motion of about $\frac{1}{14}$th of a second of arc per annum, the Sirian stars largely predominate. In the group of stars known as the Hyades, however, the components of which have a common proper motion both in amount and direction, stars of the first and second types appear to be mixed, and Professor Kapteyn assumes that the two types represent different phases of evolution, and that as the brightest stars of the group are chiefly of the solar type, these stars must be the largest of the group. From this fact he concludes the solar type stars are in a less advanced stage of evolution than those of the Sirian type. This does not agree with the generally accepted view. Professor Vogel considers the Sirian stars to represent an earlier stage of stellar evolution. Mr. Proctor held the same opinion, and in Professor Lockyer's hypothesis of increasing and decreasing temperatures in stars of various types, he

[1] A full discussion of Struve's views will be found in Chapter XVI. of "The Visible Universe," by the present writer.

places the Sirian stars at the summit of the evolution curve, and the sun and solar stars just below them on the descending branch of the curve.[1] These hypotheses are in conformity also with the current opinion that the sun is a cooling body. The discrepancy may perhaps be explained by supposing that the *brighter* stars of the Hyades form a connected group, and that some, at least, of the fainter stars do not belong to the group, but lie at a great distance behind it. In the case of the Pleiades, which form a more evident cluster, I find from the Draper "Catalogue of Stellar Spectra" that the great majority of the brighter stars have spectra of the Sirian type. Most of the stars in the Pleiades have a very similar proper motion, both in amount and in direction, and there can be no doubt that most of the brighter stars, at least, form a connected system. As already stated, it seems highly probable that the fainter stars in the Pleiades lie far beyond the brighter components, and have merely an optical connexion with them, and the same may be the case in the Hyades. The superior brilliancy of the stars composing the Hyades would suggest that they are nearer to the earth than the Pleiades group, and they may possibly form members of Gould's "solar cluster."

Assuming that the distances are inversely proportional to the proper motions, Professor Kapteyn computes the relative volumes of the spherical shells which contain the stars with different proper motions (from one-tenth of a second to one second of arc and more). Comparing these volumes with the corresponding number of stars, we arrive at an estimate of the density of star distribution at various distances. The result of this calculation shows that the distribution of stars of the Sirian type approaches uniformity when a large number of the faint stars (ninth magnitude) are considered. With reference to the stars of the second type, however, the larger the proper motion the greater the number of the stars; or, in other words, the second type, or solar stars, are crowded together in the sun's vicinity. Evidence in favour of this conclusion is

[1] "The Meteoritic Theory," pp. 380, 381.

afforded by the fact that, of eight stars having the largest measured parallax (and whose spectrum has been determined), I find that seven have spectra of the solar type. The exception is Sirius, which is evidently an exceptional star with reference to its brightness and comparative proximity to the earth, no other star of the first magnitude having nearly so large a parallax. Indeed, the average distance of all the first magnitude stars is about forty times the distance of Sirius.

Professor Kapteyn finds that the centre of greatest condensation of the solar type stars lies near a point situated about ten degrees to the west of the great nebula in Andromeda, and that this centre nearly coincides with the point which, according to Struve and Herschel, represents the apparent centre of the Milky Way considered as a ring. This would indicate that the sun and solar system lie a little to the north of the Milky Way, and towards a point situated in the northern portion of the constellation of the Centaur. The fact is worth noting, that the nearest fixed star to the earth, Alpha Centauri, lies not very far from this point. Possibly there may be other stars in this direction having a measured parallax, as the southern portion of the heavens has not yet been thoroughly explored.

Professor Kapteyn finds that for stars of equal brightness, those of the Sirian type are, on an average, about two and three-quarter times farther from the earth than those of the solar type. Now, as light varies inversely as the square of the distance, this would imply that the Sirian stars are intrinsically brighter than those of the solar type. This conclusion is confirmed by the great brilliancy of Sirius and other stars of the same type in proportion to their mass. I have shown in Chapter IV. that Sirius is about ten times brighter than the sun would be if placed at the same distance, although its mass is only twice the sun's mass, as computed from the orbit of its satellite.

The general conclusions to be derived from the above results seems to be that the sun is a member of a cluster of stars, possibly distributed in the form of a ring, and that outside this ring, at a much greater distance from us than the stars of the solar cluster, lies a considerably richer ring-shaped cluster, the light of which, reduced to nebulosity by immensity of distance, produces the Milky Way gleam of our midnight skies.

THE END.

INDEX

A

B

D

E

F

G

H

I

J

K

N

O

P

Q

R

S

T

U

V

W

Y

Z

Printed by Cowan & Co., Limited, Perth.

Printed in the United States
By Bookmasters